NEUROSCIENCE-INFORMED COUNSELING

Brain-Based Clinical Approaches,
Second Edition

edited by
Thomas A. Field • Laura K. Jones • Lori A. Russell-Chapin

AMERICAN COUNSELING
ASSOCIATION
2461 Eisenhower Avenue • Suite 300
Alexandria, VA 22314
www.counseling.org

NEUROSCIENCE-INFORMED COUNSELING:
Brain-Based Clinical Approaches
Second Edition

Copyright © 2024 by the American Counseling Association. All rights reserved. Printed in the United States of America. Except as permitted under the United States Copyright Act of 1976, no part of this publication may be reproduced or distributed in any form or by any means, or stored in a database or retrieval system, without the written permission of the publisher.

American Counseling Association
2461 Eisenhower Avenue, Suite 300
Alexandria, VA 22314

Library of Congress Cataloging-in-Publication Data

Names: Field, Thomas A., editor. | Jones, Laura K., editor. | Russell-Chapin, Lori A., editor.
Title: Neuroscience-informed counseling: Brain-based clinical approaches | edited by Thomas A. Field, Laura K. Jones, Lori A. Russell-Chapin.
Description: Second edition. | Alexandria, VA : American Counseling Association, [2024] | Includes bibliographical references and index. Includes bibliographical references and index.
Identifiers: 2023042589 (print) | LCCN 2023042590 (ebook) | ISBN 9781556204142 (paperback) | ISBN 9781394222902 (epub) | ISBN 9781394222926 (pdf)
Subjects: LCSH: Counseling psychology. | Brain. | Neurophysiology.
Classification: LCC BF636.6 .N487 2024 (print) | LCC BF636.6 (ebook) | DDC 158.3--dc23/eng/20240131
LC record available at https://lccn.loc.gov/2023042589
LC ebook record available at https://lccn.loc.gov/2023042590

We dedicate this book to everyone who has been working toward integrating neuroscience into the counseling field and to the next generation of counselors, eager to understand the connections between brain, mind, body, and behavior.

REVIEWS

Thomas A. Field, Laura K. Jones, and Lori A. Russell-Chapin have individually written some of the most inspiring and practical textbooks that integrated neuroscience and neuroscientific concepts into the counseling field. Here in the second edition of their edited book, *Neuroscience-Informed Counseling: Brain-Based Clinical Approaches,* they contribute to the continued significance of enhancing the relational work that counseling professionals and counseling students do by being informed by neuroscience research and literature. This book not only speaks to the dialogues between brain, mind, body, and behavior in a deficit-focused way but also helps counseling professionals conceptualize biological, neurological, and physical factors that play a role in from human development over the life span to overall wellness. Additionally, the organization of the book's chapters aligns with the step-by-step approach addressed in a training model established for development of neuroscience competencies in counseling professionals.

— **Yoon Suh Moh, PhD, LPC, CRC, NCC, BC-TMH, BCN,** author of *Neurobiology of Stress-Informed Counseling: Healing and Prevention Practices for the Helping Professions*

The latest edition of *Neuroscience-Informed Counseling* successfully bridges the gap between neuroscience research and practical clinical application. Packed with insightful case illustrations, this invaluable book translates emerging brain science into clear interventions clinicians can employ immediately for improved treatment outcomes. I am thrilled to recommend the newest edition of this essential guide to brain-based therapies. The authors have vastly expanded the clinical utility for frontline counselors through enhanced psychopharmacology content, lifespan neurodevelopmental context, and demonstrating neuroscience-aligned skills via rich case scenarios. Connecting neurophysiological processes with counseling best practices, this multidimensional resource truly empowers therapist capabilities. From illuminating how neurobiology interacts with substance abuse to providing neuroscience interventions across specialty areas, practitioners now have an accessible framework for harnessing this knowledge toward evidence-based care.

— **Carl Sheperis, PhD, NCC, CCMHC, MAC**

An excellent and accessible read to get therapists up to speed on the brain. Practicing counselors have limited time to stay on top of current research, and we all know how mentally draining it can be trying to piece together real world applications from abstract publications. This updated edition of *Neuroscience-Informed Counseling: Brain-Based Clinical Approaches* does the heavy lifting for you, summarizing thousands of hours of reading into key observations and practical suggestions specifically relevant to the modern clinician. As a licensed therapist who has spent 20 years designing clinical applications for neurotechnology, I have seen many deeply curious counselors who want answers but don't know where to begin. This book is the solution: the perfect starting place for busy, overwhelmed clinicians to learn how to counsel mind, body, *and* brain.

— **Penijean Gracefire, LMHC, BCN, qEEG-D**

TABLE OF CONTENTS

Preface — xi
About the Editors — xxv
About the Contributors — xxvii

PART I FOUNDATIONS OF CASE CONCEPTUALIZATION

Chapter 1 Anatomy and Brain Development — 3
Laura K. Jones

Chapter 2 Neurophysiological Development Across the Life Span — 33
Laura K. Jones

Chapter 3 Biology of Marginality: A Neurophysiological Exploration of the Social and Cultural Foundations of Psychological Health — 55
Kathryn Z. Douthit and Justin Russotti

Chapter 4 Neurophysiology of Traumatic Stress — 77
Laura K. Jones

Chapter 5 Clinical Neuroscience of Substance Use Disorders — 101
Sean B. Hall, Laura K. Jones, and Kiera Walker

Chapter 6 Psychopharmacology Basics — 123
Nancy E. Sherman and Thomas A. Field

PART II COUNSELING ASSESSMENTS, RELATIONSHIPS, AND INTERVENTIONS

Chapter 7 Neurocounseling Assessment 143
Lori A. Russell-Chapin

Chapter 8 Neurocounseling Approaches to Wellness and Optimal Performance 161
Theodore J. Chapin

Chapter 9 Neuroscience of Attention: Empathy and Counseling Skills 181
Carlos P. Zalaquett, Ravza N. Aksoy, Allen E. Ivey, Mary Bradford Ivey, and Thomas Daniels

Chapter 10 Leveraging the Neuroeducation Process to Enhance Outcomes 199
Eric T. Beeson, Thomas A. Field, Chad Luke, Raissa Miller, Laura K. Jones, and Isaac Burt

Chapter 11 Neuroscience-Informed Counseling Theory 219
Carlos P. Zalaquett, SeriaShia Chatters, Ravza N. Aksoy, and Allen E. Ivey

Chapter 12 Neuro-Informed Career-Focused Counseling 237
Chad Luke and Thomas A. Field

Chapter 13 Neuro-Informed Group Work 251
Chad Luke, Joel F. Diambra, and Christine J. Schimmel

PART III ADVANCED APPLICATIONS

Chapter 14 Enhancing Counseling Practice With Neuroscience-Informed Research 269
G. Michael Russo, Eric T. Beeson, and Isaac Burt

Chapter 15 Neuroscience-Informed Clinical Supervision:
           An Emerging Transtheoretical Approach   287
              *Theodore J. Chapin, Lori A. Russell-Chapin,
              and Raissa Miller*

Chapter 16 Ten Guidelines for Integrating Neuroscience
           Into Your Practice   305
              *Lori A. Russell-Chapin, Thomas A. Field,
              and Laura K. Jones*

Glossary   317

Appendix   327

Index   337

PREFACE

Many therapeutic fields are embracing principles of neuroscience into their work, with such principles rapidly influencing best practices. For over 10 years, the counseling field has also been exploring how neuroscience, neurobiology, and related physiology (e.g., endocrine, immune, and gastrointestinal functioning) can best be used to inform, explain, and enhance the theory and practice of counseling while still honoring its humanistic roots.

In 2012, leaders in the counseling field articulated that neuroscientific findings were becoming the "practice standards of the future" (Myers & Young, 2012, p. 21). Recognizing the growing influence of neuroscience on counseling practice, the American Counseling Association (ACA), the Association for Counselor Education and Supervision, and the American Mental Health Counselors Association (AMHCA) established interest networks in which members have worked collaboratively to present a unified vision of how neuroscience can be used to explain and enhance counseling practice. National groups such as AMHCA have also developed specific training standards in the biological basis of behavior. In 2015, ACA's magazine *Counseling Today* began featuring a monthly column on neurocounseling.

In 2016, the Council for Accreditation of Counseling and Related Educational Programs (CACREP) began requiring counselor training programs to address the biological, neurological, and physiological factors that affect human development, functioning, and behavior. In 2023, reflecting further development of knowledge in the field, CACREP broadened the 2016 training mandate to encompass "biological, neurological, and physiological factors that affect lifespan development, functioning, behavior, resilience, and overall wellness" (CACREP, 2024; Section 3, Standard C.10.). The inclusion of resilience and overall wellness in this CACREP standard reflects how neuroscience helps counselors understand the adaptive nature of the nervous system, rather than using neuroscience to merely support deficit-based models such as diagnostic psychopathology.

Since 2017, the *Journal of Mental Health Counseling* has devoted a section to neuroscience-informed counseling. In addition, a growing number of counseling texts and national, regional, and state conference presentations have highlighted the enduring integration of neuroscience into counselor practice. New or revised textbooks have emerged addressing counseling theories from the neuroscience perspective (Chad Luke), neuroscience-informed child and adolescent counseling (Thomas Field and Michelle Ghoston), and neuro-informed clinical supervision (Lori Russell-Chapin and Ted Chapin). Other recent publications include a neuroeducation toolbox (Raissa Miller and Eric Beeson) and a practical neurocounseling text showing computerized LORETA brain maps written by counseling students and faculty (Lori Russell-Chapin, Nicole Pacheo, and Jason DeFord). In this preface, we underscore how integrating neuroscientific principles into the practice of counseling can support and advance the profession (Beeson & Field, 2017) and how this second edition builds on this knowledge.

2024 CACREP Standards

With that background in mind, this preface addresses the 2024 CACREP Standards that are pertinent to the foundational curriculum area of Professional Counseling Orientation and Ethical Practice (Section 3, Standard A):

- History and philosophy of the counseling profession and its specialized practice areas (Standard A.1.)
- The role and process of the professional counselor advocating on behalf of the profession (Standard A.5.)

■ ■ ■

Definition of Counseling

Who are counselors? What do they do? What does it mean to be a counselor? How are we similar to and different from other mental health professionals? How do advances in the field, such as neuroscience and neurobiology, pertain to counseling?

All of these questions are important to ponder throughout one's career, from new graduate student to experienced practitioner. As counselors consider who they are as professionals, neuroscience and related physiology provide the information and tools to support their belief in certain core principles as foundational to counseling practice.

The field of counseling is unique among the mental health professions in its historical beliefs about the human condition and how to enhance optimal living. In 2010, a consensus definition of *counseling* was agreed on by multiple stakeholders in what was formally titled "20/20: A Vision for the Future of Counseling" (Kaplan et al., 2014). The stakeholders distilled into one sentence how the counseling profession could be defined: "Counseling is a professional relationship that empowers diverse individuals, families, and groups to accomplish mental health, wellness, education, and career goals" (Kaplan et al., 2014, p. 368). This definition identifies as foundational to counseling practice several important concepts: (a) the professional relationship takes precedence; (b) the goal of counseling is to empower people, especially those from diverse backgrounds, and address systemic barriers that prevent optimal health; and (c) counseling assists people to not only achieve optimal mental health and wellness but also alleviate distress and mental disorders.

Neuroscience supports and provides models for implementing these concepts into counseling practice. As the chapters in this text will elucidate, neuroscience can help counselors to understand how relationships are forged, leading to deeper and more meaningful working relationships with clients; to recognize the persisting impact of systemic barriers such as oppression, marginalization, and trauma on clients' ability to achieve their goals; and to take a wellness and strengths-based perspective that serves to empower clients and increase optimal performance. In other words, the use of neuroscience in counseling is consistent with the orientation and identity of the counseling profession.

Neurocounseling and Neuroscience-Informed Counseling Defined

Mental health fields, including counseling, are rapidly evolving. One of the most important emerging trends in counseling has been the integration of neuroscience into practice (Beeson & Field, 2017), often referred to as neurocounseling (Montes, 2013; Russell-Chapin, 2016) or as neuroscience-informed counseling (Duenyas & Luke, 2019; Field & Ghoston, 2020; Field et al., 2022; Russo et al., 2021; Schauss et al., 2019). *Neurocounseling* has been defined as "the integration of neuroscience into the practice of counseling, by teaching and illustrating the physiological underpinnings of many of our mental health concerns" (Russell-Chapin, 2016, p. 93). *Neuroscience-informed counseling* is commonly understood to mean the same as neurocounseling, but the use of different terms has become a source of friendly discussion among leaders in this area, and the terms may mean different things to different practitioners, including the editors of this text. Both neurocounseling and neuroscience-informed counseling reflect a focus on the brain-mind-body and neurological-physiological factors that influence cognition, behavior, and emotion.

Lori prefers to use the term neurocounseling because it reflects a thorough psycho/social/medical assessment that customizes and prioritizes client interventions for neurological and physiological dysregulation. For chronic or severe client symptoms, a more thorough assessment than the typical paper-and-pencil instruments and psycho/social/medical history may be necessary. A 5- or 19-channel electroencephalogram (EEG) could be used. Turning these EEG brain wave recordings into visual brain maps assists the counselor and client to better understand where the areas of dysregulation occur. Then, symptom priority and goal setting are personalized for each client. Using both qualitative and quantitative assessments creates an easy method for successful goal obtainment and research (Russell-Chapin et al., 2021).

Conversely, Laura and Thom fear that the term neurocounseling implies that a counselor who integrates neuroscience into their work is doing a different form of counseling altogether. There are growing concerns within the counseling field that the counseling profession is overemphasizing specialization (e.g., school counseling, couples and family counseling, career counseling) at the expense of its unified identity and straying away from its humanistic roots. Laura and Thom prefer the term neuroscience-informed counseling because it represents a unified identity, and the integration of neuroscience and related physiology can be applied to any and all work in which counselors are engaging. Furthermore, Laura and Thom agree that using a term that suggests integration, rather than using a wholly new term, emphasizes that they are still honoring the core traditions of the field while also incorporating knowledge from other fields that may further strengthen our field and the work of counselors within it. Perhaps reflecting the shared valuing of a unified counselor identity, the term neuroscience-informed counseling is also becoming more prevalent within the counseling literature.

Honoring all views, we all agreed as editors to use the term neuroscience-informed counseling in the title of this second edition but to allow the authors of each chapter to use whatever term they preferred. The terms are roughly equivalent but sometimes differentiated in focus.

The integration of principles of neuroscience and related physiology into counseling has a variety of uses. For example, it can be used by counselors to

- understand how and why psychotherapy changes the brain (Russell-Chapin, 2016) and, thus, why clients begin to feel, think, and behave in healthier and more adaptive ways;
- better understand client concerns, conceptualize cases, and plan treatment by using a brain-based perspective;

- help clients understand their experience through brain-based psychoeducation; and
- take a more holistic, wellness-based, and mind-body integrative approach to client work.

In addition, counselors can use technical approaches such as biofeedback and neurofeedback to determine the physiological and neurological underpinnings of a client's distress and dysfunction (assessment) and to help clients modify physiology and brain waves to enhance their functioning and reduce distress and dysfunction (intervention). Approaches such as biofeedback and neurofeedback can also be used to improve optimal performance, not just to modify distress and dysfunction.

For some counselors, integrating principles of neuroscience and related physiology is part of what they are already doing and is another tool in the toolbox (i.e., an adjunctive strategy to psychotherapy). However, it can also entirely change the way they conceptualize client cases, conduct assessments, and select interventions. For example, as Lori once wrote, "For decades, my goal was to assist clients in changing their unwanted thoughts, feelings and behaviors . . . but the overarching goal of all my counseling [today] is to help clients to improve their emotional and physiological self-regulation" (Russell-Chapin, 2016, p. 94).

Purpose of This Text

As counselors learn more about neuroscience and related physiology, they need guidance regarding how to integrate this new brain- and body-based knowledge into counseling practice with clients. The ability to translate complex knowledge to clients is a separate skill set that requires the ability to distill rather than dilute information. Counselors are not immune from believing inaccurate pop psychology information about neuroscience, or *neuromyths* (Kim & Zalaquett, 2019), such as about learning styles and hemispheric dominance ("right or left brained"). Counselors whose case conceptualizations are becoming informed by neuroscientific knowledge also require guidelines regarding how to apply these concepts in clinical practice.

The purpose of this text is to provide a resource for how neuroscientific concepts can be translated and applied to the counseling field, with the objective of both explaining and enhancing the theory and practice of counseling. In doing so, we hope to provide guidance and facilitate learning about how counselors are integrating neuroscience into their work, with the hope of better understanding and identifying methods for effectively and responsibly incorporating key principles of neuroscience into the profession. To advance this effort, we use the 2024 CACREP Standards as our markers of learning to ensure that CACREP-accredited (and all) programs have the information needed to apply neuroscientific concepts to all major areas of counseling practice.

While writing and editing this text, we also understood that for some counselors, especially those for whom science and research are not strength areas, neuroscience can be an overwhelming and frightening concept. The scientific terminology, complex anatomy, and technology-based brain measurements may seem irrelevant to daily counseling practice with clients who bring forth deep existential human struggles that cannot easily be quantified. The specialized knowledge required to be a neuroscience-savvy practitioner may also seem outside the scope of counseling practice.

With that in mind, the purpose of this text is to provide counselors with guidelines, ideas, and tips on how to become effective and skillful neuroscience-informed counselors. We have purposefully asked each author to convey these concepts in a way that is understandable yet retains important information (distill, not dilute; Field & Ghoston, 2020). The chapters are organized so that you will understand foundational neuroscience concepts that inform client case conceptualization (e.g., human development, social and cultural background) before learning how to approach assessment and intervention from a neuroscience-informed perspective.

We hope that this text will be useful not only to counseling practitioners but also to master's- and doctoral-level students in counseling programs. In that regard, it addresses the eight core areas in the Foundational Counseling Curriculum (Section 3) of the 2024 CACREP Standards: professional counseling orientation and ethical practice, social and cultural identities and experiences, lifespan development, career development, counseling practice and relationships, group counseling and group work, assessment and diagnostic processes, and research and program evaluation. We also address several 2024 CACREP Standards that are integrated into the eight core areas, such as the impact of crises, disaster, and traumatic events; the neurobiology of addictions; wellness and optimal performance; and psychopharmacology. Some chapters also address the 2024 CACREP Doctoral Standards for Counselor Education and Supervision (Section 6). We are proud that the first edition of this text was the first publication to discuss the application of neurocounseling and neuroscience to the CACREP standards specifically. Furthermore, the first edition of this text represents the first publication to broadly address the application of neurocounseling and neuroscientific concepts across the core counseling curriculum, an approach that provides a practical, comprehensive model for the integration of neuroscience into counseling practice.

In addition to being an adjunctive text for all foundational courses in the master's-level counseling curriculum, this text can serve as a primary resource for counseling students (both master's- and doctoral-level students) who are taking specialization courses in neuroscience-informed counseling, neurocounseling, brain and behavior, the biological basis of behavior, and so forth. Finally, the text could also be a resource for counselor educators and supervisors who want to learn more about neuroscientific applications to counseling practice. Thus, it is broadly designed for practicing counselors in the field, counselor education students-in-training, and counselor educators and supervisors.

Text Organization and Chapters

From 2017 to 2020, leaders in the neuroscience integration movement worked together on an AMHCA-sponsored task force to create a training model for the development of neuroscience competencies (Field et al., 2022). Their training model (see Figure) highlighted a step-by-step learning process whereby counselors first are provided foundational knowledge in neuroanatomy and physiology (Step 1); then learn about neuroscience applications to clinical presentations (Step 2); and last move up to counseling practice, interventions, and techniques (Step 3).

In this text, the organization of the chapters corresponds to the steps in the training model, so that the reader learns basic neuroanatomy and physiology (Chapters 1 to 3) and clinical presentations (Chapters 4 to 6) before exploring applications to counseling practice (Chapters 7 to 13). The last three chapters (Chapters 14 to 16) are about advanced applications pertaining to research, supervision, and 10 principles for holistic integration. Thus, the text is divided into three parts.

Part I reviews foundational information about neuroanatomy and neurophysiological development across the life span before exploring the effects of social and cultural issues such as marginalization and oppression. This part also covers clinical presentations, such as the neurophysiology of traumatic stress, substance use, and indications for psychopharmacological intervention.

Part II applies the foundational knowledge learned in Part I to counseling assessments and relationships from the intake process through case closure. These chapters describe an approach to completing a comprehensive neurocounseling assessment, assessing for client wellness and enhancing optimal performance, using attentional processes and empathic understanding, leveraging the neuroeducation process to enhance client outcomes, applying neuroscience-informed counseling theory, career counseling, and group work.

Part III addresses advanced applications of a brain- and body-based approach to conducting research and clinical supervision. It also includes 10 guidelines for integrating neuroscience into counseling practice using the clinical case study of Muna.

Specializations
technological interventions
• bio- and neurofeedback

Theories and Applications
case conceptualization • assessment • wellness and TLCs • critiques of neuro-informed theories and interventions • consultation • advocacy

Clinical Presentations
myelination • pruning • LTD, LTP • neurological basis of mental disorders • neurological conditions • medications • impacts to relationships • integrative neuro-informed counseling theories

Foundations
central and peripheral nervous system structures and functions • endocrine system • neurons and neurotransmission • epigenetics • biological components of relationships • neuro-informed understanding of existing counseling practice

Locate and appraise literature

FIGURE
Training Model for the Development of Neuroscience-Informed Counseling Competencies

Note. TLC = therapeutic lifestyle change; LTD = long-term depression; LTP = long-term potentiation. Reprinted from "A Training Model for the Development of Neuroscience-Informed Counseling Competencies," by T. A. Field, Y. S. Moh, C. Luke, P. Gracefire, E. T. Beeson, and G. M. Russo, 2022, *Journal of Mental Health Counseling, 44*(3), 266–281. Copyright © 2022 by American Mental Health Counselors Association. All rights reserved. Reprinted with permission.

Text Features

As editors, we sought to ensure that each chapter made direct connections between the content and clinical practice. As an anchor for the content knowledge, each chapter references a fictional case study to ensure the material is relevant to client work. This preface presents a case study with reflective questions that are further explored in the final chapter. Reflective questions are integrated throughout each chapter so that readers can pause and consider how the content knowledge that has been covered could be relevant to the client case being discussed. We encouraged chapter authors to share their own brain-based approach to their case study so you, as the reader, can consider how to use the information presented with clients in your own unique way. A few quiz questions are included at the conclusion of each chapter so that you can test your knowledge. The quiz answers are located at the back of the text. A glossary is also provided at the conclusion of the text to help you understand the concepts taught in the chapters. You are encouraged to return to sections of the chapter in which those terms are described if you are not confident in your knowledge.

Changes to the Second Edition

This revision builds upon the first edition by (a) updating content to highlight changes in the scientific literature since 2016; (b) reflecting new research within the counseling field regarding the empirical understandings of integration, the ethics of integration, cultural considerations when integrating neuroscience into counseling practice, and best practices for training counselors in neuroscience; and (c) addressing pertinent 2024 CACREP Standards. Empirical findings regarding neuroscience are always emerging, and this text reflects important updates to the scientific literature.

In the years since the first edition was published, counselors also have been recognized by law as being eligible for independent Medicare provider status, have worked with increased numbers of clients struggling with addiction (especially opioid addiction), and have assisted clients through the coronavirus pandemic that began in 2019. This second edition text includes new content relevant to these current practice needs for counselors, such as (a) the neuroscience of aging, cognitive decline, and brain injury; (b) medication-assisted treatment for addiction; and (c) the role of inflammation in mental health conditions. This second edition also features two new chapters reflecting emerging themes in the counseling literature: leveraging neuroeducation to enhance client outcomes and neuroscience-informed clinical supervision.

Addressing Diversity and Social Justice

Neuroscience informs the importance of cultural competency and social justice work as counselors (Ivey et al., 2024). When used appropriately, the integration of neuroscience also can reduce the stigma associated with mental health (Lebowitz & Appelbaum, 2017) and perhaps facilitate help seeking by members of marginalized groups. Chapter authors discuss such considerations and give attention throughout the text to the different factors related to providing neuroscience-informed counseling with clients from marginalized backgrounds. For example, the biology of marginality chapter provides clear guidance for exploring a client's heritage and marginalized experiences during the counseling process and assisting clients to navigate systemic and environmental barriers to reduce the impacts of such stress.

Concluding Thoughts

As Lori likes to say, once you have learned about how the brain works in relation to physical and emotional health, you cannot go back. We are confident that this knowledge will forever change how you approach case conceptualization, assessment, and intervention in clinical practice. We hope the subsequent chapters will be your starting point on this journey.

Clinical Case Study

The client described here, Muna, was first presented in the first edition of this text. In this second edition, we provide updated case information to catch the reader up on Muna's journey.

Muna is a 42-year-old Iraqi woman living in a metropolitan area of a large U.S. city. Seven years ago, she sought counseling to address anxiety that she was experiencing at her new job in an accounting firm. Muna was also struggling with feelings of inadequacy related to her long-standing dating relationship of nearly a decade. Her family lives in Iraq, and she emigrated to attend a U.S. college in her early 20s. She lived in constant dread of her family finding out that she was living with her boyfriend outside of marriage. She drank alcohol to cope, mostly at night (4 to 5 units). Muna also struggled with sleep, usually only getting 3 to 5 hours per night. She had a past diagnosis of attention-deficit/hyperactivity disorder and twice a day took 20 milligrams of Adderall, a stimulant. Muna had experienced psychological abuse from her father throughout her childhood. She was warm and engaging during the initial interview, although her nonverbal fidgeting suggested she was somewhat anxious. Muna acknowledged holding deep-seated fears that something was deeply wrong with her.

During the first 2 years of counseling, Muna made significant progress. At the counselor's request, Muna scheduled a full physical exam with a primary care physician. Her bloodwork assessment indicated that she was anemic and vitamin D deficient, with estrogen levels below the normal range, and that her thyroid functioning was outside the normal range. These medical concerns were addressed, and Muna's anxiety abated somewhat. Muna explored her past and present relationships and realized her attachment pattern was to end relationships before the other person might call it quits.

Treatment for Muna included discussing the effects of early trauma on her physiological hyperarousal, her development, and her overall functioning. Knowing about these connections reduced Muna's presenting symptoms of anxiety and sleeplessness. She then proceeded to achieve treatment goals of reducing her alcohol consumption through attending Alcoholics Anonymous meetings. She began attending a mosque. She also received neurofeedback and eventually was weaned off the stimulant medication. When treatment concluded, Muna reported a great deal of self-efficacy from achieving her goals.

Five years later, Muna contacts your office to arrange a new intake appointment. You learn during the first meeting that the coronavirus pandemic was particularly hard on Muna. Shaking as she talks, she discloses that she has broken up with her partner and moved into her own apartment. The close proximity and "cabin fever" of quarantine had exacerbated disputes between them to the point that she feared for her safety. During the pandemic, Muna grew uncomfortable with socializing in large groups and currently has limited social activity and few friends. She also was diagnosed with ulcerative colitis and is having difficulty managing her symptoms. Muna wonders if these conditions developed following lingering symptoms of coronavirus infection. In addition, a year before the pandemic, Muna was injured in a serious car accident that left her with chronic pain she manages daily with opioids. Muna is still employed at the accounting firm but now works from home. Her goals for counseling are to better manage her medical problems, reduce her reliance on opioids, and become more socially active again.

The ensuing chapters of this text contain information that will help you conceptualize Muna's concerns. You will learn possible answers to important questions such as the following:

- How might stress and the inflammatory response be connected with mental health?
- Which areas of Muna's brain and body are being compromised?
- How does anxiety "happen" in the body?
- What is the potential extended impact of emotional abuse on the client's functioning?

- How does a person become neurophysiologically dependent on a drug?
- What role do lifestyle changes and neurofeedback have in managing health conditions?

In the final chapter of this text (Ten Guidelines for Integrating Neuroscience Into Your Practice), we review each of these questions on the basis of knowledge you will acquire from the chapters that precede it.

References

Beeson, E. T., & Field, T. A. (2017). Neurocounseling: A new section of the *Journal of Mental Health Counseling*. *Journal of Mental Health Counseling, 39*(1), 71–83.

Council for Accreditation of Counseling and Related Educational Programs. (2016). *2016 CACREP standards.* http://www.cacrep.org/for-programs/2016-cacrep-standards

Council for Accreditation of Counseling and Related Educational Programs. (2024). *2024 CACREP standards.* https://www.cacrep.org/wp-content/uploads/2023/10/2024-Standards-Combined-Version-10.09.23.pdf

Duenyas, D. L., & Luke, C. (2019). Neuroscience for counselors: Recommendations for developing and teaching a graduate course. *The Professional Counselor, 9*(4), 369–380.

Field, T. A., & Ghoston, M. R. (2020). *Neuroscience-informed counseling with children and adolescents.* American Counseling Association.

Field, T. A., Moh, Y. S., Luke, C., Gracefire, P., Beeson, E. T., & Russo, G. M. (2022). A training model for the development of neuroscience-informed counseling competencies. *Journal of Mental Health Counseling, 44*(3), 266–281.

Ivey, A., Ivey, M. B., & Zalaquett, C. P. (2024). *Essentials of intentional counseling and psychotherapy: Counseling in a multicultural world* (4th ed.). Cengage.

Kaplan, D. M., Tarvydas, V. M., & Gladding, S. T. (2014). 20/20: A vision for the future of counseling: The new consensus definition of counseling. *Journal of Counseling & Development, 92*(3), 366–372. https://doi.org/10.1002/j.1556-6676.2014.00164.x

Kim, S. R., & Zalaquett, C. (2019). An exploratory study of prevalence and predictors of neuromyths among potential mental health counselors. *Journal of Mental Health Counseling, 41*(2), 173–187.

Lebowitz, M. S., & Appelbaum, P. S. (2017). Beneficial and detrimental effects of genetic explanations for addiction. *International Journal of Social Psychiatry, 63*(8), 717–723. https://doi.org/10.1177%2F0020764017737573

Montes, S. (2013, December). The birth of the neurocounselor. *Counseling Today, 56*(6), 32–40.

Myers, J. E., & Young, S. J. (2012). Brain wave feedback: Benefits of integrating neurofeedback in counseling. *Journal of Counseling & Development, 90*(1), 20–28. https://doi.org/10.1111/j.1556-6676.2012.00003.x

Russell-Chapin, L. A. (Ed.). (2016). Integrating neurocounseling into the counseling profession: An introduction [Special issue]. *Journal of Mental Health Counseling, 38*(2), 93–102. https://doi.org/10.17744/mehc.38.2.01

Russell-Chapin, L., Pacheco, N., & DeFord, J. (Eds.). (2021). *Practical neurocounseling: Connecting brain functions to real therapy interventions.* Routledge.

Russo, G. M., Schauss, E., Naik, S., Banerjee, R., Ghoston, M., Jones, L. K., Zalaquett, C. P., Beeson, E. T., & Field, T. A. (2021). Extent of counselor training in neuroscience-informed counseling competencies. *Journal of Mental Health Counseling, 43*(1), 75–93.

Schauss, E., Horn, G., Ellmo, F., Reeves, T., Zettler, H., Bartelli, D., Cogdal, P., & West, S. (2019). Fostering intrinsic resilience: A neuroscience-informed model of conceptualizing and treating adverse childhood experiences. *Journal of Mental Health Counseling, 41*(3), 242–259. https://doi.org/10.17744/mehc.41.3.04

About the EDITORS

Thomas A. Field, PhD, LMHC (MA, WA), LPC (OR, VA), LPC-MH (SD), NCC, CCMHC, ACS, is an associate professor and head of counselor, adult, and higher education in the College of Education at Oregon State University. He was previously a faculty member at Boston University School of Medicine. Thom holds a PhD in counseling and supervision from James Madison University. His research focuses on the integration of neuroscience into counseling practice and professional and social justice advocacy. He has published numerous articles and authored two books on the topic of neuroscience integration. Thom is currently a member of a research team that is studying the development of an emerging counseling theory called neuroscience-informed cognitive behavior therapy. Since 2017, Thom has served as the associate editor of the Neuroscience-Informed Counseling section of the *Journal of Mental Health Counseling*. He is a former coeditor of the "Neurocounseling: Bridging Brain and Behavior" column in Counseling Today magazine. In addition to performing faculty responsibilities, he has actively helped clients with mental health concerns since 2006. He has provided counseling to more than 1,000 clients during his career and currently maintains a small private practice.

Laura K. Jones, PhD, MS, is an associate professor at the University of North Carolina Asheville, where she teaches coursework in both health and wellness promotion (health sciences) and neuroscience. She also serves as athletics mental health coordinator and health care administrator for the university's Division I athletics program, helping athletes and coaches better understand how mental health physiologically influences peak performance. In addition to having a PhD in counseling and counselor education from the University of North Carolina at Greensboro, she holds an MS in psychology–cognitive neuroscience from the University of Oregon. Laura has professional experience conducting functional MRI (fMRI) research and has presented at numerous conferences on the intentional and informed integration of neuroscience into the counseling field and counselor training programs. She has authored and coauthored publications and book chapters detailing the application of neuroscience and related physiology (e.g., endocrine, immune, and gastrointestinal functioning) to clinical mental health counseling and trauma and crisis intervention. Laura is a member of the American Counseling Association's Neurocounseling Interest Network. Laura previously served as the inaugural chair of the Association for Counselor Education and Supervision's Neuroscience Interest Network and was a founding coeditor of the "Neurocounseling: Bridging Brain and Behavior" column in *Counseling Today* magazine.

Lori A. Russell-Chapin, PhD, NCC, CCMHC, LCPC, BCN, is a professor of counselor education at Bradley University in Peoria, Illinois. Lori earned a PhD in counselor education from the University of Wyoming and a master's in counselor education from Eastern Montana College. Currently, Lori teaches graduate counseling courses in Bradley University's campus-based and online brain-based master's programs. She codirects the Center for Collaborative Brain Research, a partnership among Bradley University, OSF Saint Francis Medical Center, and the Illinois Neurological Institute. Board certified in neurofeedback, Lori has authored or coauthored 13 books on topics ranging from practicum internship and supervision to neurocounseling and neurofeedback. Lori is the chair of the ACA Neurocounseling Interest Network and has served as coeditor of the "Neurocounseling: Bridging Brain and Behavior" column in *Counseling Today* magazine. Lori maintains a small private practice. In addition to being an award-winning researcher and teacher at Bradley University, she is an American Counseling Association fellow and past recipient of the ACA Trailblazer Award for her work in neurocounseling.

About the CONTRIBUTORS

We are very fortunate to have been able to gather together some of the best minds (pun intended!) within the field of neurocounseling to share with you their expertise related to core counseling content areas. The authors who contributed to this book are listed here in alphabetical order.

Ravza N. Aksoy, MEd, is a doctoral student in the Department of Educational Psychology, Counseling, and Special Education at The Pennsylvania State University.

Eric T. Beeson, PhD, is a professor and department chair of counselor education at Marshall University. He is coauthor of *The Neuroeducation Toolbox: Practical Translations of Neuroscience in Counseling and Psychotherapy.*

Isaac Burt, PhD, is an associate professor in the School of Education at Johns Hopkins University. His research interests include integrating scientific principles with Africana counseling and combining neuroscience concepts (flow state) with multiculturalism.

Theodore J. Chapin, PhD, is the president and clinical director of Resource Management Services and the Neurotherapy Institute of Central Illinois, a private business consulting and counseling firm in Peoria, Illinois. He is board certified in neurofeedback.

SeriaShia Chatters, PhD, is an assistant vice provost of educational equity at The Pennsylvania State University and adjunct associate professor in the Department of Educational Psychology, Counseling, and Special Education at The Pennsylvania State University.

Thomas Daniels, PhD, is a retired professor of psychology at Memorial University of Newfoundland (Grenfell Campus). He is known internationally for his work in microcounseling and microskills.

Joel F. Diambra, PhD, is associate professor and director of graduate studies in the Educational Psychology and Counseling Department at the University of Tennessee at Knoxville. Before becoming an academic, Joel worked as an employment specialist for clients who had sustained a traumatic brain injury.

About the Contributors

Kathryn Z. Douthit, PhD, is a faculty member in counseling and human development at the University of Rochester. Before her counseling training, she earned an MA in microbiology and immunology.

Sean B. Hall, PhD, is an assistant professor of counselor education at Florida Gulf Coast University. He earned his doctorate in counseling, specializing in clinical mental health and educational research methods, from Old Dominion University.

Allen E. Ivey, EdD, is a distinguished professor (emeritus) at the University of Massachusetts, Amherst. He is a fellow of the American Counseling Association. Board certified by the American Board of Professional Psychology, he is also a fellow of the American Psychological Association.

Mary Bradford Ivey, EdD, is former vice president of Microtraining Associates, an educational publishing firm, and an independent consultant. She is a fellow of the American Counseling Association.

Chad Luke, PhD, is an associate professor in the Department of Counseling at St. Bonaventure University. Among his many books is a classic titled *Neuroscience for Counselors and Therapists: Integrating the Sciences of Brain and Mind*, now in its second edition.

Raissa Miller, PhD, is an associate professor of counselor education at Boise State University. She is coauthor of *The Neuroeducation Toolbox: Practical Translations of Neuroscience in Counseling and Psychotherapy*.

G. Michael Russo, PhD, is an assistant professor of professional counseling at the University of Oklahoma. He holds board certification in neurofeedback and a graduate minor in applied statistics.

Justin Russotti, PhD, is on the research faculty at the University of Rochester Medical College and leads a research lab examining the sequelae of childhood trauma.

Christine J. Schimmel, PhD, is an associate professor in the Department of Counseling and Well-Being at West Virginia University. She has authored numerous books, including *Applying Neuroscience to Counseling Children and Adolescents: A Guide to Brain-Based, Experiential Interventions*.

Nancy E. Sherman, PhD, is a professor emeritus in the Department of Leadership in Education, Nonprofits, and Counseling at Bradley University.

Kiera Walker, MA, is a licensed professional counselor and works as a clinical counselor at the University of Alabama at Birmingham. Kiera has over 10 years of research experience, with expertise in molecular biology techniques.

Carlos P. Zalaquett, PhD, is a professor in the Department of Educational Psychology, Counseling, and Special Education in the College of Education at The Pennsylvania State University. He is a coleader of the neurofeedback laboratory at Penn State's College of Education.

Part I
FOUNDATIONS OF CASE CONCEPTUALIZATION

This first section of the text reviews foundational knowledge needed to conceptualize client cases from a neurophysiological perspective. You will first be introduced to basic brain anatomy and systems before learning about neurophysiological development across the life cycle and the impact of neurophysiological marginality and traumatic stress on psychological health. This knowledge is considered fundamental to understanding the client's presenting problem from a neurophysiological perspective, leading to more effective counseling relationships, assessments, and interventions.

■ ■ ■

Chapter 1

Anatomy and Brain Development

Laura K. Jones

Learning about brain anatomy and the functioning of related systems may not be the first thing that comes to mind when you think about counselor education and training. You may wonder how your clinical decisions with a client, supervisor, or community would be influenced by your knowledge of the brain and body. As you read this and later chapters, you might find yourself becoming fascinated by the many wonders of the brain and body and how this information affects your work as a counselor.

Interest in the application of neuroanatomy to mental health has been long-standing. Sigmund Freud (1914/2014) suggested in his classic paper *On Narcissism* that "we must recollect that all of our provisional ideas in psychology will presumably one day be based on an organic substructure" (p. 78). Since Freud's early conjecture, limitations in technology have hampered understanding of the effect of the workings of the brain on mental health functioning. On April 2, 2013, a paradigm shift in mental health research began with the launch of the National Institutes of Health's Brain Research Through Advancing Innovative Neurotechnologies (BRAIN) Initiative. The mission of this initiative is to understand "the circuits and patterns of neural activity that give rise to mental experience and behavior" (National Institutes of Health, 2014, p. 12) and, in doing so, to cultivate an integrative understanding of brain-behavior processes. As technology and science continue to develop, so may our understanding of mental health as being a result of an intricate interplay between the brain, immune, endocrine, and gastrointestinal systems (Anisman et al., 2018; Brown et al., 2020; Hou et al., 2022; Peirce & Alviña, 2019; Skonieczna-Żydecka et al., 2018). This chapter is a first step in introducing you to the inner workings of the brain in an effort to inform your case conceptualizations, treatment plans, and clinical effectiveness with clients.

2024 CACREP Standards

This chapter addresses 2024 Council for Accreditation of Counseling and Related Educational Programs (CACREP) Standards pertinent to the Foundational Counseling Curriculum (Section 3) area of Lifespan Development (Standard C):

- Theories of learning (Standard C.3.)
- Biological, neurological, and physiological factors that affect lifespan development, functioning, behavior, resilience, and overall wellness (Standard C.10.)

Clinical Case Study: Rein

Rein is a 12-year-old female preadolescent of Native American descent. Her home environment is intact; she lives with both parents and a brother. Rein describes having an attentive but reserved family with limited communication regarding emotions, physiological changes to the body, or interpersonal relationships. Her parents scheduled an appointment because of a marked decrease in Rein's grades at school—Rein previously having been a straight A student—and a noticeable withdrawal from group activities during class. Rein reports feeling very sad and lonely and indicates that she has been feeling increasingly more distant from friends in the past 6 months. Rein denies feeling bullied or victimized, but she does not feel that anyone likes or understands her and has a hard time connecting with others. She also reports feeling lethargic. Her mother confirms that Rein often cries and isolates herself from her family. Her mother also shared that Rein recently reached menarche and was concerned that Rein may feel embarrassed about this.

The Brain: Structure, Function, and Systems

Despite incredible advances in science, the human brain in many ways remains a mystery. Part of its enigmatic nature rests in its complexity. Not only are various internal and external structures of the brain specified for certain functions, but those parts directly and indirectly influence one another and other parts of the body by way of various chemical messengers and electrochemical networks.

Furthermore, several areas of the brain can work in concert to govern other aspects of an individual's mental and physical functioning. Something as simple as reading the word "counselor" requires a remarkable succession of processes that involve virtually the entire brain.

In addition, researchers are now discovering more about how a person's physical health, and even the nature of the microbes in their gut, influences the functioning of the brain and vice versa. Some of the paradigms of mental health are now shifting as researchers begin to further investigate the reciprocal functioning of the body and brain, such as the role of inflammation (the process by which white blood cells help to protect people from infection) in depression (Miller & Raison, 2016). Thus, let us explore some of what is known about the various parts and coordinated systems of the brain and body.

External Structures

What is your first thought when you hear the word "brain"? Most people think of a gray folded mass that sits inside the skull. This folded mass is the outer layer or lateral part of the brain, called the *cerebral cortex*, or cortex for short. You may also know that in the interior of this folded cortex are other exceedingly important parts of the brain. These internal and external structures control virtually everything about you. They allow you to think, feel, behave, breathe, and survive.

The cerebral cortex, or outermost part of the brain, is actually a 2- to 4.5-millimeter-thick mass of gelatinous tissue (Fischl & Dale, 2000). Even though it is often thought of as gray (and parts of it are called *gray matter*), it is actually pink when it is healthy living tissue, much like other tissue in the body. This folded mass includes the ridges of cortex known as *gyri* (singular, *gyrus*) and the shallower grooves between the gyri known as *sulci* (singular, *sulcus*). *Fissures* are similar to sulci but are deeper and more clearly divide regions of the brain. The gyri, sulci, and fissures help to demarcate different regions of the brain.

Hemispheres

The cortex is made up of two hemispheres: one on the left and one on the right. Connecting these two hemispheres is a thick band of nerve fibers known as the *corpus callosum*. It is the largest collection of nerve fibers in the entire nervous system, containing roughly 200 million interhemispheric connections (Luders et al., 2010). This band of fibers allows the two hemispheres to communicate back and forth and integrate the information being processed on either side of the brain. A common misconception in popular culture is that individuals are either "left brained" or "right brained." It is accurate that certain functions may be predominantly controlled by brain regions in one hemisphere or another; for example, the area of the brain responsible for language production is typically located in the left hemisphere. Yet the notion that a person can be either left brained or right brained is an overgeneralization and a misrepresentation of brain functioning. More often than not, both sides of the brain are working in coordinated action to allow people to more fully perceive, respond, and adapt to their internal and external environments.

Lobes

The cerebral cortex is further divided into four sets of primary lobes, with analogous lobes in each hemisphere. Each lobe is specialized for certain functions, such as sight, somesthesis (e.g., skin senses and proprioception), hearing and language comprehension, motor control, and executive functioning. The four lobes are the occipital lobe, parietal lobe, temporal lobe, and frontal lobe. Figure 1.1 depicts the general location of each of the four lobes.

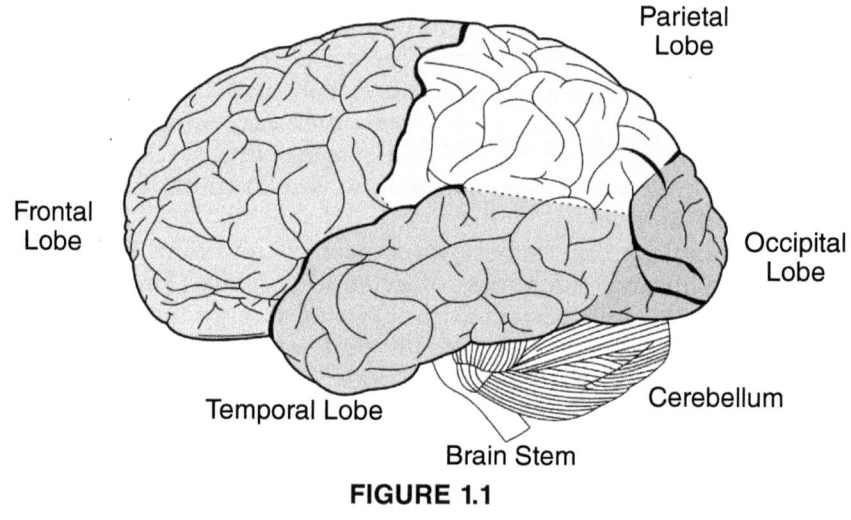

FIGURE 1.1

Lobes of the Brain

Note. From *Anatomy of the Human Body* (20th ed., Plate 728), by H. V. Carter, 1918, Lea & Febiger. In the public domain. Vectorized by Mysid, 2008, via Wikimedia Commons.

Occipital lobe. At the very back of the brain sits the occipital lobe, which is the smallest of the four lobes. This is the visual center of the brain. The occipital lobe pieces together the visual components of the surrounding world. This allows people to interpret and understand what their eyes are seeing, such as shape, color, size, depth, and motion. However, visual processing is limited, meaning that people cannot process everything in the world around them. Visual perception is determined by several factors, including how many objects are present, how long the objects are in the visual field, and to which objects people allocate their attention.

Parietal lobe. The parietal lobe is located between the frontal lobe and the occipital lobe toward the crown of the head. This part of the brain contains the primary somatosensory cortex, which regulates the sensations that are perceived by the physical body, such as touch (e.g., temperature, pressure, pain), and the awareness of bodily movement and the orientation of the body in space (i.e., proprioception). The most anterior (i.e., toward the front) gyrus of this cortex (i.e., the postcentral gyrus) extends across both hemispheres and contains a map of one's entire body. For example, regions of this gyrus represent the thumb, tongue, arm, stomach, and even pinky toe. Every part of the body that one can feel is allocated a certain area of cortex on this gyrus, with the number of sensory receptors on the skin of that area governing how much of the cortex it represents. For example, the hands, face, and tongue are all represented by very large areas of cortex because people need to have very refined sensations of touch for these areas. The body maps are also *contralateral* (as opposed to ipsilateral), meaning that the right side of the body is mapped onto the left postcentral gyrus and vice versa.

Temporal lobe. The temporal lobe is located just behind the ears, below the parietal lobe and between the frontal lobe and the occipital lobe. The primary function of the temporal lobe is hearing and language comprehension. This region of the brain allows people to put together and comprehend the various sounds that are coming into their ears. Wernicke's area is the region of the temporal lobe that allows people to process spoken language. The temporal lobes are also involved in memory. The hippocampus is located in the medial (i.e., interior) region of the temporal lobes and is responsible for the formation of long-term, explicit (i.e., consciously declared or declarative) memories.

Frontal lobe. The front section of the brain contains the largest cortical lobe—and the largest of any mammal—which covers nearly half of the entire cortex. Aptly named the frontal lobe, this area is also the most extensive in terms of the functions that it controls. The most posterior (i.e., toward the back) gyrus of the frontal cortex (neighboring the parietal lobe) contains the primary motor cortex. Like the postcentral gyrus (i.e., somatosensory cortex), this area contralaterally maps out the entire body, not for the purpose of feeling those parts but for the purpose of moving them. So as you reach to turn the page of this book, neurons (i.e., brain cells) in the upper left side of this gyrus (or the right side if you are using your left hand) are firing.

At the base of this primary motor cortex, extending down and forward and typically in the left hemisphere, is Broca's area, which is responsible for language production. Note that the area for language comprehension is different from that for language production. Broca's area is located at the base of the primary motor cortex because that is the area of the cortex that is responsible for the movement of the face and mouth, allowing individuals to move in ways that are necessary to vocalize words and sounds. The functioning of this area can be affected by experiences of extreme traumatic stress, as further discussed in Chapter 4. Given that counseling most often consists of "talk therapy," especially with adults, any altered functioning of Broca's area may have notable clinical implications.

Over and above these functions, the frontal lobe is involved in problem-solving, decision-making, planning, moral reasoning, attention, emotion regulation, and even priming in memory. The frontal lobe also plays a role in one's personality. This was famously demonstrated by the case of Phineas Gage, a railroad worker who experienced significant damage to the most anterior region of his frontal lobe. An explosion sent a metal spike diagonally through his skull, from around his ear up to the opposite top side of his skull. He not only survived the accident but was walking around and talking minutes later. The most striking consequence of this accident was his change in personality and behavior. Before the accident, Phineas was seen as a friendly, polite, and even-tempered fellow. Afterward, he was reported to be angry, cantankerous, and even violent, often using considerable profanity and lacking in impulse control (Macmillan, 2002). The portion of his frontal lobe that was most notably affected was a region called the *prefrontal cortex*. The prefrontal cortex serves as the "executive control" center of the brain and is in charge of rational thinking. Much like a corporate executive, the prefrontal cortex helps people decide or judge what is best for them, plan for the future, work toward goals, and discriminate among varying options. It also regulates their emotions, attention, and social functioning. Put simply, it allows people to think through situations and behave in a deliberate, goal-directed manner.

Just underneath the outer surface of cortex in the frontal lobe, near the confluence of the frontal, parietal, and temporal lobes, lies an area of cortex called the *insula* (insular cortex). Responsible for a host of sensory and affective functions (e.g., pain perception), this region has been scarcely explored in empirical research (Uddin et al., 2017). However, one of the most well-understood functions is in helping individuals translate the emotions that they feel in their body into their cognitive understanding of those emotions, or what is known as feelings. Yes, emotions and feelings are actually different things, processed by different areas of the brain (Damasio, 2001). The insula is the area of the brain that acts as a bridge between these two, converting emotions into feelings or helping people to understand and put words to their bodily sensations. This sense of having an awareness of one's internal bodily state is known as *interoception*. Given its ability to facilitate awareness of one's emotional state, the insula is also a key brain region involved in consciousness and empathy (Craig, 2009; Decety & Lamm, 2006) and has been implicated in pain perception, substance use disorders, and eating disorders, among other struggles (Kim et al., 2012; Lu et al., 2016; Naqvi & Bechara, 2010).

The cingulate cortex also lies underneath the outer surface of the cortex. This band of cortex extends lengthwise from the frontal to the occipital lobes of the brain, following the shape and curve of the corpus callosum. The job of the cingulate cortex is quite varied, given its role in learning, memory, reward, and social and emotional processing, with the anterior (frontal) and posterior (back) sections controlling different functions. The anterior cingulate cortex (ACC) is responsible for emotional processing and regulation, empathic responding, and socially driven interactions, with the dorsal (top) ACC (also called the *middle cingulate cortex*) participating in more cognitive aspects and the rostral (front) ACC participating in more affective aspects of such processes (Lavin et al., 2013; Stevens et al., 2011). The functioning of the posterior cingulate cortex (PCC), however, is a bit more elusive. Recent research has determined that it appears to play a role in internally directed cognition, retrieval of autobiographical memories, and planning for the future (Buckner et al., 2008; Leech & Sharp, 2014; Raichle et al., 2001). Such research has also identified the PCC as a central node of the default mode network (DMN), a system of functionally connected brain regions that become engaged when the brain is in a resting state and not involved in a specific attention-demanding, goal-oriented task. Difficulties with DMN activation are associated with challenges in regulating attention and thoughts, including being able to shift attention from intrusive (i.e., unwanted) thoughts. As a result, the functioning of the PCC and this central node of the DMN has notable implications in post-traumatic stress disorder (A. A. Nicholson et al., 2022).

Cerebellum and brain stem. In addition to these four primary lobes, two other key features of the exterior portion of the brain are the cerebellum and brain stem. The cerebellum, meaning "little brain," is the cauliflower-shaped structure at the back base of the brain. Like the cerebral cortex, the cerebellum is also divided into two hemispheres. It contains more neurons than the cortex—nearly 3.6 times as many (Herculano-Houzel, 2010). The cerebellum was originally thought to govern only motor control, such as posture and balance; fine-tuned motor learning (e.g., riding a bike); and coordination of the fluid movements of multiple muscle groups. Researchers have more recently begun to recognize the integral role of the cerebellum in a range of functions related to cognition, emotion, sensory perception, attention, threat, and pleasure (Strick et al., 2009; Turner et al., 2007).

The brain stem is the structure of the brain that extends from the base of the brain to the spinal cord. It consists of the midbrain, the pons, and the medulla oblongata. The brain stem connects the brain to the rest of the body and is vital to survival. Evolutionarily and developmentally the most primitive part of the brain, it regulates many of the nonconscious (not consciously directed) processes that keep people alive, such as breathing, heart rate, blood pressure, and circadian rhythms, including the sleep cycle.

Cranial Nerves

Extending directly from the brain out to various organs, muscles, and sensory systems are the 12 pairs of cranial nerves (CNs). These nerves function in sensory, motor, and parasympathetic control, with most

> **Reflective Question**
>
> Which cortical areas seem related to Rein's symptoms?

of the nerves controlling muscles of the face and neck or regulating visual, olfactory, and auditory sensations. These nerves allow people to chew, swallow, blink reflexively (corneal reflex), constrict or dilate the pupils, and vocalize, among other functions. CNs also connect to organs that help people to regulate or calm their autonomic nervous system (part of the peripheral nervous system that regulates involuntary physiological processes), thus helping to control nonconscious bodily functions such as heart rate, breathing, digestion, and sexual arousal. For example, the facial cranial nerve (CN VII) controls the salivary glands. Among these nerves is the 10th cranial (CN X) or *vagus* nerve, which extends down the trunk to connect with the heart, lungs, and gastrointestinal tract. It provides input to help slow breathing and heart rate, and it stimulates activity of the stomach and intestines and thus digestion. The vagus nerve has been hypothesized to play an important role in social connection, attachment, and internal experiences of safety (Porges & Furman, 2011).

Internal (Subcortical) Structures

Underneath the large outer cortex of the brain are a number of vital subcortical (below the cortex) brain structures. It is beyond the scope of this chapter to discuss every subcortical structure; however, several of the most notable are discussed later in this book. Figure 1.2 provides a useful diagram of some of these structures.

The most centrally located of the subcortical structures is the thalamus. This is the primary relay station of the brain—all of the messages from most sensory systems and the body that are relayed to the cortex, or vice versa, are sent through the thalamus. In this way, the thalamus ensures that those messages are directed to the correct place and allows for the senses to be consciously perceived (i.e., understood). It can be thought of as the Grand Central Station of the brain, through which all the trains (i.e., signals) to and from the cerebral cortex pass to get to their final destinations (Taber et al., 2004). The sensory system of *olfaction*, or sense of smell,

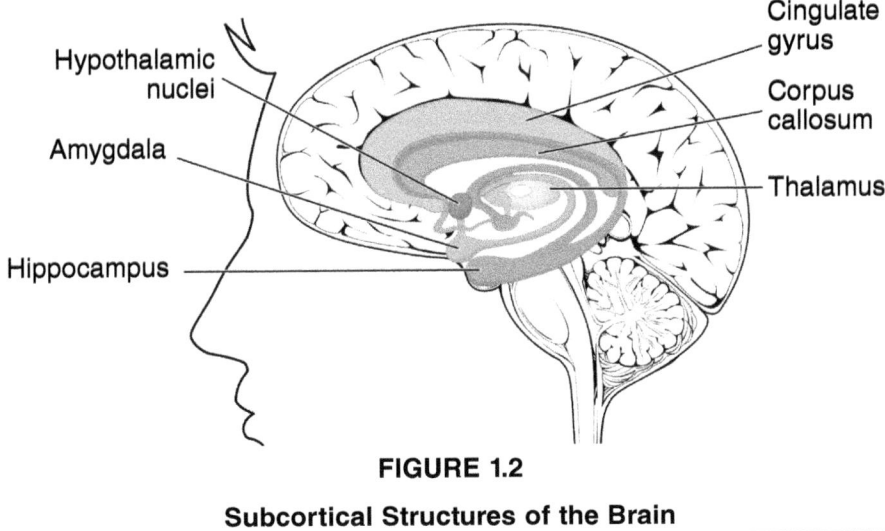

FIGURE 1.2

Subcortical Structures of the Brain

Note. From OpenStax College, 2013, via Wikimedia Commons. Used under Creative Commons Attribution 3.0 Unported License.

is the only one that does not travel through the thalamus, which has intriguing clinical implications given that the sense of smell is a particularly powerful emotional stimulus (Herz & Schooler, 2002; Shepherd, 2005). This has led clinicians and researchers to begin exploring the use of olfactory virtual reality with certain populations, such as those struggling with posttraumatic stress disorder (Herz, 2021).

Wrapping around the thalamus is a coordinated set of subcortical nuclei known as the *basal ganglia*. This set consists of the caudate nucleus and putamen (together known as the striatum), nucleus accumbens, globus pallidus, substantia nigra, and subthalamic nucleus. These structures play a considerable role in learning and memory, particularly implicit (i.e., outside of awareness) learning of automatized responses (Graybiel, 2000). The basal ganglia also contribute to motor functions, including both inhibiting and motivating movement. Over- or underactivity in this region is seen in motor disorders such as Parkinson's disease (Graybiel, 2000). Disrupted functioning of the basal ganglia area also occurs in addictions, obsessive-compulsive disorder, Tourette's disorder, and even schizophrenia (Graybiel, 2000). The nucleus accumbens, a key region of the basal ganglia, is colloquially known as the "pleasure and reward" center of the brain and is integral to reward, aversion, motivated learning, and translating motivation into goal-directed action (Klawonn & Malenka, 2018; Yang et al., 2018). Closely associated with the basal ganglia and located roughly at the bottom center of the brain is a group of neurons called the *ventral tegmental area*. It is the source of a significant number of dopamine (a neurotransmitter related to pleasure and motivation) projections and is the primary source of dopamine for the prefrontal cortex and nucleus accumbens. As a result, the ventral tegmental area functions in reward and motivation and has thus been implicated in addiction and other mental health disorders (Lammel et al., 2012).

Anterior and ventral (i.e., below, toward the base of the skull) to the thalamus and basal ganglia lies the hypothalamus. This almond-sized structure links the brain with the endocrine (collection of glands that secrete hormones) and nervous systems in the body. It also assists the body in maintaining a state of internal balance or equilibrium known as *homeostasis*; for example, it controls body temperature, food intake, and water intake. The hypothalamus is also responsible for the release of various key hormones in the body, for sexual development and physiology, and for the ability to respond to stress.

Closely connected and adjacent to the hypothalamus, located approximately behind the bridge of the nose, is the pituitary gland. This pea-sized structure is the master gland of the body, both producing and regulating the functioning of numerous hormones. It also controls other glands throughout the body, such as the thyroid and adrenal glands. The pituitary gland is highly active in the production of sex hormones and vital to the body's ability to respond and adapt to stress.

Another central subcortical brain structure is the amygdala. Although the amygdala is sometimes conceptualized as the part of the brain responsible for fear and anger, this understanding is limited and misrepresents the amygdala's role in overall functioning. LeDoux (2013) emphasized that fear and, similarly, anger are cognitive constructs that have no inherent meaning. They are evolutionarily advanced words and concepts, so to say that the amygdala is the source of specific feelings is erroneous. Instead, think of the amygdala as the principal brain region involved in detecting and responding to both innate and learned threats in one's internal or external environment (LeDoux, 2012). The amygdala also plays a role in classical conditioning related to emotional responses and nondeclarative (i.e., implicit, not consciously retrievable) memories, and it can regulate the strength of both declarative and nondeclarative memories (Squire, 2004).

> **Reflective Question**
>
> Which subcortical areas seem related to Rein's symptoms?

Just posterior to the amygdala is a seahorse-shaped structure called the hippocampus. The hippocampus is the area of the brain important for the consolidation (i.e., formation and storage) of long-term declarative (i.e., explicit, consciously retrievable) memories. Explicit memory—what is typically thought of as memory—is memory for facts and events that can be recalled.

Last, stemming from underneath the frontal and prefrontal cortices on either side are the olfactory bulbs. These structures are the sensory organs for the sense of smell. They are the only sensory organs that are directly connected to the subcortical areas of the brain, most notably the limbic system.

Coordinated Systems of the Brain and Body

Several key systems in the brain and body are essential to any discussion of neurophysiology. Each of these systems represents a set of functionally or anatomically interconnected structures that play a crucial role in mental health and healthy functioning in general. The first two systems discussed here are brain-based systems and the remaining three are systems involving the brain and body.

Limbic System

The *limbic system,* a term coined by Paul MacLean in the 1950s, is a set of functionally connected structures first thought to include the cingulate cortex and select subcortical nuclei and considered to be the emotional center of the brain (Nakano, 1998; Rajmohan & Mohandas, 2007). Researchers have since come to understand that the functioning of the limbic system extends well beyond emotion. Not only does the limbic system allow people to respond to emotionally salient cues and threats in their environment, it also plays a role in memory, social processing, sleep, appetite, motivation, addiction, and even sexual behavior (Nakano, 1998; Rajmohan & Mohandas, 2007). Some debate remains about which structures make up this functional system. Subcortical structures such as the amygdala, hippocampus, hypothalamus, and olfactory bulbs have long been accepted to be part of the limbic system. However, the thalamus and basal ganglia are also occasionally included. With regard to cortical structures, the limbic system also includes the surrounding cingulate cortex.

DMN

The next functional system or network in the brain is the DMN, which, as previously discussed in this chapter, serves as the default or resting state of the brain. The brain regions involved in the DMN include the prefrontal cortex, PCC, and regions of the parietal and temporal lobes (Buckner et al., 2008; Raichle et al., 2001). Even when not actively directed toward some task, the brain is not simply inactive. It is in a sort of idle state, ready to engage. For example, the DMN is active during self-generated thought, such as when a person is daydreaming, passively reminiscing, or getting lost in thought. These areas deactivate when one engages in active cognitive tasks. The functioning of the DMN is disrupted in disorders such as attention-deficit/hyperactivity disorder, schizophrenia, autism spectrum disorder, depression, and posttraumatic stress disorder and can be disrupted by substance use as well (Andrews-Hanna et al., 2014).

Autonomic Nervous System

The autonomic nervous system is the part of the peripheral nervous system that innervates the tissues and organs of the body, such as the heart, lungs, stomach, and intestines, over which people have little-to-no conscious awareness. The autonomic nervous system consists of two different divisions, the sympathetic and parasympathetic branches, which act as complementary systems. The sympathetic nervous system functions to engage or excite the organs of the body, including in preparation for protective action. In this way, it activates the fight-or-flight system. The heart starts beating faster, the bronchi in the lungs dilate to allow one to take in more oxygen, the pupils dilate, the stomach and intestines slow down, and the liver makes more glucose. The opposite is true for the parasympathetic nervous system. The heartbeat slows, the bronchi constrict, and the stomach and intestines begin to function more regularly. In short, the parasympathetic nervous system helps people to regulate autonomic arousal and relax. It is also thought to have two branches—namely, the dorsal and ventral branches of the vagus nerve (Porges, 2011). According to Stephen Porges's (2011) polyvagal theory, the dorsal vagal side of the vagus nerve is associated with the "freeze" response to threat and the ventral vagal side of the vagus nerve is associated more with relaxed communication and social engagement.

Hypothalamic-Pituitary-Adrenal and Sympathetic-Adrenal-Medullary Axes

The hypothalamic-pituitary-adrenal (HPA) axis is the functional connection between three endocrine glands. This axis, along with the sympathetic-adrenal-medullary (SAM) axis, allows people to adapt to both emotional and physical stress. The hypothalamus releases corticotropin-releasing hormone to the pituitary gland. This hormone stimulates the pituitary gland to release adrenocorticotropic hormone (ACTH). ACTH then travels down to the adrenal glands, which sit on top of the kidneys. The ACTH activates the adrenal cortex (which is outside of the adrenal glands and not to be confused with the cortex in the brain) to release glucocorticoids, in particular cortisol. The function of cortisol is to help restore homeostasis after stress, and it is essential to life. It also helps to control the reaction to stress by initiating what is called a *negative feedback loop* that turns off the production of ACTH. However, the HPA axis is not the only functional system involved in the stress response. It is one of two systems governing one's response to stressors. As part of the SAM axis, or the sympathetic adrenomedullary system, it serves to activate the sympathetic nervous system, thereby leading to the release of epinephrine/adrenaline and norepinephrine/noradrenaline in hormonal form from the adrenal gland. The role of the SAM axis, and what happens when the feedback process of the HPA axis breaks down, is further explained in Chapter 4 on traumatic stress.

Microbiota-Gut-Brain Axis

Did you know that microbes in your gut can have a direct impact on your mental health? A growing body of research has substantiated the direct connection between the gut and the brain, a connection termed the *microbiota-gut-brain axis*. The gastrointestinal tract has its own nervous system, called the *enteric nervous system,* which contains as many as 600 million neurons. The gut also has the highest concentration of immune cells in the body and is filled with trillions of bacteria, fungi, and viruses (Forsythe et al., 2016). It is controlled in part by the vagus nerve, which is one of the mechanisms of the direct bidirectional connection between the gut and the brain (Cryan & Dinan, 2012). Not only has the microbiota-gut-brain axis been linked to memory, pain, concentrations of brain-derived neurotrophic factor helpful in the development of brain cells and intercellular connections, and immune functioning, but recent research has linked aberrations in gut microbiota with disorders such as depression, anxiety, stress responses, and posttraumatic stress disorder (Cryan & Dinan, 2012; Cryan & O'Mahony, 2011; Forsythe et al., 2016; Leclercq et al., 2016).

Psychoneuroendocrinology and Immunology

As technology expands, so too does the potential to discover intricate multidirectional connections between the brain and other systems of the body. In addition to connections with the gastrointestinal tract, researchers have long illuminated the connections between the brain and the immune and endocrine systems and are beginning to better understand how the related functioning of these systems impacts mental health. This body of research and practice—known as psychoneuroendocrinology, psychoneuroimmunology, or psychoneuroendocrineimmunology—has been rapidly expanding (A. G. Bottaccioli et al., 2019; F. Bottaccioli et al., 2021; Fioranelli et al., 2018; Raony et al., 2020).

Having learned about the HPA and SAM axes earlier in this chapter, you can already see how the brain influences and is influenced by our hormones; however, this influence extends well beyond that of stress response and allostasis (i.e., the ongoing efforts of the body and brain to achieve homeostasis or balance). Thyroid hormones, for example, are essential to brain development, the production and growth of neurons, cell differentiation, the development of new connections between neurons, and so on. We also see the influence of thyroid hormones in mental health struggles such as depression (Ritchie & Yeap, 2015), anxiety (Weiner, 2019), bipolar disorder (Seshadri et al., 2022), schizophrenia (Misiak et al., 2021), and even the pathophysiology (i.e., disrupted processes associated with a condition) of autism (Wilson et al., 2021) and dementia (Chaker et al., 2016). In conditions such as postpartum depression, the interplay is between thyroid hormones and gonadal hormones, such as estrogen and progesterone, and mental health (Li et al., 2019). Gonadal hormones influence mental and emotional well-being, including mood, suicidality, adaptation and reactivity to stress, and substance use and abuse across all sexes, in multiple ways. For example, anabolic steroids (e.g., testosterone) are a commonly used and efficacious pharmacological treatment for adult males presenting with depressive symptoms (Walther et al., 2019).

Further still, researchers have explored the intricate relationship between the immune system and the neurological, endocrine, and gastrointestinal systems. The immune system comprises mechanisms both innate (i.e., nonspecific processes with which individuals are born, such as skin, enzymes in sweat, mucus membranes, inflammatory processes, and cells such as neutrophils) and acquired (i.e., specific immunity individuals build through exposure, by way of B cells and T cells). A comprehensive explanation of the immune system is well beyond the scope of this chapter; for an overview, see L. B. Nicholson (2016) and Parkin and Cohen (2001). The functioning of both the innate and adaptive immune systems is influenced by both internal (e.g., neurotransmitters, gut microbiome, hormones) and external (e.g., acute and chronic stress) factors and can influence numerous physiological processes (El Aidy et al., 2014; Pondeljak & Lugović-Mihić, 2020). Researchers have also substantiated the connection between immune functioning and conditions such as depression (Beurel et al., 2020; Dantzer et al., 2008), anxiety (Peirce & Alviña, 2019), and schizophrenia (Khandaker et al., 2015), to name a few.

Despite ongoing attempts to silo and reduce various body systems for ease of research and explanation, it has become overwhelmingly clear just how complex and interconnected every system of the body is. This knowledge is beginning to translate into more integrated approaches to health and well-being, both physical and mental. Although learning may seem daunting, it is essential to remember that this is an ever-evolving body of knowledge and that your learning will be ongoing.

> **Reflective Question**
>
> Which systems seem related to Rein's symptoms?

How the Brain Communicates

You may be asking yourself how all of these various parts of the brain and body communicate with one another. For example, how does the prefrontal cortex help to regulate the amygdala and associated limbic regions of the brain? How does the thalamus transmit messages from one area to another?

Neurons

Communication in the brain is carried out by electrochemical messages transmitted between specialized cells called *neurons.* In addition to neurons, neuroglia or glial cells are another type of cell that exists in the nervous system. These cells help neurons by providing insulation, nourishment, repair, and structural support. They also remove waste from the nervous system.

Neurons are cell bodies that have extensions on either side that allow them to connect to neighboring neurons. They are composed of four primary parts—namely, the cell body (soma), dendrites, the axon, and the presynaptic or axon terminal (see Figure 1.3). The soma or cell body is the core part of the neuron, containing the nucleus and other cellular structures, such as mitochondria used in energy production. Connected to the cell body are the dendrites, which are branches extending out from the cell body. These branches bring chemical messages or information into the cell body from neighboring neurons. Extending from the other side of the neuron is the axon, a long hairlike projection that carries chemical messages away from the cell body to neurons, glands, or muscles on the other side. Some axons are covered in a myelin sheath through a process called *myelination.* This fatty sheath, made out of fatty glial cells, helps electrochemical messages be transmitted faster down the axon. Rapid transmission is incredibly important to the functioning of the nervous system in general, such as when messages must be carried long distances or transmitted very quickly. Disruptions and deterioration in myelin are a central feature of the autoimmune disorder multiple sclerosis (Steinman, 1996). This impairment can lead to neurological symptoms such as vision loss, pain, involuntary muscle movements, muscle paralysis, tingling and burning sensations, and impaired coordination.

The axon terminal is the area at the end of the axon where the axon branches and extends to come in close proximity to the neighboring structure (e.g., other neurons, muscles, glands). This is also the area where the chemical messages are released. Finally, the small space (tens of nanometers in width) between two neurons is called the *synaptic cleft.* The axon terminal, the synaptic cleft, and the dendrites comprise what is called a *synapse.* There are nearly 1 billion neurons in the adult brain, and each one of these neurons connects to 10,000 others through these synapses. This means that adults have more than 100,000 trillion synaptic connections (Kolb & Gibb, 2011).

Often the terms *gray matter* and *white matter* are used to describe certain visually distinct areas of the brain. For example, if you look at a structural image of the brain generated by an MRI machine, you will see areas of the brain that distinctly look gray and some areas that distinctly look white. The gray areas are the neuronal cell bodies and the white areas are the axons—in particular, myelinated axons. The myelin gives the white matter its characteristic white color. Keep in mind that if you were to look directly at healthy brain tissue, all of it would be pink, and you would be unable to see this delineation.

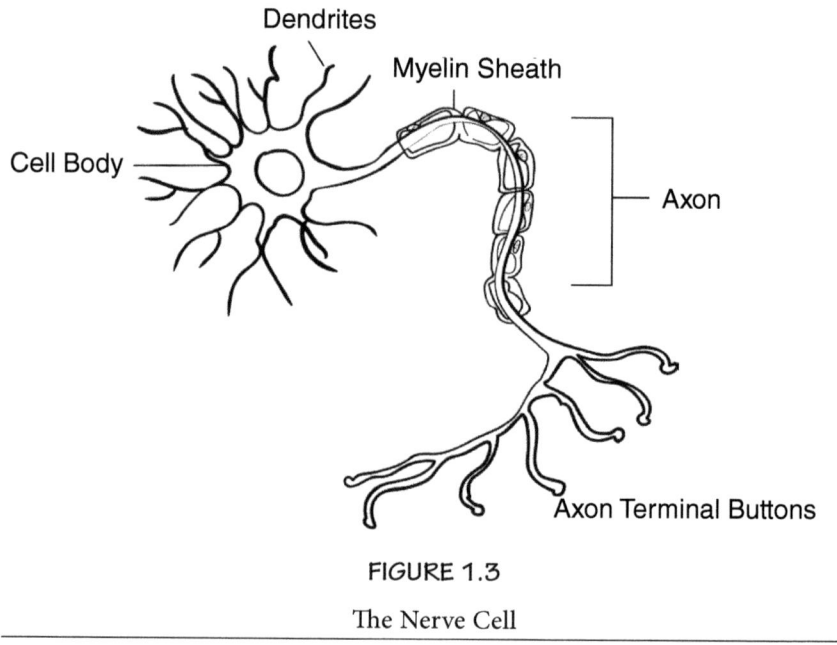

FIGURE 1.3

The Nerve Cell

Note. From National Institute on Drug Abuse, 2005. In Wikimedia Commons.

Neurotransmission and Action Potentials

So how do neurons communicate with each other? *Neurotransmission* is the communication of electrical or chemical signals from one neuron to another. During neurotransmission, a molecule, such as a hormone or neurotransmitter, binds to receptors on a cell's surface and relays external information into and along the neuron and then out to neighboring neurons on the other side. Neurons are said to "fire" when they begin transmitting these electrochemical messages via an action potential. This process starts when an external agent—for example, a neurotransmitter—approaches the dendrites of a neuron. These electrically charged chemicals bind to receptors on the dendrites and change the configuration of the ion channels, which control the flow of ions through the membrane. This binding, depending on the combination and concentration of the chemical messengers, can initiate an action potential in the postsynaptic neuron. The transmission of this action potential down the axon leads to the opening of calcium channels. The calcium ions then cause the release of neurotransmitters into the synaptic cleft at the end of the axon. The neurotransmitters that are released into the synaptic cleft and do not bind to the postsynaptic neuron are then either metabolized by enzymes or recycled back into the presynaptic neuron through a process known as *reuptake*. Psychotropic drugs, such as serotonin selective reuptake inhibitors, work by influencing this process of communication between neurons. This is just one example of how psychotropic drugs can influence the communication between neurons. For more on psychotropic drugs and neurotransmission, see Chapter 6.

Neurotransmitters and Hormones

Neurotransmitters are the chemical messengers that relay information across a synapse between two neurons. There are two basic categories of neurotransmitters: excitatory and inhibitory. Excitatory neurotransmitters enhance or increase the functioning of postsynaptic neurons, whereas inhibitory neurotransmitters inhibit or downregulate (decrease) the functioning of the postsynaptic neuron. Basically, they alter the likelihood of whether or not the postsynaptic neuron will fire or create an action potential. Some neurotransmitters can actually have both excitatory and inhibitory properties. Neurotransmitters act quickly across very short distances between neurons. Hormones, however, secreted by the endocrine system into the bloodstream, can travel longer distances to act on target cells farther away from the endocrine gland. Interestingly, some chemicals can function as both. Some key hormones and neurotransmitters are presented in Table 1.1.

> **Reflective Question**
>
> Which neurotransmitters and hormones seem related to Rein's symptoms?

TABLE 1.1

Key Neurotransmitters, Peptides, and Hormones and Their Related Functioning

Name	Abbreviations/ Other Names	Structure	Type	Function
Acetylcholine	Ach	Amine neurotransmitter	Mostly excitatory	Often excitatory with muscles; involved in muscle movement; essential in memory; also important in attention, learning, neuroplasticity, arousal, and reward
Dopamine	DA	Monoamine/ catecholamine neurotransmitter	Both inhibitory and excitatory	Motivation, reward, addiction; feelings of pleasure, and voluntary muscle movement
Epinephrine	Epi; adrenaline	Monoamine/ catecholamine; neurotransmitter; hormone	Excitatory	Involved in fight-or-flight responses to increase heart rate and blood flow and enhance awareness
Gamma-aminobutyric acid	GABA	Amino acid neurotransmitter	Inhibitory	Primary inhibitory neurotransmitter; reduces neuronal excitability; helps to reduce anxiety. As an interesting side note, it is primarily excitatory in the developing brain.
Glucocorticoids	Cortisol (one example)	Steroid hormone		Essential for life, regulates stress, assists in homeostasis and immune functioning
Glutamate		Amino acid neurotransmitter	Excitatory	Primary excitatory neurotransmitter; essential in learning and memory; important in long-term potentiation (action potential) and long-term depression
Norepinephrine	NE; noradrenaline	Monoamine/ catecholamine neurotransmitter; hormone	Mostly excitatory	Prepares the brain and body to act; fight or flight; increases restlessness and anxiety; enhances attention, vigilance, and arousal; enhances memory formation and retrieval
Oxytocin		Neuropeptide; hormone		Social recognition, bonding and attachment, trust, and reproductive behaviors
Serotonin	5-HT	Monoamine neurotransmitter	Inhibitory	Mood, sleep cycles, appetite, and temperature

Neuroplasticity

The communication process just described can be modified by internal and external experiences. Neuroplasticity is the ability of the brain to alter its structure and function in response to development, learning, memory, brain injury, and disease (Tardif et al., 2016). It occurs in response to sustained changes in the pattern of neural activity, which in turn alter one's neural connections and thus how one responds in both adaptive and maladaptive ways (Lillard & Erisir, 2011).

These new patterns of firing can result from several different sources. One can develop new neurons (i.e., *neurogenesis*), develop new connections between neurons (i.e., *synaptogenesis*), enhance or strengthen the connections between neurons, weaken the connections between neurons, or altogether prune connections between neurons (i.e., *extinction*). *Long-term potentiation* (LTP) refers to stable increases in neuronal sensitivity to stimulation and thus enhancement of the synaptic connection (Bliss & Cooke, 2011; Collingridge et al., 2010). Brain-derived neurotrophic factor is one class of proteins in the brain that is integral to neuronal and synaptic development and thus the process of LTP. As a complementary process, *long-term depression* (LTD) is a reduction in synaptic transmission or a decrease in neuronal sensitivity to a repeated stimulus (Bliss & Cooke, 2011; Collingridge et al., 2010). LTP and LTD have been hypothesized to be the basis of learning and memory. In fact, a study by Nabavi et al. (2014) demonstrated that memories for fear could be inactivated or reactivated by altering the processes of LTD and LTP, respectively.

Neurogenesis, or the birth of new neurons, was historically thought to occur only during brain development and at later critical periods, such as during adolescence, and until age 20 years. In 1998, Eriksson et al. found for the first time that neurogenesis can occur in adult humans, particularly in the olfactory bulbs (smell sensory organs) and the dentate gyrus of the hippocampus (Lillard & Erisir, 2011; Lledo et al., 2006; Snyder, 2019). Impairments in neuroplasticity are thought to underlie numerous mental health disorders, such as depression, bipolar disorder, autism, schizophrenia, and drug addiction (Collingridge et al., 2010), as well as degenerative diseases such as Alzheimer's (Babcock et al., 2021) and diseases of the central nervous system such as Parkinson's (Wakhloo et al., 2022). The brain is always adapting and changing to the world around you, as well as to your internal world. This means that change is always possible. Counselors can use this understanding of neuroplasticity to empower their clients and enhance their clinical practice.

How the Brain Evolved Over Time

Understanding how and why the brain functions as it does requires examination of how the brain evolved over time. The core functional components of the brain, meaning the proteins that produce the electrical communication signals controlling behavior and thus survival, date back hundreds of millions of years to single-cell eukaryotes called *choanoflagellates* (Burkhardt et al., 2011). Choanoflagellates existed before animals and are considered animals' closest ancestor. Their basic protein structure may be the raw material that gave rise to the more complex protein structures that function in the release of neurotransmitters in animals. Some of the first mammal and mammal-like creatures that existed nearly 200 million years ago seem to have had a prominent olfactory system, the system that governs the sense of smell. This growth in the olfactory region of the brain, followed closely by regions that control tactical sensations (i.e., touch) and muscular coordination or motor control, seems to have driven the development of the mammalian brain in terms of size and specificity (Rowe et al., 2011).

> **Reflective Question**
>
> Imagine you are sitting in session with Rein. In your own words, talk about neuroplasticity in a manner that empowers Rein.

During evolution, the brain and the nervous system seem to have become more refined from the inside out, or from the brain stem to the cortex. Research examining the development of the vertebrate brain has suggested that nearly all vertebrate species have the same number of brain divisions (Northcutt, 2002). The divisions constitute three core areas: the rhombencephalon or *hindbrain* (myelencephalon and metencephalon; brain stem), the mesencephalon or *midbrain*, and the prosencephalon or *forebrain* (telencephalon and diencephalon). However, both the size of the brain compared with the body and the size of the various divisions in relation to one another differ among the different classes of vertebrates. For example, in fish and amphibians the hindbrain, which includes the medulla oblongata, pons, and cerebellum, is much larger in comparison with the other divisions. Mammals, however, have a much larger cerebral cortex (telencephalon), which is also the largest division within their brains, than do other vertebrates. For vertebrates, the first brain region to become specialized (rhombencephalon) consists of the areas that keep them alive—namely, the spinal cord and brain regions closer to the spinal cord that govern survival functions such as breathing, heart rate, and sleep. The last to specialize (the telencephalon of the forebrain) controls such processes as advanced learning and memory, sensory perception, decision-making, speech, empathy, and social functioning. The size and complexity of the prefrontal cortex in particular appears to be what most notably differentiates humans from any other organism (Johnson et al., 2009).

Figure 1.4 depicts Siegel's (2010) hand model of the human brain, which can be used with clients. Although this model is not comprehensive, it is useful for loosely demonstrating to others how the brain was formed evolutionarily and developmentally. In reading about neuroscience, you likely will see numerous references to Paul MacLean's (1964) triune theory of brain development. However, considerable research has countered this account of cerebral development (Cesario et al., 2020). Thus, although Siegel's model is useful in depicting how our survival instincts may supersede our cognitive processing in certain situations, it should not be used as a precise depiction of brain development.

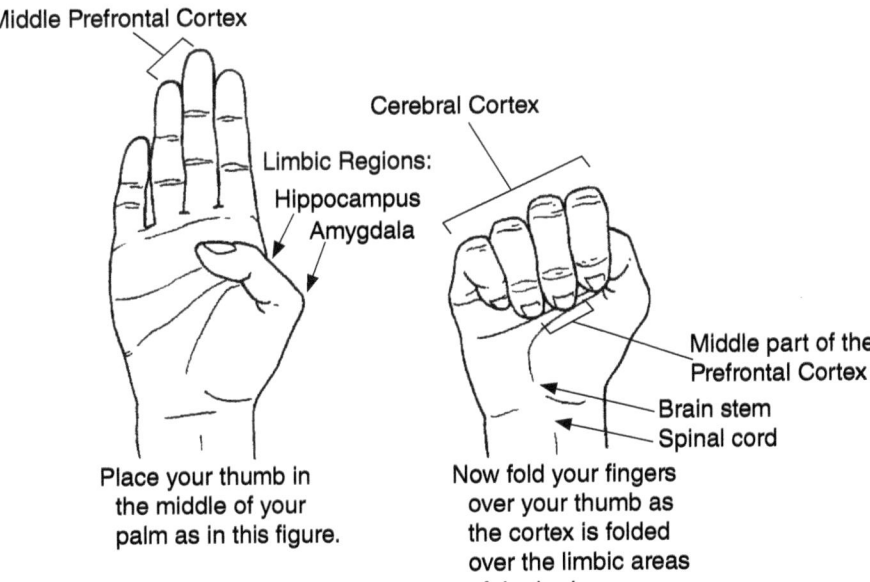

FIGURE 1.4

The Brain in the Palm of the Hand

Note. The hand model depicts the major regions of the brain: cerebral cortex in the fingers, limbic area in the thumb, and brain stem in the palm. The spinal cord is represented by the wrist. From Daniel J. Siegel, MD, *Mindsight: The New Science of Personal Transformation*, copyright © 2010 by Mind Your Brain, Inc.

Polyvagal Theory

Porges (2011) asserted that evolution has led to a functional neural organization of the brain that regulates autonomic states to best support social behavior. This burgeoning theory, termed the *polyvagal theory*, proposes to describe the development and function of two branches of the parasympathetic nervous system, as mentioned earlier. Porges contended that when in a balanced state of autonomic arousal, relaxed and engaged, an individual functions from the social engagement system, known as the *ventral vagal system*. This system promotes behaviors that enhance social bonds by regulating the ready control of muscles that function in eye gaze, facial expressions, voice tone, social gestures, and even the ability to extract the human voice from background noises. Porges also suggested that people have a physiological sensation of safety only when they are functioning within this zone of ventral vagal autonomic responding (Geller & Porges, 2014). This assertion could have considerable implications not only for social interactions but also for the counseling relationship. As a result, references to the polyvagal theory have become commonplace in counseling practice, particularly when working with trauma. However, mounting research is challenging some of the proposed developmental underpinnings of the polyvagal theory (Grossman & Taylor, 2007; Neuhuber & Berthoud, 2022; Taylor et al., 2022). The polyvagal theory provides a general, digestible, and visual description of autonomic functioning in response to stress and threats to safety, but the foundation of this theory—specifically concerning a ventral vagal complex—may not stand up to continued scientific scrutiny (Neuhuber & Berthoud, 2022). It will be important to follow developments in this area of research.

My Brain-Based Approach to the Case of Rein

When working with Rein, I would consider what is going on just below the surface in her brain and body. For example, Rein's feelings of loneliness and withdrawal from social supports immediately make me consider the functioning of her autonomic nervous system. I might talk with her about feelings of safety and markers of autonomic reactivity (e.g., rapid heart rate or breathing, feelings of detachment). I would also speak with her about the difference between emotions and feelings and help her to talk through some of what she has been experiencing in her body (i.e., emotions) and what language she is using to understand those sensations (i.e., feelings). In turn, I would discuss how this interpretation of her sensations is affecting her. I would then intentionally blend aspects of gestalt and narrative approaches to help Rein gain further insight into her emotional and interpersonal experiences. In addition, I would also want to better understand her stage of brain development, which is explored further in Chapter 2.

Understanding brain development during adolescence will provide considerably more insight into how best to work with Rein. Further still, given reports of lethargy and the onset of menstruation, I am aware that disruptions in endocrine functioning could be a possibility. Thus, I would want to make sure that Rein had been recently seen by a primary care provider and there were no acute or developmental physical health concerns. I would also ensure, from an affirming and empowering standpoint, that Rein had a sufficient understanding of the ways that the menstrual cycle can, for some individuals, impact thoughts, feelings, behaviors, social initiation and receptivity, and overall wellness.

Conclusion

To better understand the process of integrating neuroscience into professional counseling, you must first understand the physiology of the brain and body. This foundational knowledge in neuroscience serves as a metatheory of counseling because the functioning of the brain and body underlies everything counselors do. Understanding how brain structures, systems, and related connections function and change in response to internal and external events helps counselors understand symptom development and how clients can change over time. The foundational knowledge you have gained from this chapter will help you to better understand how the brain and related systems change over the life cycle. This chapter covered a vast amount of important terminology and concepts, and I recommend ensuring you understand those terms before proceeding to subsequent chapters.

Quiz

1. Which of the following lobes of the brain is in charge of executive functioning?
    a. Frontal.
    b. Parietal.
    c. Temporal.
    d. Occipital.
2. Which functional system of the brain is primarily known for helping people to respond to emotionally salient cues and threats in their environment but also plays a role in memory, social processing, motivation, addiction, and sexual behavior?
    a. Default mode network.
    b. HPA axis.
    c. Limbic system.
    d. Autonomic nervous system.

3. The study of the interconnectedness of various systems of the body is called:
    a. Craniospinalgastroenterology.
    b. Integrated care.
    c. Psychoneuroendocrineimmunology.
    d. Thalamoneuralendocrineimmunology.

References

Andrews-Hanna, J. R., Smallwood, J., & Spreng, R. N. (2014). The default network and self-generated thought: Component processes, dynamic control, and clinical relevance. *Annals of the New York Academy of Sciences, 1316*(1), 29–52.

Anisman, H., Hayley, S., & Kusnecov, A. W. (2018). *The immune system and mental health*. Academic Press.

Babcock, K. R., Page, J. S., Fallon, J. R., & Webb, A. E. (2021). Adult hippocampal neurogenesis in aging and Alzheimer's disease. *Stem Cell Reports, 16*(4), 681–693.

Beurel, E., Toups, M., & Nemeroff, C. B. (2020). The bidirectional relationship of depression and inflammation: Double trouble. *Neuron, 107*(2), 234–256.

Bliss, T. V., & Cooke, S. F. (2011). Long-term potentiation and long-term depression: A clinical perspective. *Clinics, 66*(1), 3–17.

Bottaccioli, A. G., Bottaccioli, F., & Minelli, A. (2019). Stress and the psyche–brain–immune network in psychiatric diseases based on psychoneuroendocrineimmunology: A concise review. *Annals of the New York Academy of Sciences, 1437*(1), 31–42.

Bottaccioli, F., Bottaccioli, A. G., Marzola, E., Longo, P., Minelli, A., & Abbate-Daga, G. (2021). Nutrition, exercise, and stress management for treatment and prevention of psychiatric disorders. A narrative review psychoneuroendocrineimmunology-based. *Endocrines, 2*(3), 226–240.

Brown, L. C., Murphy, A. R., Lalonde, C. S., Subhedar, P. D., Miller, A. H., & Stevens, J. S. (2020). Posttraumatic stress disorder and breast cancer: Risk factors and the role of inflammation and endocrine function. *Cancer, 126*(14), 3181–3191. https://doi.org/10.1002/cncr.32934

Buckner, R. L., Andrews-Hanna, J. R., & Schacter, D. L. (2008). The brain's default network: Anatomy, function, and relevance to disease. *Annals of the New York Academy of Sciences, 1124*(1), 1–38.

Burkhardt, P., Stegmann, C. M., Cooper, B., Kloepper, T. H., Imig, C., Varoqueaux, F., Wahl, M. C., & Fasshauer, D. (2011). Primordial neurosecretory apparatus identified in the choanoflagellate *Monosiga brevicollis*. *Proceedings of the National Academy of Sciences, USA, 108*(37), 15264–15269.

Carter, H. V. (Illus.). (1918). *Principal fissures and lobes of the cerebrum viewed laterally* [Digital image]. In W. H. Lewis (Ed.), *Anatomy of the human body* (20th ed., Fig. 728). Lea & Febiger. https://commons.wikimedia.org/wiki/File:Gray728.svg

Cesario, J., Johnson, D. J., & Eisthen, H. L. (2020). Your brain is not an onion with a tiny reptile inside. *Current Directions in Psychological Science, 29*(3), 255–260.

Chaker, L., Wolters, F. J., Bos, D., Korevaar, T. I., Hofman, A., van der Lugt, A., Koudstaal, P. J., Franco, O. H., Dehghan, A., Vernooij, M. W., Peeters, R. P., & Ikram, M. A. (2016). Thyroid function and the risk of dementia: The Rotterdam Study. *Neurology, 87*(16), 1688–1695.

Collingridge, G. L., Peineau, S., Howland, J. G., & Wang, Y. T. (2010). Long-term depression in the CNS. *Nature Reviews Neuroscience, 11*(7), 459–473.

Council for Accreditation of Counseling and Related Educational Programs. (2024). *2024 CACREP standards.* https://www.cacrep.org/wp-content/uploads/2023/06/2024-Standards-Combined-Version-6.27.23.pdf

Craig, A. (2009). How do you feel—now? The anterior insula and human awareness. *Nature Reviews Neuroscience, 10*(1), 59–70.

Cryan, J. F., & Dinan, T. G. (2012). Mind-altering microorganisms: The impact of the gut microbiota on brain and behaviour. *Nature Reviews Neuroscience, 13*(10), 701–712.

Cryan, J. F., & O'Mahony, S. M. (2011). The microbiome-gut-brain axis: From bowel to behavior. *Neurogastroenterology & Motility, 23*(3), 187–192.

Damasio, A. (2001). Fundamental feelings. *Nature, 413,* 781. https://doi.org/10.1038/35101669

Dantzer, R., O'Connor, J. C., Freund, G. G., Johnson, R. W., & Kelley, K. W. (2008). From inflammation to sickness and depression: When the immune system subjugates the brain. *Nature Reviews Neuroscience, 9*(1), 46–56.

Decety, J., & Lamm, C. (2006). Human empathy through the lens of social neuroscience. *Scientific World Journal, 6,* 1146–1163.

El Aidy, S., Dinan, T. G., & Cryan, J. F. (2014). Immune modulation of the brain-gut-microbe axis. *Frontiers in Microbiology, 5,* Article 146.

Eriksson, P. S., Perfilieva, E., Björk-Eriksson, T., Alborn, A. M., Nordborg, C., Peterson, D. A., & Gage, F. H. (1998). Neurogenesis in the adult human hippocampus. *Nature Medicine, 4*(11), 1313–1317.

Fioranelli, M., Bottaccioli, A. G., Bottaccioli, F., Bianchi, M., Rovesti, M., & Roccia, M. G. (2018). Stress and inflammation in coronary artery disease: A review psychoneuroendocrineimmunology-based. *Frontiers in Immunology, 9,* Article 2031. https://doi.org/10.3389/fimmu.2018.02031

Fischl, B., & Dale, A. M. (2000). Measuring the thickness of the human cerebral cortex from magnetic resonance images. *Proceedings of the National Academy of Sciences, USA, 97*(20), 11050–11055.

Forsythe, P., Kunze, W., & Bienenstock, J. (2016). Moody microbes or fecal phrenology: What do we know about the microbiota-gut-brain axis? *BMC Medicine, 14*(1), Article 58. https://doi.org/10.1186/s12916-016-0604-8

Freud, S. (2014). *On narcissism: An introduction.* Read & Co Great Essays. (Original work published 1914)

Geller, S. M., & Porges, S. W. (2014). Therapeutic presence: Neurophysiological mechanisms mediating feeling safe in therapeutic relationships. *Journal of Psychotherapy Integration, 24*(3), 178–192.

Graybiel, A. M. (2000). The basal ganglia. *Current Biology, 10*(14), R509–R511.

Grossman, P., & Taylor, E. W. (2007). Toward understanding respiratory sinus arrhythmia: Relations to cardiac vagal tone, evolution, and biobehavioral functions. *Biological Psychology, 74*(2), 263–285. https://doi.org/10.1016/j.biopsycho.2005.11.014

Herculano-Houzel, S. (2010). Coordinated scaling of cortical and cerebellar numbers of neurons. *Frontiers in Neuroanatomy, 4,* Article 12. https://doi.org/10.3389/fnana.2010.00012

Herz, R. S. (2021). Olfactory virtual reality: A new frontier in the treatment and prevention of posttraumatic stress disorder. *Brain Sciences, 11*(8), Article 1070. https://doi.org/10.3390/brainsci11081070

Herz, R. S., & Schooler, J. W. (2002). A naturalistic study of autobiographical memories evoked by olfactory and visual cues: Testing the Proustian hypothesis. *American Journal of Psychology, 115*(1), 21–32.

Hou, K., Wu, Z.-X., Chen, X.-Y., Wang, J.-Q., Zhang, D., Xiao, C., Zhu, D., Koya, J. B., Wei, L., Li, J., & Chen, Z.-S. (2022). Microbiota in health and diseases. *Signal Transduction and Targeted Therapy, 7*(1), Article 135. https://doi.org/10.1038/s41392-022-00974-4

Johnson, M. B., Kawasawa, Y. I., Mason, C. E., Krsnik, Ž., Coppola, G., Bogdanović, D., Geschwind, D. H., Mane, S. M., State, M. W., & Šestan, N. (2009). Functional and evolutionary insights into human brain development through global transcriptome analysis. *Neuron, 62*(4), 494–509.

Khandaker, G. M., Cousins, L., Deakin, J., Lennox, B. R., Yolken, R., & Jones, P. B. (2015). Inflammation and immunity in schizophrenia: Implications for pathophysiology and treatment. *The Lancet Psychiatry, 2*(3), 258–270.

Kim, K. R., Ku, J., Lee, J. H., Lee, H., & Jung, Y. C. (2012). Functional and effective connectivity of anterior insula in anorexia nervosa and bulimia nervosa. *Neuroscience Letters, 521*(2), 152–157.

Klawonn, A. M., & Malenka, R. C. (2018). Nucleus accumbens modulation in reward and aversion. *Cold Spring Harbor Symposia on Quantitative Biology, 83,* 119–129. https://doi.org/10.1101/sqb.2018.83.037457

Kolb, B., & Gibb, R. (2011). Brain plasticity and behaviour in the developing brain. *Journal of Canadian Academy of Child and Adolescent Psychiatry, 20*(4), 265–276.

Lammel, S., Lim, B. K., Ran, C., Huang, K. W., Betley, M. J., Tye, K. M., Deisseroth, K., & Malenka, R. C. (2012). Input-specific control of reward and aversion in the ventral tegmental area. *Nature, 491,* 212–217.

Lavin, C., Melis, C., Mikulan, E. P., Gelormini, C., Huepe, D., & Ibanez, A. (2013). The anterior cingulate cortex: An integrative hub for human socially-driven interactions. *Frontiers in Neuroscience, 7,* Article 64.

Leclercq, S., Forsythe, P., & Bienenstock, J. (2016). Posttraumatic stress disorder: Does the gut microbiome hold the key? *Canadian Journal of Psychiatry, 61*(4), 204–213. https://doi.org/10.1177/0706743716635535

LeDoux, J. (2012). Rethinking the emotional brain. *Neuron, 73*(4), 653–676.
LeDoux, J. E. (2013). The slippery slope of fear. *Trends in Cognitive Sciences, 17*(4), 155–156.
Leech, R., & Sharp, D. J. (2014). The role of the posterior cingulate cortex in cognition and disease. *Brain, 137*(1), 12–32.
Li, D., Li, Y., Chen, Y., Li, H., She, Y., Zhang, X., Chen, S., Chen, W., Qiu, G., Huang, H., & Zhang, S. (2019). Neuroprotection of reduced thyroid hormone with increased estrogen and progestogen in postpartum depression. *Bioscience Reports, 39*(9), BSR20182382.
Lillard, A. S., & Erisir, A. (2011). Old dogs learning new tricks: Neuroplasticity beyond the juvenile period. *Developmental Review, 31*(4), 207–239.
Lledo, P. M., Alonso, M., & Grubb, M. S. (2006). Adult neurogenesis and functional plasticity in neuronal circuits. *Nature Reviews Neuroscience, 7*(3), 179–193.
Lu, C., Yang, T., Zhao, H., Zhang, M., Meng, F., Fu, H., Xie, Y., & Xu, H. (2016). Insular cortex is critical for the perception, modulation, and chronification of pain. *Neuroscience Bulletin, 32*(2), 191–201.
Luders, E., Thompson, P. M., & Toga, A. W. (2010). The development of the corpus callosum in the healthy human brain. *Journal of Neuroscience, 30*(33), 10985–10990.
MacLean, P. D. (1964). Man and his animal brains. *Modern Medicine, 32*, 95–106.
Macmillan, M. (2002). *An odd kind of fame: Stories of Phineas Gage*. MIT Press.
Miller, A. H., & Raison, C. L. (2016). The role of inflammation in depression: From evolutionary imperative to modern treatment target. *Nature Reviews Immunology, 16*(1), 22–34.
Misiak, B., Stańczykiewicz, B., Wiśniewski, M., Bartoli, F., Carra, G., Cavaleri, D., Samochowiec, J., Jarosz, K., Rosińczuk, J., & Frydecka, D. (2021). Thyroid hormones in persons with schizophrenia: A systematic review and meta-analysis. *Progress in Neuro-Psychopharmacology and Biological Psychiatry, 111*, Article 110402.
Nabavi, S., Fox, R., Prolux, C. D., Lin, J., Tsien, R. Y., & Malinow, R. (2014). Engineering a memory with LTD and LTP. *Nature, 511*(7509), 348–352. https://doi.org/10.1038/nature13294
Nakano, I. (1998). The limbic system: An outline and brief history of its concept. *Neuropathology, 18*(2), 211–214.
Naqvi, N. H., & Bechara, A. (2010). The insula and drug addiction: An interoceptive view of pleasure, urges, and decision-making. *Brain Structure and Function, 214*(5), 435–450.
National Institutes of Health. (2014). *Brain 2025: A scientific vision.* https://braininitiative.nih.gov/vision/nih-brain-initiative-reports/brain-2025-scientific-vision
Neuhuber, W. L., & Berthoud, H. R. (2022). Functional anatomy of the vagus system: How does the polyvagal theory comply? *Biological Psychology, 174*, Article 108425.

Nicholson, A. A., Rabellino, D., Densmore, M., Frewen, P. A., Steyrl, D., Scharnowski, F., Théberge, J., Neufeld, R. W. J., Schmahl, C., Jetly, R., & Lanius, R. A. (2022). Differential mechanisms of posterior cingulate cortex downregulation and symptom decreases in posttraumatic stress disorder and healthy individuals using real-time fMRI neurofeedback. *Brain and Behavior, 12*(1), e2441.

Nicholson, L. B. (2016). The immune system. *Essays in Biochemistry, 60*(3), 275–301.

Northcutt, R. G. (2002). Understanding vertebrate brain evolution. *Integrative and Comparative Biology, 42*(4), 743–756.

OpenStax College. (2013). *The limbic lobe* [Digital image]. https://commons.wikimedia.org/wiki/File:1511_The_Limbic_Lobe.jpg

Parkin, J., & Cohen, B. (2001). An overview of the immune system. *The Lancet, 357*(9270), 1777–1789.

Peirce, J. M., & Alviña, K. (2019). The role of inflammation and the gut microbiome in depression and anxiety. *Journal of Neuroscience Research, 97*(10), 1223–1241.

Pondeljak, N., & Lugović-Mihić, L. (2020). Stress-induced interaction of skin immune cells, hormones, and neurotransmitters. *Clinical Therapeutics, 42*(5), 757–770.

Porges, S. W. (2011). *The polyvagal theory: Neurophysiological foundations of emotions, attachment, communication, and self-regulation.* Norton.

Porges, S. W., & Furman, S. A. (2011). The early development of the autonomic nervous system provides a neural platform for social behaviour: A polyvagal perspective. *Infant and Child Development, 20*(1), 106–118.

Raichle, M. E., MacLeod, A. M., Snyder, A. Z., Powers, W. J., Gusnard, D. A., & Shulman, G. L. (2001). A default mode of brain function. *Proceedings of the National Academy of Sciences, USA, 98*(2), 676–682.

Rajmohan, V., & Mohandas, E. (2007). The limbic system. *Indian Journal of Psychiatry, 49*(2), 132–139.

Raony, Í., de Figueiredo, C. S., Pandolfo, P., Giestal-de-Araujo, E., Oliveira-Silva Bomfim, P., & Savino, W. (2020). Psycho-neuroendocrine-immune interactions in COVID-19: Potential impacts on mental health. *Frontiers in Immunology, 11*, Article 1170. https://doi.org/10.3389/fimmu.2020.01170

Ritchie, M., & Yeap, B. B. (2015). Thyroid hormone: Influences on mood and cognition in adults. *Maturitas, 81*(2), 266–275. https://doi.org/10.1016/j.maturitas.2015.03.016

Rowe, T. B., Macrini, T. E., & Luo, Z. X. (2011). Fossil evidence on origin of the mammalian brain. *Science, 332*(6032), 955–957.

Seshadri, A., Sundaresh, V., Prokop, L. J., & Singh, B. (2022). Thyroid hormone augmentation for bipolar disorder: A systematic review. *Brain Sciences, 12*(11), Article 1540. https://doi.org/10.3390/brainsci12111540

Shepherd, G. M. (2005). Perception without a thalamus: How does olfaction do it? *Neuron, 46*(2), 166–168.

Siegel, D. J. (2010). *Mindsight: The new science of personal transformation.* Bantam Books.

Skonieczna-Żydecka, K., Marlicz, W., Misera, A., Koulaouzidis, A., & Łoniewski, I. (2018). Microbiome—the missing link in the gut-brain axis: Focus on its role in gastrointestinal and mental health. *Journal of Clinical Medicine, 7*(12), Article 521. https://doi.org/10.3390/jcm7120521

Snyder, J. S. (2019). Recalibrating the relevance of adult neurogenesis. *Trends in Neurosciences, 42*(3), 164–178.

Squire, L. R. (2004). Memory systems of the brain: A brief history and current perspective. *Neurobiology of Learning and Memory, 82*(3), 171–177.

Steinman, L. (1996). Multiple sclerosis: A coordinated immunological attack against myelin in the central nervous system. *Cell, 85*(3), 299–302.

Stevens, F. L., Hurley, R. A., & Taber, K. H. (2011). Anterior cingulate cortex: Unique role in cognition and emotion. *Journal of Neuropsychiatry and Clinical Neurosciences, 23*(2), 121–125.

Strick, P. L., Dum, R. P., & Fiez, J. A. (2009). Cerebellum and nonmotor function. *Annual Review of Neuroscience, 32*, 413–434.

Taber, K. H., Wen, C., Khan, A., & Hurley, R. A. (2004). The limbic thalamus. *Journal of Neuropsychiatry and Clinical Neurosciences, 16*(2), 127–132.

Tardif, C. L., Gauthier, C. J., Steele, C. J., Bazin, P. L., Schäfer, A., Schaefer, A., Turner, R., & Villringer, A. (2016). Advanced MRI techniques to improve our understanding of experience-induced neuroplasticity. *NeuroImage, 131*, 55–72. https://doi.org/10.1016/j.neuroimage.2015.08.047

Taylor, E. W., Wang, T., & Leite, C. A. (2022). An overview of the phylogeny of cardiorespiratory control in vertebrates with some reflections on the "polyvagal theory." *Biological Psychology, 172*, Article 108382.

Turner, B. M., Paradiso, S., Marvel, C. L., Pierson, R., Ponto, L. L. B., Hichwa, R. D., & Robinson, R. G. (2007). The cerebellum and emotional experience. *Neuropsychologia, 45*(6), 1331–1341.

Uddin, L. Q., Nomi, J. S., Hébert-Seropian, B., Ghaziri, J., & Boucher, O. (2017). Structure and function of the human insula. *Journal of Clinical Neurophysiology, 34*(4), 300–306. https://doi.org/10.1097/WNP.0000000000000377

Wakhloo, D., Oberhauser, J., Madira, A., & Mahajani, S. (2022). From cradle to grave: Neurogenesis, neuroregeneration and neurodegeneration in Alzheimer's and Parkinson's diseases. *Neural Regeneration Research, 17*(12), 2606–2614.

Walther, A., Breidenstein, J., & Miller, R. (2019). Association of testosterone treatment with alleviation of depressive symptoms in men: A systematic review and meta-analysis. *JAMA Psychiatry, 76*(1), 31–40.

Weiner, H. (Ed.). (2019). The psychobiology and pathophysiology of anxiety and fear. In A. H. Tuma & J. D. Maser (Eds.), *Anxiety and the anxiety disorders* (pp. 333–354). Routledge.

Wilson, H. A., Creighton, C., Scharfman, H., Choleris, E., & MacLusky, N. J. (2021). Endocrine insights into the pathophysiology of autism spectrum disorder. *The Neuroscientist, 27*(6), 650–667.

Yang, H., De Jong, J. W., Tak, Y., Peck, J., Bateup, H., & Lammel, S. (2018). Nucleus accumbens subnuclei regulate motivated behavior via direct inhibition and disinhibition of VTA dopamine subpopulations. *Neuron, 97*(2), 434–449. https://doi.org/10.1016/j.neuron.2017.12.022

Chapter 2

Neurophysiological Development Across the Life Span

Laura K. Jones

Understanding and appreciating the client's developmental stage is a core tenet of the counseling profession. With the rapid expansion of technology and associated research in neuroscience and neurophysiology, our understanding of development is now being revolutionized. This chapter builds on Chapter 1 (Anatomy and Brain Development) to provide an overview of individual human growth and development with an emphasis on the neurophysiological changes that occur across the life span. This chapter not only discusses the unobstructed development of the brain and body across significant life stages but also touches on difficulties that can arise during these sensitive periods in brain development. The implications of such knowledge for case conceptualization, treatment planning, and the intentional selection of effective clinical interventions are also discussed and illustrated using the case example of Rein from Chapter 1.

2024 CACREP Standards

This chapter addresses 2024 Council for Accreditation of Counseling and Related Educational Programs (CACREP) Standards pertinent to the Foundational Counseling Curriculum (Section 3) area of Lifespan Development (Standard C):

- Theories of learning (Standard C.3.)
- Models of resilience, optimal development, and wellness in individuals and families across the life span (Standard C.7.)

- Biological, neurological, and physiological factors that affect lifespan development, functioning, behavior, resilience, and overall wellness (Standard C.10.)

This chapter also addresses the following Entry-Level Specialized Practice Area (Section 5) standard for Marriage, Couple, and Family Counseling (Standard F):

- Aging and intergenerational influences and related family concerns (Standard F.2.)

Brain Development Over the Life Span

Each individual brain follows a sequence similar to the evolutionary course of development discussed in Chapter 1. As with evolution, the individual brain develops from the inside out, or from the brain stem to the prefrontal cortex. This process of brain development is slow and gradual. Although the base structure of the brain is formed prenatally, starting during the 3rd week of gestation, the brain continues to become more refined, in both structure and function, into the late teens and early 20s. Let us look at this process in detail.

Prenatal Brain Development

First Trimester

The brain starts to develop during the 3rd week of gestation, with such development and changes lasting virtually across the life span (Stiles & Jernigan, 2010). From approximately Day 17 to Day 20, the embryo develops a *neural plate*, or the structure that will eventually become the nervous system, including the brain and spinal cord. Then during the 3rd gestational week, a groove begins to form on the neural plate. This is the neural groove, which will later become the brain. Soon thereafter, the two sides of the groove begin to curl, folding in on themselves and becoming a tubelike structure known as the *neural tube*. Neural stem cells line the neural tube and eventually give rise to the many specialized cells in the nervous system, including neurons and glia. The cranial nerves begin to develop around Week 5 or 6 postconception. Soon thereafter, at around 50 days of gestation, begins *neurogenesis*. At its peak, neurons develop at the astonishing rate of 250,000 new neurons per minute (Cowan, 1979). This equates to more than 4,000 new neurons per second. This process of neurogenesis lasts until around the 5th month of gestation.

Researchers are trying to understand what drives stem cells to become neurons rather than glia or vice versa. Proteins called neurotrophic factors constitute one such mechanism for differentiation.

Neurotrophic factors support the growth and survival of neurons. Brain-derived neurotrophic factor (BDNF) is one class of neurotrophic factor proteins that specifically supports the growth, survival, and differentiation of neurons in the central and peripheral nervous systems. BDNF also supports the development and strength of synapses between neurons. Events such as sustained or traumatic stress, head trauma, drug use, hypoglycemia, and even the microbes in the gut can influence the activity of BDNF, which may have mental health implications (Pitts et al., 2016; Tapia-Arancibia et al., 2004; Zhang et al., 2006).

At approximately 60 days postconception, the presence of sex hormones, notably androgens (i.e., testosterone) or the absence thereof, initiates sexual differentiation of both the body and the brain. Exposure to sex hormones during development can also affect the functioning of the brain postnatally and throughout life. The importance of sex hormones in the brain and associated development is further discussed later in the chapter.

Second Trimester

By approximately the start of the second trimester, at 3 months postconception, the telencephalon, or cerebral cortex, becomes the largest structure of the prenatal brain. The cerebellum follows a similar process of regionalized development beginning in the second trimester (Liu et al., 2011). Neurogenesis starts to slow by the 5th month of gestation but continues until early postnatal stages (Urbán & Guillemot, 2013). Neurogenesis was historically thought not to extend beyond this period of early development, meaning that the neurons that develop during prenatal and early postnatal development would constitute all of the neurons that one would have in life. However, groundbreaking research in the late 1990s revealed that the human brain continues to produce new neurons throughout adulthood (Eriksson et al., 1998).

Looking beyond the brain, researchers have debated whether microbiota begin colonizing the intestinal tract in utero—specifically, during the second trimester (Kennedy et al., 2023; Li et al., 2020). Although this line of research is still nascent, if development of the microbiome does indeed begin in utero, it could challenge the long-standing belief that the womb is a sterile environment. Furthermore, if this is the case, researchers likely will begin to explore whether and how this burgeoning microbiome may influence brain development in utero, or whether the influence of the microbiome begins neonatally (Gars et al., 2021).

Third Trimester

During the third trimester, synapses begin developing between neurons at a rate of approximately 40,000 per minute (Bourgeois, 1997). This also means that the brain is increasing in size, almost tripling in volume, during this last prenatal period. Consequently, during the 7th month of development, the gyri (folds) and sulci (grooves) of the cerebral cortex, which give it its characteristic wrinkled look, begin to emerge. Myelin begins to form around axons at about the 7th month of gestation and continues until around 9 months postnatally, with this process generally moving from the back to the front of the brain. The frontal region of the brain is the last to become myelinated. This means that basic survival functions (e.g., crying for food) precede rational thinking, planning, decision-making, and emotion regulation. The cerebellum also increases dramatically in surface area. Given this rapid and expansive development of the brain during the third trimester, this period is also the most vulnerable to injury by both internal and external factors.

Brain Development During Infancy and Childhood

The prenatal period is only the first step in brain development, which is a long and extensive process that lasts through late adolescence and into early adulthood (Stiles & Jernigan, 2010). Development progresses from the bottom to the top and from the back to the front, meaning that the prefrontal cortex, maturing in the early to mid-20s, is one of the last areas to develop. However, it has been argued that the architecture of the brain continues to change throughout life and that the last areas to develop are some of the first to experience decline with increasing age. These developmental stages of the brain and the changes occurring during each stage may have marked influences on the counseling process.

After birth, the brain continues to rapidly grow and becomes more tailored in its functioning, increasing fourfold and reaching nearly 90% of adult brain volume by age 6 years (Stiles & Jernigan, 2010). However, this does not always happen in a smooth and gradual process. People experience growth spurts in brain development when their brain rapidly undergoes changes. These sensitive or optimum periods of development occur in the brain but are represented as periods of behavior change, when children (and, later, adolescents) are able to take on new tasks and regulate their emotions and behaviors accordingly. Unlike what are traditionally considered critical periods with definitive end points, sensitive or optimal periods are thought to be dependent on the learning process and are not inherently limited by time (Johnson, 2005; Werker & Tees, 2005).

The first rapid period of development, which starts during the third trimester and encompasses rapid neuronal growth and myelination of axons, continues until roughly the 3rd year of life (Jernigan et al., 2011). Changes to the brain during this time are more clearly seen on structural scans than are changes that occur later in childhood and adulthood (Jernigan et al., 2011). Gray matter increases nearly 150% during the 1st year of life, with the cerebellum growing a whopping

240% (Knickmeyer et al., 2008). By age 3 years, a child has twice as many neuronal connections as he or she does as an adult. This may seem counterintuitive at first, given that the brain and head continue to grow in size well beyond age 3. However, during this early period the brain produces many more neuronal connections than one could ever actually use. This abundant growth in neurons and connections primes the child for rapid learning. Through learning (i.e., both structured learning and experiences), certain connections become reinforced, and others that are not used are pruned. This synaptic pruning, or thinning process, actually represents a fine-tuning of cerebral functioning that occurs in later childhood and through adolescence. It is a process of refining and engaging higher order cortical areas over the more evolutionarily primitive subcortical regions. This process of refinement, which may actually represent an increased myelination of axons, is generally first seen in the primary sensory and sensorimotor cortex and then moves out to multimodal areas (Jernigan et al., 2011; Sowell et al., 2004).

These first few years are considered the most critical to development and can have a marked impact on development later in life (Knickmeyer et al., 2008). To use an analogy, the first few years are similar to constructing the scaffolding on which later learning will build. During the first few years, humans move from primarily using motor reflexes that are key to survival (e.g., crying and the rooting-sucking reflex) to using their senses and vocalizations to communicate (e.g., head turning, cooing, and reaching for objects). They begin to crawl and then walk. They learn how to communicate through a symbolic system, most often language. Humans form interpersonal attachments with a primary caregiver and later learn social cognitive skills useful in cooperative play. These developments are rooted in changes taking place in the brain.

Attachment

Infants' survival and primary well-being are contingent on their attachment to a primary caregiver. Classic attachment theory (Bowlby, 1982) suggests that through these early relationships, people develop internal working models of human bonding that set the stage for later neurological and socioemotional development. These relationships help people to feel safe and accepted (Porges, 2011; Sullivan, 2003). However, finding a neurological system for attachment has been elusive. Eloquently paraphrasing Wittgenstein, Coan (2008) likened trying to find a functional system of attachment to "trying to find the real artichoke by peeling away all its leaves" (p. 242). In other words, attachment is a complex neurological system made up of a delicate balance of myriad intricate parts.

Much of the early research on infant attachment was in the area of imprinting, or the formation of an enduring cognitive representation of and behavioral response to a caregiver or caregiver-like figure soon after birth (Insel & Young, 2001). This process involves a predominance of different sensory systems in varying species. In birds, for example, imprinting is largely facilitated using more visual and auditory cues, whereas mammals (e.g., rats) use more olfactory cues (Insel & Young, 2001). In humans, the neurophysiological process of attachment seems to involve all three sensory systems, olfactory, auditory, and visual. The neuropeptide oxytocin seems to facilitate attachment bonding (Insel & Young, 2001). Oxytocin, developed in the hypothalamus, enhances learning and emotional memory and is thought to be transmitted from mother to child during early infant attachment experiences (Feldman et al., 2010). Strong correlations have been found between salivary (spit) concentrations of oxytocin in rat mothers and offspring both before and after social intervening (e.g., grooming behaviors), with the higher oxytocin levels being associated with higher emotional synchrony and social engagement in infants (Feldman et al., 2010). In addition, the functioning of the amygdala helps humans to understand emotions on faces, and concentrations of both dopamine from the nucleus accumbens and norepinephrine from the locus coeruleus facilitate reward-driven behavior toward attachment figures (Coan, 2008). These attachment relationships support brain development. For example, physical touch and connection appear to attenuate the stress response, because physical touch is a primary release mechanism for oxytocin. In other words, attachment relationships help humans to regulate their autonomic nervous systems and the hypothalamic-pituitary-adrenal (HPA) axis (Coan, 2008; Gunnar & Hostinar, 2015). Secure attachment relationships have also been found to influence adversity-related epigenetics (see the discussion later in this chapter), meaning that having safety and security in one's attachment figures early in life may decrease the degree to which adverse events can lead to deleterious changes in the way one's genes are read and expressed later in life (Provenzi et al., 2020).

Language

The ability to develop a representative form of communication is a remarkable feat and one that unfolds over the first few years of life. At around 6 to 8 months of age, a baby can vocalize by canonical babbling or repeated consonant vowel sounds (Bates & Dick, 2002). As the child reaches toddlerhood, strings of meaningful words are juxtaposed to interact with adults and other children. All children follow a remarkably similar developmental course for language learning (Kuhl & Rivera-Gaxiola, 2008). Children acquire language in part through imitating or mirroring adults in the production of speech. Language development is thought to be assisted by mirror neurons

(Bates & Dick, 2002; Kuhl & Rivera-Gaxiola, 2008). Located in the premotor and primary motor cortex, frontal lobe, Broca's area, and inferior parietal lobe (Kilner & Lemon, 2013), mirror neurons fire in the same way when one is perceiving the action of another as when one is conducting the action oneself. Some researchers consider mirror neurons to also be associated with emotional contagion, a rudimentary form of empathy.

Numerous factors may influence the process of language development. For instance, social interaction is clearly essential for language development. As a result, it is important to understand how context and environment can influence this process, even at the physiological level, as they do with other development processes (e.g., cognitive, motor; Rowe & Weisleder, 2020). Further still, research suggests that steroid hormones (e.g., testosterone, estradiol) early in life may impact various aspects of language development, such as variance in phonemic discrimination, expressive vocabulary, and phonological working memory (Kung et al., 2016; Schaadt et al., 2014). As researchers continue to discover how the systems of the body interact developmentally, we will be able to better understand this very intricate process across the life span.

Memory

The hippocampus and prefrontal cortex do not fully develop until adolescence and early adulthood, respectively. This protracted development has marked implications for children's memory systems. According to Squire (2004), the brain has two primary memory systems. Declarative or *explicit* memories are conscious, language-based memories of events that rely on the hippocampus and frontal lobes. Nondeclarative or *implicit* memories are memories largely outside of conscious awareness (i.e., preconscious), such as conditioned emotional and sensory learning, reflexes, and procedural learning. Implicit memories are formed with the aid of the amygdala and cerebellum (Squire, 2004). Research has suggested that children younger than the age of 18 months have not yet developed their declarative memory system (Fishbane, 2007) and that this system is still being solidified through the first 5 years of life (Siegel, 2006). Counselors should consider these limitations when working with young children.

> **Reflective Questions**
>
> Recall Rein's presenting concerns and broader context.
>
> Given her current age, what milestones has Rein encountered with respect to her brain development?
>
> How might Rein's environment have affected her brain development in childhood?

Brain Development During Adolescence

Spanning from roughly age 11 years to the mid-20s, adolescence represents another notably sensitive period of brain development and is considered by some to be perhaps the most critical period in development (Paus et al., 2008). It is also a period during which the onset of numerous mental health disorders is seen. A number of factors may contribute to this vulnerability and have implications for clinical practice, as summarized by Jones (2015).

During this time, individuals experience volumetric changes in both gray matter and white matter in the brain that represent an increased refinement in brain functioning (Casey et al., 2008). For example, following a bell-shaped curve, the volume of gray matter in the frontal, and particularly the prefrontal, cortex peaks in early adolescence and gradually decreases into late adolescence. Along with such alterations in gray matter, the volume of white matter follows a linear course of development, increasing across adolescence. Such changes in the prefrontal cortex parallel improvements in cognitive abilities and logical reasoning skills.

Emotion Regulation

Along with these changes in cortical structures, concurrent changes are also taking place in the limbic and paralimbic regions of the brain. The amygdala, hippocampus, anterior cingulate cortex, and nucleus accumbens also demonstrate progressive synaptic pruning during adolescence. In general, the limbic structures of the brain mature earlier than the prefrontal cortex (Casey et al., 2008). This asymmetrical development in part explains the emotional lability, self-consciousness, impulsiveness, and risk-taking behavior often seen during adolescence. The limbic, or more emotionally responsive and instinctual, parts of the brain are in overdrive, but the more cognitive, decision-making, and problem-solving areas of the brain have not yet caught up. Consequently, adolescents are not yet able to optimally regulate their limbic regions in emotionally charged situations. Adolescents appear to demonstrate a capacity for self-control and impulse regulation in nonemotional situations similar to that of adults (Casey & Caudle, 2013). As the adolescent matures and the volume of white matter increases between the prefrontal cortex and limbic regions, top-down control of limbic cortices becomes more firmly established, and the individual becomes better able to self-regulate in emotional contexts. In this way, adolescence also presents an opportunity to optimize brain development and functioning, potentially leading to enhanced cognitive, social, and emotional function and overall mental health in adulthood.

The HPA axis similarly goes through another sensitive period of development during adolescence (Gunnar et al., 2009; Gunnar & Hostinar, 2015; Spear, 2000). Adolescents on average experience higher levels of cortisol in their systems in response to both psychological and social stressors. Thus, chronic or acute stress during adolescence may have pronounced effects on the emergence of mental health disorders during adolescence as well as an individual's ability to self-regulate and cope with stress as an adult. During adolescence, the body also experiences an influx of neurotransmitters and hormones. Dopamine levels are particularly high during adolescence, and this increased dopaminergic activity can help to explain why adolescents engage in more sensation-seeking and risk-taking behaviors, including substance use (Sturman & Moghaddam, 2011).

The microbiota-gut-brain axis follows a similar developmental time course as that of the brain. Adolescence represents one of the most dynamic periods of change for gut microbiota (Borre et al., 2014). Such changes in the microbiota-gut-brain axis during adolescence may even influence the development of anxiety (Foster & Neufeld, 2013). Also, Hoban et al. (2016) found that a lack of proper colonization of microbes in the gut could influence the development of the prefrontal cortex through the overproduction of myelin. These findings and the growing interest in the gut-brain axis may have significant implications for the development of mental health disorders, such as schizophrenia and mood disorders, that present during this time.

Social Development

Adolescence is a period of differentiating from parents and developing stronger interpersonal connections with friends. It is also, in many cultures, the onset of developing amorous interests and relationships with potential partners. Blakemore and Mills (2014) emphasized that adolescence is a decisive period in the development of *social cognition*, which is the ability to recognize and understand the mental states of others. Social cognition is governed by a set of brain structures that includes the medial prefrontal cortex, anterior cingulate cortex, inferior frontal gyrus, posterior superior temporal sulcus, temporo-parietal junction, amygdala, and anterior insula (Sebastian et al., 2010). Changes going on in these regions underlie the process of theory of mind, or the ability to attribute mental states to others, and can lead adolescents to be more sensitive to social cues, especially facial expressions (L. A. Thomas et al., 2007). Adolescents can thus become more sensitive to bullying and social ostracization, feeling more emotional effects than adults (Sebastian et al., 2010). K. M. Thomas et al. (2001) examined how adolescents and adults perceive facial expressions differently, examining the brain activity underlying behavioral responses. The adolescents in their study frequently perceived more emotions incorrectly. Their amygdalae were also more active than those of adults, even to neutral faces. This pattern of brain development in adolescence may help to explain why individuals feel increasingly self-conscious during this time.

Just as they start to develop the capacity to grasp the perspective of another (theory of mind), adolescents' emotional response may be exaggerated and more negative than adults'. However, during early and mid-adolescence, they may not yet have the cortical capacity to optimally regulate that emotional response.

The reciprocal importance of social functioning and brain development during this time also leads to questions of what impact a reliance on technology during this time has on the brain. Although research is ongoing in this area and many questions remain to be answered, one study has suggested the potential effects of technology on empathy. Uhls et al. (2014) found that tweens who abstained from television, computer, and smartphone use for 5 days significantly improved their ability to read nonverbal emotion cues compared with peers with technology access.

Influence of Sex Hormones

Nearly every change in the brain that occurs during adolescence is in some manner influenced by the deluge of sex hormones initiated during this period (Goddings et al., 2019; Jones, 2016). This is also when more of the broad sex-related differences in mental health begin to emerge in terms of the prevalence of certain disorders. Cerebral blood flow, HPA axis functioning, and levels of neurotransmitters are all regulated by levels of gonadal steroids, such as estrogen and testosterone. Estrogen not only enhances cerebral blood flow but influences levels of serotonin, dopamine, and oxytocin. Increased levels of estrogen and testosterone both enhance the functioning of the HPA axis and related corticosterone release and negative feedback loops. During the menstrual cycle, levels of gonadal hormones fluctuate. Early evidence has suggested that this may affect susceptibility to addiction, relapse, posttrauma pathology, and even suicidality. For example, Baca-Garcia et al. (2010) found that female suicide attempts increased when levels of estrogen and progesterone were low (early follicular phase), that these attempts were more severe, and that suicidal ideation increased in female adolescents when progesterone levels were low. The influx of hormones furthermore leads to a sex differentiation in the gut microbiome, which again influences mental health (Jašarević et al., 2016). Researchers such as Sisk and Zehr (2005) have suggested that it is this influx of hormones coupled with rapid brain development that increases the risk for psychopathology during adolescence.

Epigenetics: The Influence of Environment

You are likely familiar with the long-standing debate over the influence of nature versus nurture when considering the course of human development. As counselors, we understand that a person's environment (social and physical) has a strong influence on development and overall wellness. Feminist theories (e.g., the person is political; Crethar et al., 2008), multicultural theories (e.g., understanding and appreciating the real differences among racial, ethnic, and cultural groups; Singh et al., 2020; Speight et al., 1991), ecological systems theories (Bronfenbrenner & Ceci, 1994), and theories concerning the social determinants of mental health (Compton & Shim, 2015; Lund et al., 2018; Pester et al., 2023) all describe the role of the environment and social context in one's well-being. As humans, we never exist outside of an environment. Even our prenatal development is in part dictated by our environment within the womb (Raja et al., 2022) and, at times, even by the environment of those whose DNA created our own (Sharma, 2019; Yehuda & Lehrner, 2018). However, we rarely discuss how our environment and social context influence our well-being. How does our environment "get under our skin" so to speak?

Waddington (1942/2012) was the first to introduce the concept of epigenetics as an explanation for how our behaviors and environment can affect the way genes (segments of our DNA) work (Centers for Disease Control and Prevention, n.d.). Technically speaking, *epigenetics* is any process that alters gene activity without changing the DNA sequence itself (Weinhold, 2006). There are many ways in which epigenetic changes can happen, such as through DNA methylation. During this process, a methyl group (one carbon and three hydrogen atoms) attaches to a DNA sequence and, essentially, turns a gene "off," preventing proteins from reading the gene and thus blocking gene expression. The role of epigenetics in our lives is essential. The processes are often very adaptive and advantageous to our well-being, allowing us to better function within our environment. In this way, the process of epigenetics is central to learning and behavioral control and is at the core of how one's environment can lead to changes in physical and mental health across the life span and across generations (Lester et al., 2012; McGowan & Szyf, 2010; Mews et al., 2021; Toyokawa et al., 2012). However, at times those changes can be maladaptive and lead to both physical and mental health challenges. It is important to remember that not all epigenetic changes are permanent, which leaves space not only for challenges but also for resilience. This finding further underscores counseling's vital role in social justice and advocacy work with our clients and within communities.

> **Reflective Questions**
>
> Reflect back on the case of Rein and consider the following questions:
>
> How might adolescent brain development be influencing Rein's symptoms?
>
> What other factors might you want to consider in addition to these influences?

The Aging Brain

According to the United Nations Department of Economic and Social Affairs (2019), 1 in 6 people worldwide will be older than age 65 by 2050. It is estimated that over 50 million individuals are currently living with Alzheimer's disease (Patterson, 2018). Further still, nearly 15% of adults who are age 60 or over experience mental health challenges such as depression, anxiety, and substance use disorders (World Health Organization, 2017). The mental health needs of these older adults are not currently being met (Karel et al., 2012). Various aspects of cognitive functioning, including processing speed, executive functioning, and episodic declarative memory (i.e., conscious memory of events), have also been found to decline with age (Anderson, 2019; Grady, 2012). Other cognitive functions, such as emotion regulation, and crystallized tasks, such as vocabulary, appear to stay intact (Woodhead & Yochim, 2022; Wright & Díaz, 2014). Understanding the neuroscience of aging elucidates how counselors can best support older adult clients.

Gray Matter

As individuals age, they often experience a loss of brain volume, with different brain areas displaying different aging trajectories (Sele et al., 2020). Generally speaking, the last brain areas to develop are often the first to experience age-related declines (Tamnes et al., 2013). The most notable volumetric loss has been reported in the frontal and prefrontal cortices, temporal lobes and hippocampus, thalamus, and nucleus accumbens (Fjell & Walhovd, 2010). Rather than being the result of necrosis (neuronal death), gray matter changes appear to be related to a shrinkage of neurons, reduction of synaptic spines, decrease in the number of synapses, and loss of glial cells (Fjell & Walhovd, 2010).

White Matter

Changes in the volume and integrity of white matter are also seen in the healthy aging brain. After the increase and peak in white matter during adolescence and subsequent volumetric plateau, white matter volume seems to start to decline gradually around age 50 and more steeply after age 70 (Fjell & Walhovd, 2010). The majority of this decline is seen in the frontal lobe, with notable loss also occurring in the corpus callosum (Geerligs et al., 2015; Gunning-Dixon et al., 2009). Decreases in myelinated axons within and across various brain regions may underscore much of this volumetric change, with a reduction in the length of myelinated axons of nearly 50% (Fjell & Walhovd, 2010). This is also thought to be indicative of decreased connectivity between brain regions. For example, functional connectivity of the default mode network declines with age, which may have implications for working memory (Tomasi & Volkow, 2012).

Increased Bilateral Brain Activity

One other interesting phenomenon documented in healthy aging brains is an increase in bilateral activation of the frontal and prefrontal cortices during tasks that used to be unilateral in young adults (Grady, 2012). Researchers have provided several explanations for such seemingly counterintuitive changes. Initially, such increased activation was thought to represent compensatory activity in the brain. In other words, greater activation and more brain regions are needed to perform the same task. Another possible explanation is a concept called *dedifferentiation.* Dedifferentiation can be defined as "a process by which structures . . . that were specialized for a given function lose their specialization and become simplified, less distinct or common to different functions" (Baltes & Lindenberger, 1997, as cited in Sleimen-Malkoun et al., 2014, p. 2). Thus, multiple different tasks can now activate similar brain regions. For example, brain regions that support sensory, motor, and cognitive tasks are more correlated in older adults, as are different types of visual processing and memory functions (Sleimen-Malkoun et al., 2014).

Alzheimer's Disease

Significant research has also examined changes in the brain related to various degenerative disorders, such as dementia and Alzheimer's disease. Alzheimer's disease is the most common form of dementia that progressively affects memory, thought processes, and consequently behavior. The characteristic markers of Alzheimer's disease include reduced brain weight, cortical atrophy, and associated ventricular enlargements in addition to neurofibrillary tangles of tau protein filaments typically occurring in the hippocampus and related limbic structures and amyloid plaques typically found throughout the cortex (Rossini et al., 2007). These changes can be associated with synapse elimination and necrosis. Alzheimer's-related impairments in episodic memory have also been documented as being noticeable more than 6 years before diagnosis was made (Bäckman et al., 2001). Grady (2012) suggested that, given this window, providers should be assessing aging clients for Alzheimer's disease risk factors, such as mild cognitive impairment.

Developmentally Informed Interventions

Understanding brain development over the life span enriches case conceptualization and treatment planning. In clinical work, child and adolescent clients are often grouped together into a single special population. From this review of brain development, it is clear that children and adolescents are two separate groups, each warranting a unique approach. Best practices for working with children, such as play therapy, honor the neurodevelopmental stage of young clients by providing a symbolic framework that does not rely on linguistic production and declarative memory systems. The child does not have to tell a story in words but symbolically represents the feelings of the body and the implicit memory of events in an environment that is safe and secure, fostering relationship and attachment. According to Gaskill and Perry (2014), the somatosensory experiences involved in play can help to create the necessary neurological foundations for "advanced mental skills, such as creativity, abstract thought, prosocial behavior, and expressive language" (p. 180). However, neuroscience-informed best practices for adolescents alone are not as well established.

The importance of assessing for depression in aging clients, in addition to assessing for mild cognitive impairment, was emphasized by Wright and Díaz (2014). Depression coupled with volumetric changes in the hippocampus can speed cognitive decline (Sawyer et al., 2012). *Cognitive decline* refers to the loss of neurons and/or synaptic connections that occurs with age, resulting in challenges with memory loss and attention. Common in older adults, mild cognitive decline is differentiated from neurocognitive disorders. Research has examined the rehabilitative effects of training on the aging brain. Older adults who practiced active tasks related to divided attention, episodic memory, and working memory for 2.5 to 10 hours had cortical activation similar to that of younger adults in their prefrontal, frontal, and temporal cortices (Grady, 2012). Further still, researchers have explored the efficacy of interventions targeting nutrition (Roy et al., 2021), music (Ito et al., 2022), exercise (Sanders et al., 2019), and social engagement (Evans et al., 2019) as a means of stemming the progression of mild cognitive decline.

My Brain-Based Approach to the Case of Rein

Rein's developmental stage of early adolescence is central to understanding her presenting symptoms. During this period, the subcortical and limbic regions of her brain and associated HPA axis are developing at a faster rate than the cortical regions of her brain, such as the prefrontal cortex. When emotional or under stress, she is likely to have limited cognitive control and may respond in an impulsive and emotionally driven manner. Rein is beginning to develop the capacity for theory of mind and mentalizing; thus, she may be more self-conscious about how others view her. Because Rein is just starting to develop these cognitive aspects of empathy, she is likely more sensitive to social cues, particularly facial expressions. She recently initiated menarche, and the considerable influx of gonadal hormones such as estrogen is altering the functioning of neurotransmitters, hormones, and the HPA axis—all of which can have significant implications for her mental health.

In working with Rein, I would first provide psychoeducation about how the changes she is experiencing in her brain and body affect how she is presently feeling. I would explore with Rein the emotions and physical sensations she is experiencing in her body. Verbally labeling an emotional stimulus can assist Rein with emotion regulation by decreasing cerebral blood flow (i.e., activity) in the amygdala and increasing activation of the prefrontal cortex (Hariri et al., 2000). Grounding and breathing exercises, mindfulness practices, biofeedback, and neurofeedback have all been used to enhance optimal brain development among adolescents. Brain-based guidance curriculums are also available for school settings. Last, I would work collaboratively with a primary care provider or nutritionist to develop a nutrition and exercise program. Among a host of other advantages, exercise promotes the release of BDNF from the hippocampus and cortex (Rasmussen et al., 2009).

Conclusion

The principle of working from a developmental perspective represents one of the pillars of the counseling profession. Understanding brain development across the life span enriches the ability to conceptualize client concerns within a developmental context and address clients' developmental needs. This chapter provided a brief overview of notable milestones, sensitive periods, and considerations in the development of the human brain over the life span as related to cognitive, emotional, behavioral, and interpersonal functioning. As research in this area continues to burgeon—especially research examining the nuances of neurophysiological functioning and shifts in adolescent and aging populations—the importance of considering clients' developmental stage will become even more clear, and counselors will be better equipped to support the cognitive, behavioral, emotional, interpersonal, and neurophysiological needs of their clients.

Quiz

1. During which stage of development do synapses begin developing at a rate of approximately 40,000 per minute?
    a. First trimester.
    b. Second trimester.
    c. Third trimester.
    d. First 3 months of postnatal development.

2. During adolescent brain development, which of the following is true?
   a. Subcortical limbic regions of the brain develop before the prefrontal cortical areas.
   b. The HPA axis no longer changes.
   c. Neurotransmitter expression is the lowest than at any other point in development.
   d. Prefrontal cortical areas of the brain develop before subcortical limbic regions.
3. In older adults, which of the following *does not* seem to be impaired by healthy aging?
   a. Memory.
   b. Emotion regulation.
   c. Processing speed.
   d. All of the above.

References

Anderson, N. D. (2019). State of the science on mild cognitive impairment (MCI). *CNS Spectrums, 24*(1), 78–87.

Baca-Garcia, E., Diaz-Sastre, C., Ceverino, A., Perez-Rodriguez, M. M., Navarro-Jimenez, R., Lopez-Castroman, J., Saiz-Ruiz, J., de Leon, J., & Oquendo, M. A. (2010). Suicide attempts among women during low estradiol/low progesterone states. *Journal of Psychiatric Research, 44*(4), 209–214.

Bäckman, L., Small, B. J., & Fratiglioni, L. (2001). Stability of the preclinical episodic memory deficit in Alzheimer's disease. *Brain, 124*(1), 96–102.

Bates, E., & Dick, F. (2002). Language, gesture, and the developing brain. *Developmental Psychobiology, 40*(3), 293–310.

Blakemore, S. J., & Mills, K. L. (2014). Is adolescence a sensitive period for sociocultural processing? *Annual Review of Psychology, 65,* 187–207. https://doi.org/10.1146/annurev-psych-010213-115202

Borre, Y. E., O'Keeffe, G. W., Clarke, G., Stanton, C., Dinan, T. G., & Cryan, J. F. (2014). Microbiota and neurodevelopmental windows: Implications for brain disorders. *Trends in Molecular Medicine, 20*(9), 509–518.

Bourgeois, J. P. (1997). Synaptogenesis, heterochrony and epigenesis in the mammalian neocortex. *Acta Paediatrica, 86*(S422), 27–33.

Bowlby, J. (1982). *Attachment and loss* (Vol. 1). Basic Books.

Bronfenbrenner, U., & Ceci, S. J. (1994). Nature-nurture reconceptualized in developmental perspective: A bioecological model. *Psychological Review, 101*(4), 568–586.

Casey, B. J., & Caudle, K. (2013). The teenage brain: Self control. *Current Directions in Psychological Science, 22*(2), 82–87.

Casey, B. J., Jones, R. M., & Hare, T. A. (2008). The adolescent brain. *Developmental Review, 28*(1), 62–67.

Centers for Disease Control and Prevention. (n.d.). *Genomics and precision health: What is epigenetics?* https://www.cdc.gov/genomics/disease/epigenetics.htm

Coan, J. A. (2008). Toward a neuroscience of attachment. In J. Cassidy & P. R. Shaver (Eds.), *Handbook of attachment: Theory, research, and clinical applications* (2nd ed., pp. 241–265). Guilford Press.

Compton, M. T., & Shim, R. S. (2015). The social determinants of mental health. *Focus, 13*(4), 419–425.

Council for Accreditation of Counseling and Related Educational Programs. (2024). *2024 CACREP standards.* https://www.cacrep.org/wp-content/uploads/2023/06/2024-Standards-Combined-Version-6.27.23.pdf

Cowan, W. M. (1979). The development of the brain. *Scientific American, 241*(3), 113–133.

Crethar, H. C., Rivera, E. T., & Nash, S. (2008). In search of common threads: Linking multicultural, feminist, and social justice counseling paradigms. *Journal of Counseling & Development, 86*(3), 269–278. https://doi.org/10.1002/j.1556-6678.2008.tb00509.x

Eriksson, P. S., Perfilieva, E., Björk-Eriksson, T., Alborn, A. M., Nordborg, C., Peterson, D. A., & Gage, F. H. (1998). Neurogenesis in the adult human hippocampus. *Nature Medicine, 4*(11), 1313–1317.

Evans, I. E., Martyr, A., Collins, R., Brayne, C., & Clare, L. (2019). Social isolation and cognitive function in later life: A systematic review and meta-analysis. *Journal of Alzheimer's Disease, 70*(s1), S119–S144.

Feldman, R., Gordon, I., & Zagoory-Sharon, O. (2010). The cross-generation transmission of oxytocin in humans. *Hormones and Behavior, 58*(4), 669–676.

Fishbane, M. D. (2007). Wired to connect: Neuroscience, relationships, and therapy. *Family Process, 46*(3), 395–412.

Fjell, A. M., & Walhovd, K. B. (2010). Structural brain changes in aging: Courses, causes and cognitive consequences. *Reviews in the Neurosciences, 21*(3), 187–222.

Foster, J. A., & Neufeld, K. A. M. (2013). Gut–brain axis: How the microbiome influences anxiety and depression. *Trends in Neurosciences, 36*(5), 305–312.

Gars, A., Ronczkowski, N. M., Chassaing, B., Castillo-Ruiz, A., & Forger, N. G. (2021). First encounters: Effects of the microbiota on neonatal brain development. *Frontiers in Cellular Neuroscience, 15,* Article 682505. https://doi.org/10.3389/fncel.2021.682505

Gaskill, R. L., & Perry, B. D. (2014). The neurobiological power of play. In C. A. Malchiodi & D. A. Crenshaw (Eds.), *Creative arts and play therapy for attachment problems* (pp. 178–194). Guilford Press.

Geerligs, L., Renken, R. J., Saliasi, E., Maurits, N. M., & Lorist, M. M. (2015). A brain-wide study of age-related changes in functional connectivity. *Cerebral Cortex, 25*(7), 1987–1999.

Goddings, A. L., Beltz, A., Peper, J. S., Crone, E. A., & Braams, B. R. (2019). Understanding the role of puberty in structural and functional development of the adolescent brain. *Journal of Research on Adolescence, 29*(1), 32–53.

Grady, C. (2012). The cognitive neuroscience of ageing. *Nature Reviews Neuroscience, 13*(7), 491–505.

Gunnar, M. R., & Hostinar, C. E. (2015). The social buffering of the hypothalamic–pituitary–adrenocortical axis in humans: Developmental and experiential determinants. *Social Neuroscience, 10*(5), 479–488.

Gunnar, M. R., Wewerka, S., Frenn, K., Long, J. D., & Griggs, C. (2009). Developmental changes in hypothalamus–pituitary–adrenal activity over the transition to adolescence: Normative changes and associations with puberty. *Development and Psychopathology, 21*(1), 69–85.

Gunning-Dixon, F. M., Brickman, A. M., Cheng, J. C., & Alexopoulos, G. S. (2009). Aging of cerebral white matter: A review of MRI findings. *International Journal of Geriatric Psychiatry, 24*(2), 109–117.

Hariri, A. R., Bookheimer, S. Y., & Mazziotta, J. C. (2000). Modulating emotional responses: Effects of a neocortical network on the limbic system. *NeuroReport, 11*(1), 43–48.

Hoban, A. E., Stilling, R. M., Ryan, F. J., Shanahan, F., Dinan, T. G., Claesson, M. J., Clarke, G., & Cryan, J. F. (2016). Regulation of prefrontal cortex myelination by the microbiota. *Translational Psychiatry, 6*(4), Article e774.

Insel, T. R., & Young, L. J. (2001). The neurobiology of attachment. *Nature Reviews Neuroscience, 2*(2), 129–136.

Ito, E., Nouchi, R., Dinet, J., Cheng, C. H., & Husebø, B. S. (2022). The effect of music-based intervention on general cognitive and executive functions, and episodic memory in people with mild cognitive impairment and dementia: A systematic review and meta-analysis of recent randomized controlled trials. *Healthcare (Switzerland), 10*(8), 1462.

Jašarević, E., Morrison, K. E., & Bale, T. L. (2016). Sex differences in the gut microbiome–brain axis across the lifespan. *Philosophical Transactions of the Royal Society B: Biological Sciences, 371*(1688), Article 20150122.

Jernigan, T. L., Baaré, W. F., Stiles, J., & Madsen, K. S. (2011). Postnatal brain development: Structural imaging of dynamic neurodevelopmental processes. *Progress in Brain Research, 189,* 77–92.

Johnson, M. H. (2005). Sensitive periods in functional brain development: Problems and prospects. *Developmental Psychobiology, 46*(3), 287–292.

Jones, L. K. (2015). The evolving adolescent brain. *Counseling Today, 57*(7), 14–17.

Jones, L. K. (2016). Sex-related variations in neuroscience and endocrinology and their effects on mental health. *Counseling Today, 58*(10), 16–19.

Karel, M. J., Gatz, M., & Smyer, M. A. (2012). Aging and mental health in the decade ahead: What psychologists need to know. *American Psychologist, 67*(3), 184–198.

Kennedy, K. M., de Goffau, M. C., Perez-Muñoz, M. E., Arrieta, M. C., Bäckhed, F., Bork, P., Braun, T., Bushman, F. D., Dore, J., de Vos, W. M., Earl, A. M., Eisen, J. A., Elovitz, M. A., Ganal-Vonarburg, S. C., Gänzle, M. G., Garrett, W. S., Hall, L. J., Hornef, M. W., Huttenhower, . . . Walter, J. (2023). Questioning the fetal microbiome illustrates pitfalls of low-biomass microbial studies. *Nature, 613*(7945), 639–649.

Kilner, J. M., & Lemon, R. N. (2013). What we know currently about mirror neurons. *Current Biology, 23*(23), 1057–1062.

Knickmeyer, R. C., Gouttard, S., Kang, C., Evans, D., Wilber, K., Smith, J. K., Hamer, R. M., Lin, W., Gerig, G., & Gilmore, J. H. (2008). A structural MRI study of human brain development from birth to 2 years. *Journal of Neuroscience, 28*(47), 12176–12182.

Kuhl, P., & Rivera-Gaxiola, M. (2008). Neural substrates of language acquisition. *Annual Review of Neuroscience, 31*, 511–534.

Kung, K. T., Browne, W. V., Constantinescu, M., Noorderhaven, R. M., & Hines, M. (2016). Early postnatal testosterone predicts sex-related differences in early expressive vocabulary. *Psychoneuroendocrinology, 68*, 111–116.

Lester, B. M., Marsit, C. J., Conradt, E., Bromer, C., & Padbury, J. F. (2012). Behavioral epigenetics and the developmental origins of child mental health disorders. *Journal of Developmental Origins of Health and Disease, 3*(6), 395–408.

Li, Y., Toothaker, J. M., Ben-Simon, S., Ozeri, L., Schweitzer, R., McCourt, B. T., McCourt, C. T., Werner, L., Snapper, S., Shouval, D. S., Khatib, S., Koren, O., Agnihorti, S., Tseng, G., & Konnikova, L. (2020). In utero human intestine harbors unique metabolome, including bacterial metabolites. *JCI Insight, 5*(21), Article e138751.

Liu, F., Zhang, Z., Lin, X., Teng, G., Meng, H., Yu, T., Fang, F., Zang, F., Li, Z., & Liu, S. (2011). Development of the human fetal cerebellum in the second trimester: A post mortem magnetic resonance imaging evaluation. *Journal of Anatomy, 219*(5), 582–588.

Lund, C., Brooke-Sumner, C., Baingana, F., Baron, E. C., Breuer, E., Chandra, P., Haushofer, J., Herrman, H., Jordans, M., Kieling, C., Medina-Mora, M. E., Morgan, E., Omigbodun, O., Tol, W., Patel, V., & Saxena, S. (2018). Social determinants of mental disorders and the sustainable development goals: A systematic review of reviews. *The Lancet Psychiatry, 5*(4), 357–369.

McGowan, P. O., & Szyf, M. (2010). The epigenetics of social adversity in early life: Implications for mental health outcomes. *Neurobiology of Disease, 39*(1), 66–72.

Mews, P., Calipari, E. S., Day, J., Lobo, M. K., Bredy, T., & Abel, T. (2021). From circuits to chromatin: The emerging role of epigenetics in mental health. *Journal of Neuroscience, 41*(5), 873–882.

Patterson, C. (2018). *World Alzheimer report 2018: The state of the art of dementia research: New frontiers.* Alzheimer's Disease International. https://www.alzint.org/u/WorldAlzheimerReport2018.pdf

Paus, T., Keshavan, M., & Giedd, J. N. (2008). Why do many psychiatric disorders emerge during adolescence? *Nature Reviews Neuroscience, 9*(12), 947–957.

Pester, D. A., Jones, L. K., & Talib, Z. (2023). Social determinants of mental health: Informing counseling practice and professional identity. *Journal of Counseling & Development, 101*(4), 392–401. https://doi.org/10.1002/jcad.12473

Pitts, E. G., Taylor, J. R., & Gourley, S. L. (2016). Prefrontal cortical BDNF: A regulatory key in cocaine- and food-reinforced behaviors. *Neurobiology of Disease, 91*, 326–335.

Porges, S. W. (2011). *The polyvagal theory: Neurophysiological foundations of emotions, attachment, communication, and self-regulation*. Norton.

Provenzi, L., Brambilla, M., Scotto di Minico, G., Montirosso, R., & Borgatti, R. (2020). Maternal caregiving and DNA methylation in human infants and children: Systematic review. *Genes, Brain and Behavior, 19*(3), Article e12616.

Raja, G. L., Subhashree, K. D., & Kantayya, K. E. (2022). In utero exposure to endocrine disruptors and developmental neurotoxicity: Implications for behavioural and neurological disorders in adult life. *Environmental Research, 203*, Article 111829.

Rasmussen, P., Brassard, P., Adser, H., Pedersen, M. V., Leick, L., Hart, E., Secher, N. H., Pedersen, B. K., & Pilegaard, H. (2009). Evidence for a release of brain-derived neurotrophic factor from the brain during exercise. *Experimental Physiology, 94*(10), 1062–1069.

Rossini, P. M., Rossi, S., Babiloni, C., & Polich, J. (2007). Clinical neurophysiology of aging brain: From normal aging to neurodegeneration. *Progress in Neurobiology, 83*(6), 375–400.

Rowe, M. L., & Weisleder, A. (2020). Language development in context. *Annual Review of Developmental Psychology, 2*, 201–223.

Roy, M., Fortier, M., Rheault, F., Edde, M., Croteau, E., Castellano, C. A., Langlois, F., St-Pierre, V., Cuenoud, B., Bocti, C., Fulop, T., Descoteaux, M., & Cunnane, S. C. (2021). A ketogenic supplement improves white matter energy supply and processing speed in mild cognitive impairment. *Alzheimer's & Dementia: Translational Research & Clinical Interventions, 7*(1), Article e12217.

Sanders, L. M., Hortobagyi, T., la Bastide-van Gemert, S., van der Zee, E. A., & van Heuvelen, M. J. (2019). Dose-response relationship between exercise and cognitive function in older adults with and without cognitive impairment: A systematic review and meta-analysis. *PloS One, 14*(1), Article e0210036.

Sawyer, K., Corsentino, E., Sachs-Ericsson, N., & Steffens, D. C. (2012). Depression, hippocampal volume changes, and cognitive decline in a clinical sample of older depressed outpatients and non-depressed controls. *Aging & Mental Health, 16*(6), 753–762.

Schaadt, G., Hesse, V., & Friederici, A. D. (2014). Sex hormones in early infancy seem to predict aspects of later language development. *Brain and Language, 141*, 70–76.

Sebastian, C., Viding, E., Williams, K. D., & Blakemore, S. J. (2010). Social brain development and the affective consequences of ostracism in adolescence. *Brain and Cognition, 72*(1), 134–145.

Sele, S., Liem, F., Mérillat, S., & Jäncke, L. (2020). Decline variability of cortical and subcortical regions in aging: A longitudinal study. *Frontiers in Human Neuroscience, 14*, Article 363.

Sharma, U. (2019). Paternal contributions to offspring health: Role of sperm small RNAs in intergenerational transmission of epigenetic information. *Frontiers in Cell and Developmental Biology, 7*, Article 215.

Siegel, D. J. (2006). An interpersonal neurobiology approach to psychotherapy. *Psychiatric Annals, 36*(4), 248–256.

Singh, A. A., Appling, B., & Trepal, H. (2020). Using the multicultural and social justice counseling competencies to decolonize counseling practice: The important roles of theory, power, and action. *Journal of Counseling & Development, 98*(3), 261–271. https://doi.org/10.1002/jcad.12321

Sisk, C. L., & Zehr, J. L. (2005). Pubertal hormones organize the adolescent brain and behavior. *Frontiers in Neuroendocrinology, 26,* 163–174.

Sleimen-Malkoun, R., Temprado, J. J., & Hong, S. L. (2014). Aging induced loss of complexity and dedifferentiation: Consequences for coordination dynamics within and between brain, muscular and behavioral levels. *Frontiers in Aging Neuroscience, 6,* Article 140.

Sowell, E. R., Thompson, P. M., Leonard, C. M., Welcome, S. E., Kan, E., & Toga, A. W. (2004). Longitudinal mapping of cortical thickness and brain growth in normal children. *Journal of Neuroscience, 24*(38), 8223–8231.

Spear, L. P. (2000). The adolescent brain and age-related behavioral manifestations. *Neuroscience and Biobehavioral Reviews, 24*(4), 417–463.

Speight, S. L., Myers, L. J., Cox, C. I., & Highlen, P. S. (1991). A redefinition of multicultural counseling. *Journal of Counseling & Development, 70*(1), 29–36. https://doi.org/10.1002/j.1556-6676.1991.tb01558.x

Squire, L. R. (2004). Memory systems of the brain: A brief history and current perspective. *Neurobiology of Learning and Memory, 82*(3), 171–177.

Stiles, J., & Jernigan, T. L. (2010). The basics of brain development. *Neuropsychology Review, 20*(4), 327–348.

Sturman, D. A., & Moghaddam, B. (2011). The neurobiology of adolescence: Changes in brain architecture, functional dynamics, and behavioral tendencies. *Neuroscience and Biobehavioral Reviews, 35*(8), 1704–1712.

Sullivan, R. M. (2003). Developing a sense of safety. *Annals of the New York Academy of Sciences, 1008*(1), 122–131.

Tamnes, C. K., Walhovd, K. B., Dale, A. M., Østby, Y., Grydeland, H., Richardson, G., Westlye, L. T., Roddey, J. C., Hagler, D. J., Jr., Due-Tønnessen Holland, P., Holland, D., & Fjell, A. M. (2013). Brain development and aging: Overlapping and unique patterns of change. *NeuroImage, 68,* 63–74.

Tapia-Arancibia, L., Rage, F., Givalois, L., & Arancibia, S. (2004). Physiology of BDNF: Focus on hypothalamic function. *Frontiers in Neuroendocrinology, 25,* 77–107.

Thomas, K. M., Drevets, W. C., Whalen, P. J., Eccard, C. H., Dahl, R. E., Ryan, N. D., & Casey, B. J. (2001). Amygdala response to facial expressions in children and adults. *Biological Psychiatry, 49*(4), 309–316.

Thomas, L. A., De Bellis, M. D., Graham, R., & LaBar, K. S. (2007). Development of emotional facial recognition in late childhood and adolescence. *Developmental Science, 10*(5), 547–558.

Tomasi, D., & Volkow, N. D. (2012). Aging and functional brain networks. *Molecular Psychiatry, 17*(5), 549–558.

Toyokawa, S., Uddin, M., Koenen, K. C., & Galea, S. (2012). How does the social environment 'get into the mind'? Epigenetics at the intersection of social and psychiatric epidemiology. *Social Science & Medicine, 74*(1), 67–74. https://doi.org/10.1016/j.socscimed.2011.09.036

Uhls, Y. T., Michikyan, M., Morris, J., Garcia, D., Small, G. W., Zgourou, E., & Greenfield, P. M. (2014). Five days at outdoor education camp without screens improves preteen skills with nonverbal emotion cues. *Computers in Human Behavior, 39,* 387–392.

United Nations Department of Economic and Social Affairs. (2019, October 10). *Our world is growing older: UN DESA releases new report on ageing.* https://www.un.org/development/desa/en/news/population/our-world-is-growing-older.html

Urbán, N., & Guillemot, F. (2013). Neurogenesis in the embryonic and adult brain: Same regulators, different roles. *Frontiers in Cellular Neuroscience, 8,* 396.

Waddington, C. H. (2012). The epigenotype. *International Journal of Epidemiology, 41*(1), 10–13. (Original work published 1942)

Weinhold, B. (2006). Epigenetics: The science of change. *Environmental Health Perspectives, 114*(3), A160–A167.

Werker, J. F., & Tees, R. C. (2005). Speech perception as a window for understanding plasticity and commitment in language systems of the brain. *Developmental Psychobiology, 46*(3), 233–251.

Woodhead, E. L., & Yochim, B. (2022). Adult development and aging: A foundational geropsychology knowledge competency. *Clinical Psychology: Science and Practice, 29*(1), 16–27.

World Health Organization. (2017). *Mental health of older adults* [Fact sheet]. https://www.who.int/news-room/fact-sheets/detail/mental-health-of-older-adults

Wright, S. L., & Díaz, F. (2014). Neuroscience research on aging and implications for counseling psychology. *Journal of Counseling Psychology, 61*(4), 534–540.

Yehuda, R., & Lehrner, A. (2018). Intergenerational transmission of trauma effects: Putative role of epigenetic mechanisms. *World Psychiatry, 17*(3), 243–257.

Zhang, H., Ozbay, F., Lappalainen, J., Kranzler, H. R., van Dyck, C. H., Charney, D. S., Price, L. H., Southwick, S., Yang, B. Z., Rasmussen, A., & Gelernter, J. (2006). Brain derived neurotrophic factor (BDNF) gene variants and Alzheimer's disease, affective disorders, posttraumatic stress disorder, schizophrenia, and substance dependence. *American Journal of Medical Genetics Part B: Neuropsychiatric Genetics, 141*(4), 387–39

Chapter 3

Biology of Marginality: A Neurophysiological Exploration of the Social and Cultural Foundations of Psychological Health

Kathryn Z. Douthit and Justin Russotti

Honoring social and cultural diversity in counseling practice has been a cornerstone of the profession for the past half century. Counselors have long grasped the ethical imperative in honoring how culture and ethnicity provide a framework for the ways in which clients make meaning of their world. With an eye toward social justice, the profession is also deeply concerned by the suffering incurred by those whose social or cultural status places them in the margins of the social world. This chapter challenges counselors to expand their understanding of those living in the margins by grappling with some of the biological mechanisms that are affected by a life of marginality. The implications of the neurobiological literature related to marginality are profound; injustices in the social world become imprinted in flesh-and-blood realities that have an impact on mental and physical health. Whether it is the chronic stress wrought by bigotry, dehumanization, and disdain or the environmental scars of poverty that deprive whole communities of quality food, uncontaminated water, safe housing, green spaces, and satisfactory educational experiences, marginality leaves in its wake psychological suffering, life-threatening diseases, and, for some, premature death.

Multiple dimensions of human biology inform understanding of the physiological responses to social injustice. This chapter focuses on three of these dimensions, the first being dysregulation of the hypothalamic-pituitary-adrenal (HPA) axis, the second being the injury wrought by mechanisms identified in the field of psychoneuroimmunology (PNI), and the third being the field of epigenetics. These three highly interdependent dimensions are spotlighted primarily because of their inextricable association with the biologically driven mental health struggles of marginalized clients and their potential for counseling prevention and intervention—approaches that are pivotal in navigating the case of Henrietta, the client portrayed in this chapter's case study.

2024 CACREP Standards

This chapter addresses 2024 Council for Accreditation of Counseling and Related Educational Programs (CACREP) Standards pertinent to the Foundational Counseling Curriculum (Section 3) areas of Social and Cultural Identities and Experiences (Standard B) and Lifespan Development (Standard C):

- The effects of historical events, multigenerational trauma, and current issues on diverse cultural groups in the United States and globally (Standard B.4.)
- The effects of stereotypes, overt and covert discrimination, racism, power, oppression, privilege, marginalization, microaggressions, and violence on counselors and clients (Standard B.5.)
- The effects of various sociocultural influences, including public policies, social movements, and cultural values, on mental and physical health and wellness (Standard B.6.)
- Disproportional effects of poverty, income disparities, and health disparities toward people with marginalized identities (Standard B.7.)
- Models of psychosocial adjustment and adaptation to illness and disability (Standard C.8.)
- Biological, neurological, and physiological factors that affect lifespan development, functioning, behavior, resilience, and overall wellness (Standard C.10.)
- Systemic, cultural, and environmental factors that affect lifespan development, functioning, behavior, resilience, and overall wellness (Standard C.11.)

Clinical Case Study: Henrietta

Henrietta is a 48-year-old African American woman who looks many years older than her age. She is living in an urban Rust Belt location and has had diabetes, high blood pressure, and problems with weight control for the past decade. Subsisting on an income that she pieces together from two physically taxing jobs, she resides in the second-floor unit of a poorly maintained 1940s-era house situated in a section of a medium-sized city that has gained national recognition for its deep and enduring poverty. In addition to struggling daily to keep a roof over her head, pay for her medications, and put food on the table, Henrietta assumes much of the responsibility for raising her three male grandchildren, who are ages 11, 12, and 14. The children are all enrolled in the notoriously troubled city school system among a large contingent of struggling students. The 11-year-old boy stands out for his particularly severe conduct problems, including a propensity for interpersonal physical altercation. Although the youngest is most at risk for active participation in a gang culture, all three boys are at an age when peers are being lured by local gang recruitment activity. The boys' mother is currently serving time for substance-related offenses.

Henrietta, who has had lifelong struggles with anxiety and depression, has relied heavily on her faith to give her the strength to meet the environmental challenges she encounters on a daily basis. Unfortunately, she now feels like she is losing her battle. As her grandchildren have gotten older and more defiant and the infrastructure in her neighborhood has steadily declined, her anxiety has become more pervasive and intense, and she finds herself plagued by an accompanying sense of despair, isolation, and worthlessness. Enduring two long bus rides in frigid weather and absorbing the lost income from the time that it takes to make the trip, Henrietta seeks counseling services in the behavioral health clinic of a large, urban teaching hospital. Prompted by her counselor, Henrietta describes the overwhelming feelings of trying to cope with the immense odds that she faces:

> I always thought that if I prayed hard enough that the Lord would see me through. But I think the Lord has grown weary of listening to me. I worry all day and all night—I worry that the boys are going to turn bad. I worry that I won't be able to feed them. I worry that my health is going to get worse, and I won't be able to take care of them.

Inquiring about Henrietta's psychological history, the counselor learns about her growing struggle with depression.

> I could always rely on Jesus to take my hand when things got real bad. He helped me dig deep inside and feel strong. But Jesus doesn't seem to hear me anymore. I feel so alone, and when I look in the mirror, I feel like I am looking at a ghost. I see someone who is old and tired who hasn't done nothin' good in their life. If I were a better person, Jesus would have shown me the way. But I don't deserve his love and I don't have the strength to carry on.

Viewing Henrietta's Case Through a Social Determinants of Health Lens

Many of the chronic conditions plaguing Henrietta are conditions that take a disproportionate toll on marginalized communities, such as those in which African American/Black, Latinx, Indigenous, and immigrant/refugee populations often reside. Obesity, diabetes, hypertension, heart disease, stroke, dementia, certain cancers, substance use, depression, and anxiety are a few of the physical and psychological challenges that have a higher incidence among individuals who live at the intersection of racial and economic marginality (Bailey et al., 2017; Furman et al., 2019; Marmot & Wilkerson, 2005). Much of what is known about the biology of marginality emerged from two seminal British studies that set the course for several generations of transformational research on what became known more broadly as "the social determinants of health" (Marmot & Wilkerson, 2005). The now famous Whitehall I and Whitehall II studies, launched by Sir Michael Marmot and a large team of researchers, opened a window into the close relationship between human health and the surrounding social world. The alarming message emerging from these studies was that lack of status, authority, and control affects physical and mental health in ways that can have life-or-death consequences (Marmot et al., 1991; Marmot & Wilkerson, 2005).

Although the Whitehall studies focused primarily on social class, other major studies in the social determinants of health literature have focused specifically on issues of health and mental health in relation to racial discrimination (Bailey et al., 2017; Pieterse et al., 2012; Williams et al., 2019). A groundbreaking meta-analysis targeting research on discrimination against adult Black Americans included 66 of these studies and showed a compelling link between the experience of race among Black Americans and anxiety, depression, and a variety of other psychiatric conditions (Pieterse et al., 2012).

For counselors working in the mental health field with a client base that consists of individuals struggling with the daily assaults of a marginal status or a multiplicity of intersecting marginal statuses, understanding how the daily experience of marginality can become a flesh-and-blood reality that compromises the health of mind and body can help to foster an informed intervention strategy targeting psychological struggles at their source. In this spirit, the remainder of this chapter aims first to clarify some of the main biological mechanisms that are catalyzed by experiences of marginality and foster processes injurious to psychological and physical health, and second to suggest intervention strategies that disrupt the cycles of harm fueled by day-to-day experiences of marginality.

Chronic Stress: Multiple Pathways to Harm

For clients living in the margins, the barrage of assaults they experience translates into a relentless exposure to stress. Whether it is the fear of becoming the next crime victim, the inability to access adequate health care, perpetual fear of being racially profiled by law enforcement, or chronic unemployment, repeated exposure to stress-inducing events is a reality when one is marginalized and a springboard for an array of pathways that lead to mental and physical injury.

Being able to mount the occasional stress response in the face of impending harm is an important survival tool. It is the body's way of garnering peak physical performance and sharpened senses when one is confronted with life-threatening conditions. Situations in which the body needs to "pull out all the stops" in the interest of survival are not difficult to imagine. Relying on primitive instincts, the amygdala, a part of the brain that perceives threat, sounds an alarm that puts into motion a neuroendocrine cascade that helps the body either stand its ground and enter into a battle, freeze in place, or flee to a secure safe space (Sapolsky, 2004). Chapter 4 has an extended discussion of this fight-or-flight acute stress response and the related functions of the cortisol-producing HPA axis and the epinephrine-producing sympathetic-adrenal-medullary axis (Sapolsky, 2004).

As Chapter 4 details, a key to maintaining physical and psychological health rests on one's ability to return, by way of a negative feedback system, to an inactivated state (McEwen, 2007; Sapolsky, 2004). What happens, however, in an environment in which threats to psychological and physical safety and security are an everyday affair—in which feelings of aggression and anger are repeatedly triggered by despair, defeat, insult, and a lack of fairness, respect, or dignity? In short, the feedback systems that normally function to return cortisol to normal levels become increasingly less efficient, doing less and less to maintain a healthy, regulated stress response. The acute stress response that can be so lifesaving is thus transformed into a state of chronic stress, which is a threat to mental and physical well-being and, if unabated over time, can ultimately be life-threatening (McEwen, 2007).

Since the publication of Whitehall I in 1978, countless studies have underscored the toxicity of chronic stress (see, e.g., Juster et al., 2010). The myriad physical health outcomes of chronic stress include diabetes, heart disease, stroke, asthma, cancers, and dementia, and the psychological outcomes can include depression, anxiety, and exacerbation of a long list of psychiatric problems, including bipolar disorder, schizophrenia, and posttraumatic stress disorder (Juster et al., 2010; McEwen, 2003; Syed & Nemeroff, 2017; Wilson et al., 2002). Although numerous biological pathways threaten wellness under conditions of chronic stress, two particularly salient and highly interrelated pathways to ill health can be found in the fields of PNI and epigenetics, both of which are discussed at length in this chapter.

Early Childhood Programming of the Stress Response

To this point in the chapter, our discussion of stress response dysregulation has been generally within the context of an adult, Henrietta, who finds herself in a position of trying to manage a sustained assault by a myriad of challenges. To fully grasp her situation, however, it is essential to consider the impact of conditions imposed by marginality on HPA function prenatally or in early childhood. It is now well established in the literature that the HPA axis is a powerful mediator channeling early adversity into later development of compromised psychological coping. Understanding the potential impact of early life adversity may indeed provide a plausible explanation for some of Henrietta's current psychological challenges as well as insight into the behaviors of her grandchildren.

Although the literature in this area is quite complex and empirical work often produces varied outcomes, one thing is certain regarding HPA regulation and early/prenatal childhood: A young child or fetus developing in an environment filled with adversity and few social safety nets is at risk for HPA dysregulation that persists in later life, thus increasing their chances of developing depression, anxiety, attention-deficit/hyperactivity disorder, or substance use, and in special cases of developing externalizing psychopathology in the form of conduct disorder or antisocial personality disorder (Furman et al., 2019; Koss & Gunnar, 2018; Tarullo & Gunnar, 2006). Conduct disorder and antisocial personality disorder are most commonly seen in a subset of mostly male, maltreated children in which stress responsivity to their toxic environment is initially ramped up, making them more vigilant and ready to respond to threats in their surroundings. However, this increase in reactivity is replaced by a switch to a more blunted or attenuated response (Del Giudice et al., 2011; van Goozen et al., 2007). The turn to a blunted stress response allows these individuals to face threats, reprisals, and assaults with little fear and can result in a tendency toward interpersonal violence and clashes with school authorities and law enforcement. In the context of Henrietta's current constellation of challenges, her youngest grandson may well be the victim of a blunted HPA axis.

Although this portrayal of early childhood HPA programming is accurate in its description, it is fair to say that it is tainted by a somewhat pathologizing lens. In the interest of a socially just interpretation of this programming process, it is helpful to frame it in evolutionary terms. A particular evolutionary lens, the adaptive calibration model (Del Giudice et al., 2011), achieves this alternative view by underscoring the essential life-preserving features of HPA adaptation. This model posits that in the earliest phases of the life course, the fetus, developing infant, and young child are able to instinctively take stock of the environmental forces in which they are immersed and lock into place a stress response that in evolutionary terms will give them the best chance of surviving through their reproductive years (Del Giudice et al., 2011; Koss & Gunnar, 2018). For example, a young child immersed in an environment with conditions of abuse and neglect will likely prepare for a world where they need to defend themselves against violence, aggression, and other threats to safety—a world in which vigilance and increased stress reactivity is a key to survival. By contrast, a child in a nurturing and safe environment, who has likely developed a well-regulated stress response, can focus more of their attention on relationship building and cognitive enrichment. A key concept that emerges from this drive toward adaptive advantage is that individuals who have adapted to a tumultuous environment will often find their adaptations a mismatch in spaces where those adaptations are no longer advantageous, such as classroom settings, and run the risk of receiving labels such as conduct disorder that reflect psychopathology or, worse, criminal deviance (Liu et al., 2018). Thus far, research focused on prenatal and early postnatal childhood has proven to be most successful at identifying sensitive periods in early life that can lead to enduring dysregulated HPA activity patterns (Koss & Gunnar, 2018). That being said, promising interventions to address HPA dysregulation during early childhood are discussed in Chapter 4.

PNI Pipeline

Although a myriad of pathways can threaten physical or psychological wellness under sustained challenging conditions, the connection of the HPA system to immunity provides a now well-established route that is the focus of much PNI research. This PNI pipeline involves a complex array of biological mechanisms that collectively comprise some of the most startling evidence that humans are inextricably one with their social world. Emerging in the 1970s, the field of PNI brought into sharp relief how nonmaterial perceptions of one's social context can be transformed into material flesh-and-blood realities and shed light on the ways in which the work of counselors can have an impact on both mental and physical health (Ader & Cohen, 1993).

More broadly, PNI describes the complex associations between the psychological state of a given individual, the neurological and hormonal processes that respond to that state, and the immunological mechanisms that communicate with those neurological and hormonal processes (Ader et al., 1995; Bower & Kuhlman, 2023). Although these systems can work in perfect harmony to maintain health, certain challenges can wreak havoc on how the systems interrelate, thus resulting in threats to health and well-being. To understand how the immune system, which is so vital in protecting people from potentially harmful invaders such as bacteria, viruses, and cancer cells, can threaten physical and psychological wellness, it is helpful to start with some basic information on the immune system more generally.

Immune function is generally divided into two types: acquired and innate. *Acquired immunity* refers to the ability to target and destroy specific disease-producing entities (e.g., specific bacteria, specific viruses, cancer cells; Janeway et al., 2001; Kaufman, 2019). *Specificity* in this case means that an acquired immune response has a very specific target (e.g., an acquired immune response to a cold virus does not help one's ability to fight salmonella from contaminated food or a staphylococcus infection from an injury).

In contrast to the very targeted mechanism of acquired immunity, *innate immunity*, which is more commonly known as inflammation, mounts a defense that is nonspecific (Cuevas et al., 2020; Janeway et al., 2001; Kaufman, 2019). Innate immunity involves a generalized rallying of cells (i.e., leukocytes or white blood cells) that travel to a site of tissue damage. Regardless of how the tissue became injured, the damage signals a generalized response that attempts to destroy and clear any invading microorganisms and/or foreign debris and initiates tissue repair (Janeway et al., 2001).

This introduction to immunology, although brief, provides a window into the deeply rooted relationship between people's perceptions of the world around them and the health of their minds and bodies. For those who, like Henrietta, live under conditions of chronic stress, the life-sustaining mechanisms of acquired and innate immunity are each compromised in distinctive ways. The dysregulation of the stress response that occurs with chronic stress causes the neuroendocrine system to communicate with the immune system in a manner that suppresses the cells responsible for acquired immune mechanisms and activates the cells responsible for innate immunity (Kendall-Tackett, 2009; Marshall, 2011). The consequences of this ill-fated neuroendocrine-immune dialogue are twofold. In the case of a dampened acquired immune response, one sees increased susceptibility to infections and various cancers. In contrast, the problems related to intensification of the innate immune response—that is, inflammation—are connected to a multitude of physical and psychological challenges, including heart disease, cerebrovascular disease, dementia, depression, arthritis, autoimmune diseases, chronic fatigue, and fibromyalgia (Furman et al., 2019; Hänsel et al., 2010; Kendall-Tackett, 2009; Kiecolt-Glaser et al., 2005; Majd et al., 2020; Marshall, 2011; Wilson et al., 2002). Specific substances implicated in these destructive inflammatory processes are called *proinflammatory cytokines*. These proinflammatory cytokines are released by cells in the innate immune system and are commonly used in research settings to study inflammation in human subjects. Among the most commonly cited cytokines in the psychopathology literature are interleukin-1 (IL-1), interleukin-6 (IL-6), and tumor necrosis factor (TNF) alpha (Shariq et al., 2018).

What is most relevant for counselors to understand from this description of the chronic stress–immune dysregulation relationship is that within this mind-body communication there are two major ways through which mental health is affected. Both pathways can be understood through a concept that is well known to many counselors—namely, a diathesis–stress model (D-SM) of mental health. This model draws the logical conclusion that the mental health of any given individual is determined by their degree of predisposing vulnerability and the environmental stresses to which they are subjected (Zuckerman, 1999). In the D-SM, individuals have a threshold for managing environmental stress, after which they no longer have the capacity to "bounce back," thus generating the conditions for psychological suffering.

In the context of PNI, both the environmental stress and the forces of predisposing vulnerability, which together are the foundation of the D-SM, are affected. It is not difficult to see how the environmental stress aspect of the D-SM might be intensified when PNI mechanisms are left unchecked. Individuals such as Henrietta may be faced with the prospect of living with pain and disability in a neighborhood in which good medical care is nonexistent and transportation to geographic locations with available health care is dismal. When one considers the inherent vulnerability component of the D-SM, the role of PNI in translating chronic environmental stress into psychological susceptibility is quite clear. Strong evidence exists for a distinct connection between inflammation and depression, which occurs by way of two different but related pathways (Glassman & Miller, 2007; Shariq et al., 2018; Surtees et al., 2008). The proinflammatory cytokines mediate both of these pathways. In one pathway, the cytokines trigger a series of reactions that end in a disruption of the production of the neurotransmitter serotonin, which has an important role in mood (Dantzer et al., 2011). In the other pathway, a cytokine participates in a series of reactions that slow the production of brain-derived neurotrophic factor (BDNF), which in turn suppresses neuroplasticity and results in neuron death and brain tissue shrinkage, leading to depression, anxiety, anhedonia, and social withdrawal (Brunoni et al., 2008; Dowlati et al., 2010). Thus, there is a "a vicious cycle in which chronic stress begets immunologically mediated mood disturbance, which then compromises resilience and further fuels the chronic stress condition" (Douthit, 2015b, p. 14).

It is important for counselors to be aware that these PNI processes are not reserved for clients who are exhibiting observable psychological suffering. Newer research suggests that individuals who "beat the odds" and achieve success (e.g., going to college, earning good incomes, having positive mental health) despite significant adversity, hardship, or marginalization may suffer greater inflammation—a pattern termed *skin-deep resilience* (Chen et al., 2020, 2022; Russotti et al., 2020). In essence, a tradeoff may occur such that psychosocial resilience following marginalization comes at the expense of physical health (see also *John Henryism*, as discussed in James, 1994).

Figure 3.1 depicts the complex dynamic relationship between stress and illness through the lens of marginality, mediated by variables such as epigenetics and mental health challenges.

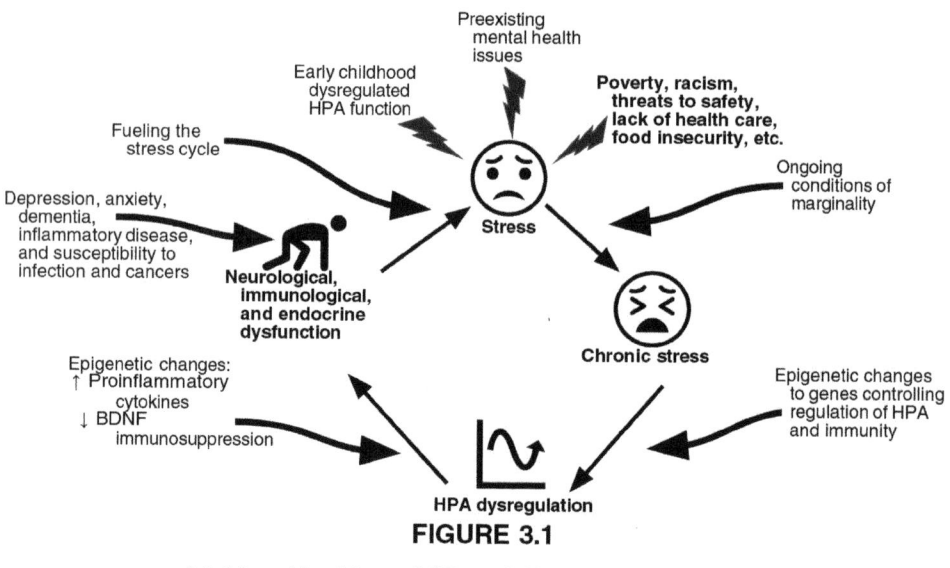

FIGURE 3.1

Linking Health and Mental Health to Marginality

Note. HPA = hypothalamic-pituitary-adrenal; BDNF = brain-derived neurotrophic factor.

> **Reflective Question**
>
> As a counselor with an understanding of the relationship between the stresses of extreme poverty and immunologically mediated changes in mental health, how might you help Henrietta to reframe her fears that God has grown weary of her need for support and has consequently abandoned her?

If one returns to the D-SM and looks more closely at the notion of predisposing vulnerability, the biology of marginality becomes even more complex than just discussed. The science of vulnerability now shows that social forces can insert themselves at the level of gene function to actually shape predisposing vulnerability. In particular, work in the field of epigenetics has done much to elucidate the mechanisms that allow contextual hardships to genetically reconfigure people's capacity for resilience. Although we focus on epigenetics in relation to chronic stress, suffice it to say that epigenetics can intersect with marginality in other injurious ways beyond the scope of this chapter. In contrast, epigenetic mechanisms can also positively participate in aiding cognitive and emotional development.

Epigenetics

Any understanding of the mechanisms driving HPA plasticity and PNI would be incomplete without attention to the field of epigenetics. The term *epigenetics* is used to describe an array of mechanisms in which aspects of the environment are able to control how genes are expressed. Genes are composed of DNA that carries the instructions or codes that direct the synthesis of proteins needed for growth, development, and other life-sustaining functions (McGowan & Szyf, 2010). Epigenetics shows that environmental forces, including those related to structural inequality, can alter the way genes are expressed both quantitatively and qualitatively without altering the code embedded in the gene. The most widely studied epigenetic mechanism in human research is called *methylation*. In the process of methylation, tiny molecules called *methyl groups* attach to various parts of the gene and through this attachment are able to physically control gene output and guide qualitative changes in the final gene product (Jones & Takai, 2001). In some cases, these epigenetic changes allow organisms (including humans) to rapidly adapt to new and challenging environments. Often, these adaptations are life-sustaining, but in some cases, as in the case of chronic stress, they can also be toxic (McGowan & Szyf, 2010; Nestler, 2014).

So, how do epigenetic mechanisms relate to the D-SM and predisposing vulnerability? To answer this question, we need to return to early childhood stress and adult chronic stress to see how epigenetic mechanisms, as a response to HPA activity, can create biological outcomes that promote dysregulated stress reactivity, physical disability, and psychological challenges. Studies related to early childhood have suggested that fetal exposure to poverty-related prenatal stressors, including poor nutrition, smoking, maternal depression, maternal anxiety, and interpersonal violence, is associated with epigenetic changes to genes involved in the regulation of the stress response of the neonate. These epigenetic outcomes, as described in the previous section on early childhood programming of the stress response, then result in increased stress reactivity that feeds the stress cycle in the newborn child and can result in dysregulation of the HPA axis across the life course (Aristizabal et al., 2020; Conradt et al., 2013; Knopik et al., 2012; Lester et al., 2016; Oberlander et al., 2008). Although there is clear evidence for prenatal programming, there is also a large literature on epigenetic programming during the early postnatal environment, particularly in the presence of stressful early life experiences such as poor maternal sensitivity, maternal separations, and threatening environmental conditions (Conradt et al., 2016). Overall, the dysregulated stress response that is the outcome of this potential series of epigenetic modifications to genes involved in stress regulation creates a greater predisposing vulnerability to psychological outcomes of chronic stress such as anxiety, depression, and manifestations of preexisting psychiatric conditions (McEwen, 2003; Walker & Diforio, 1997).

Another type of epigenetic change that occurs in the face of extreme childhood stress results in decreased production of BDNF. A detailed description of the emotional and cognitive outcomes of dampened BDNF production (Radtke et al., 2015) can be found in Chapter 4. Suffice it to say that the decrease in BDNF resulting from epigenetic changes wrought by early childhood trauma is suspected in the etiology of emotional, behavioral, and cognitive challenges experienced across the life course, including mood disorders, suicidality, disruptive behavioral and attention problems, and struggles with learning (McClelland et al., 2011; McGowan & Szyf, 2010; Miao et al., 2020; Mill & Petronis, 2008; Nestler, 2014; Szyf, 2011).

In the adult realm, research targeting chronically stressed women in midlife has shown a direct connection between epigenetics and PNI. More specifically, epigenetic alterations of genes associated with TNF-alpha and IL-6 production have been widely documented, thus implicating epigenetics in the increased inflammatory activity seen in chronic stress (Hong et al., 2016; Palma-Gudiel et al., 2021). Tragically, inflammation wrought through these epigenetic alterations of cytokine-producing genes puts individuals at risk for stress-inducing physical illnesses, such as heart and autoimmune disease, and also increases vulnerability to depression, anxiety, and dementia (Douthit, 2015b).

> **Reflective Question**
>
> Based on what you have learned thus far about the impact of extreme stress on body function, what kinds of prevention and intervention strategies do you think would be helpful for Henrietta's current mental health concerns?

Our Brain-Based Approach to the Case of Henrietta

Using the case of Henrietta as a central reference point, this section describes how knowledge of PNI and epigenetics can inform a counseling intervention. When one views Henrietta's physiological, psychological, and spiritual health as inextricably intertwined, an integrative intervention strategy is called for that addresses the multiple factors contributing to her distress. Learning and practicing techniques that foster a sympathetic-parasympathetic shift could address physiological concerns, and the many sociocultural dimensions of her current circumstances could be addressed through culturally informed, multitargeted counseling. The progress that is made through these two counseling approaches could be reinforced through preventive wellness counseling. As stated by Douthit (2015b), "These interventions, taken as a whole, provide an ecological approach to intervention that includes individual biology, the self-in-context, and social-structural affordances" (p. 14). Each is discussed at length below.

Sympathetic-Parasympathetic Shift

Although we do not know with certainty that Henrietta has endured epigenetic changes that have affected her stress reactivity, an intervention targeting her current reactive state is key to her compromised immunological, cardiovascular, and mental health. Henrietta would benefit from tools enabling her to foster a sympathetic-parasympathetic shift, achievable through techniques that support a shift away from a state in which the sympathetic nervous system is dominant to one in which the parasympathetic nervous system prevails (Benson, 2011). The sympathetic nervous system, a major player in the stress response, increases heart rate, promotes high blood pressure, disrupts sleep, causes digestive problems such as irritable bowel, fuels anxiety, and plays a role in dysregulation of the immune response (Sapolsky, 2004). In stark contrast, the parasympathetic nervous system lowers heart rate and blood pressure, creates a sense of calm, and supports healthy immune function. Techniques that could be used to achieve a sympathetic-parasympathetic shift for Henrietta include mindfulness and other meditation modalities, breathwork, neurofeedback, biofeedback, guided imagery, sand tray work, and creative arts (Benson, 2011; McEwen, 2016). Outside of counseling, Henrietta could be encouraged to engage in low-cost, parasympathetic nervous system–promoting activities, such as crossword puzzles, sudoku, yoga, tai chi, progressive muscle relaxation, and needlework (e.g., knitting, crocheting; Benson, 2011). There is growing evidence that a shift to a parasympathetic nervous system–dominant state might reverse epigenetic effects that favor stress reactivity (Kaliman et al., 2014).

Culturally Informed, Multitargeted Counseling

Culturally informed, multitargeted counseling can be fruitfully designed to focus on the long-term psychological effects of challenging environmental conditions while attending to cultural sensitivities, internalized oppression, and the need for a range of community and government resources (Douthit, 2015a, 2015b). In many cases, the psychological sequelae of conditions of marginality can be successfully targeted by empirically supported treatments such as trauma-focused cognitive behavior therapy for intensely aversive experiences (Ponniah & Hollon, 2009) and cognitive behavior therapy for anxiety and depression (Butler et al., 2006). Counselors can also use therapeutic interventions, such as interpersonal psychotherapy, that emphasize client education and would aim to inform clients such as Henrietta about potential biological mechanisms related to marginality (Swartz, 1999).

An intervention with an eye toward social justice, such as narrative therapy, is particularly well designed to address experiences of marginality by helping clients to challenge their own experiences of internalized oppression and to recreate or, in the language of narrative therapy, *restory* their identities, thus developing narratives that transcend self-injurious identities and construct a sound base for bolstering personal empowerment (Akinyela, 2002).

Cultural congruence is the cornerstone of successful intervention with marginalized clients (Sue & Sue, 2016). Any intervention strategy that fails to attend to cultural sensitivities has the potential to revictimize clients, thus reinforcing perceptions of marginality and exacerbating the stress cycle (Ridley, 2005).

In cases in which poverty intersects with ethnicity, gender, ability, and other categories of cultural difference, it is important for counselors to familiarize themselves with community and government resources that will help them to address their clients' most fundamental and urgent material needs (Douthit, 2015b; Summers, 2016). One concrete step in addressing Henrietta's transportation issues would be to consider the option of telemental health. This alternative is possible, however, only if she has an adequate internet connection, computer/hardware access, and the ability to navigate computer operation—all of which might be possible through case management strategies to familiarize her with community resources. Another possible option, which could impact her grandchildren, would be school-based, multisystemic initiatives to successfully serve children with HPA activation in the school environment (see August et al., 2018, and Chafouleas et al., 2019, for an extended discussion of school-based strategies). Briefly, such efforts include teacher training in trauma-informed classroom management with an eye on providing safe classroom environments that do not trigger HPA-activating responses in students (Chafouleas et al., 2019), extracurricular before- and after-school programming promoting sympathetic-parasympathetic shift (e.g., mindfulness meditation and exercise programs; Magan & Yadav, 2022), and student-caregiver-teacher intervention programs (August et al., 2018). Also, the psychological and physiological stresses among school-age children of food insecurity could be partially addressed through intensified 12-month nutrition programs designed to provide a psychological sense of food security and the fuel needed for optimal brain functioning. Many additional examples of potential intervention could be cited here, but all point to more attention being focused on addressing the needs of marginalized communities at all levels of government.

Preventive Wellness Counseling and Early Interventions

In the face of chronic stress, PNI, and epigenetic challenges, Henrietta would benefit from engaging in healthy lifestyle practices that prevent or disrupt cycles of chronic stress. Chapter 8 addresses many dimensions of wellness—including spirituality, which is a central aspect of Henrietta's counseling—that would intervene in her disconnection from her internal spiritual resources. Many elements of the wellness model (Myers et al., 2000) favor a parasympathetic state and could bolster resiliency in the face of sociocultural adversity.

It is vital to remember the critical importance of the early prenatal and postnatal environment in epigenetic changes to stress reactivity, in prevention, and in early interventions targeting prenatal experiences and early parent-child interaction patterns. Several of these interventions are described in Chapter 4.

Conclusion

As science progresses, we will likely have a clearer understanding of the impact of conditions of oppression on emotional and physical well-being. Animal and human research has opened vistas of intervention to address the toxic outcomes of chronic stress, giving counselors new tools to provide aid for victims of social injustice. Such research is very promising for those who are putting their hopes in the future of neurocounseling and the ways in which it can inform intervention by building on an understanding of the relationships among mind, body, and the social world. Most important in the context of this chapter is the notion that emerging mind-body science affirms the belief, central to the counseling profession, that intervention that addresses social justice issues ultimately has the power to transform (Douthit, 2015a).

Reflective Questions

Drawing on both your understanding of the biology of marginality and the many sociocultural and psychological challenges that Henrietta faces, how might you construct a multifaceted model explaining her current depressed mood?

On the basis of the model constructed in Question 1, how might you conceptualize a comprehensive intervention strategy to address Henrietta's depressed mood?

Argue for or against the following statement: "The most effective interventions for addressing the psychological outcomes of epigenetic and PNI mechanisms must necessarily involve psychopharmacological or other medical interventions."

Quiz

1. Which of the following statements does not accurately characterize what is known about psychoneuroimmunology?
   a. Chronic stress can affect both innate and acquired immunity.
   b. Chronic stress causes inflammatory shutdown.
   c. Inflammatory pathways can lead to depression, anxiety, and social withdrawal.
   d. Chronic stress can lead to numerous debilitating chronic illnesses.
2. Which of the following statements is not a characteristic of chronic stress?
   a. It can exacerbate existing psychiatric conditions.
   b. It causes physical problems, generating additional chronic stress.
   c. It can result in psychological struggles that reduce resilience.
   d. It mobilizes negative feedback systems that help to regulate the stress response.
3. Epigenetic changes can explain which of the following?
   a. A decrease in neuroplasticity.
   b. The decreased stress reactivity seen in chronic stress.
   c. Enhanced cognitive and emotional development.
   d. All of the above.

4. Which of the following best describes the notion of predisposing vulnerability?
    a. People's level of vulnerability remains constant throughout their lifetime.
    b. It is determined by inheritance alone.
    c. It can change over time as a result of the impact of environmental forces.
    d. People are all born with the same basic vulnerability.

References

Ader, R., & Cohen, N. (1993). Psychoneuroimmunology: Conditioning and stress. *Annual Review of Psychology, 44*(1), 53–85. https://doi.org/10.1146/annurev.ps.44.020193.000413

Ader, R., Cohen, N., & Felten, D. (1995). Psychoneuroimmunology: Interactions between the nervous system and the immune system. *Lancet, 345*(8942), 99–103. https://doi.org/10.1016/S0140-6736(95)90066-7

Akinyela, M. (2002). De-colonizing our lives: Divining a post-colonial therapy. *International Journal of Narrative Therapy and Community Work, 2002*(2), 32–43.

Aristizabal, M. J., Anreiter, I., Halldorsdottir, T., Odgers, C. L., McDade, T. W., Goldenberg, A., Mostafavi, S., Kobor, M. S., Binder, E. B., Sokolowski, M. B., & O'Donnell, K. J. (2020). Biological embedding of experience: A primer on epigenetics. *Proceedings of the National Academy of Sciences, USA, 117*(38), 23261–23269. https://doi.org/10.1073/pnas.1820838116

August, G. J., Piehler, T. F., & Miller, F. G. (2018). Getting "SMART" about implementing multi-tiered system support to promote school mental health. *Journal of School Psychology, 66*, 85–96. https://doi.org/10.1016/j.jsp.2017.10.001

Bailey, Z. D., Krieger, N., Agénor, M., Graves, J., Linos, N., & Bassett, M. T. (2017). Structural racism and health inequities in the USA: Evidence and interventions. *Lancet, 389*(10077), 1453–1463. https://doi.org/10.1016/S0140-6736(17)30569-X

Benson, H. (2011). *The relaxation revolution: The science and genetics of mind–body healing*. Scribner.

Bower, J. E., & Kuhlman, K. R. (2023). Psychoneuroimmunology: An introduction to immune-to-brain communication and its implications for clinical psychology. *Annual Review of Clinical Psychology, 19*(1), 331–359. https://doi.org/10.1146/annurev-clinpsy-080621-045153

Brunoni, A. R., Lopes, M., & Fregni, F. (2008). A systematic review and meta-analysis of clinical studies on major depression and BDNF levels: Implications for the role of neuroplasticity in depression. *International Journal of Neuropsychopharmacology, 11*(8), 1169–1180. https://doi.org/10.1017/S1461145708009309

Butler, A. C., Chapman, J. E., Forman, E. M., & Beck, A. T. (2006). The empirical status of cognitive-behavioral therapy: A review of meta-analyses. *Clinical Psychology Review, 26*(1), 17–31. https://doi.org/10.1016/j.cpr.2005.07.003

Chafouleas, S. M., Koriakin, T. A., Roundfield, K. D., & Overstreet, S. (2019). Addressing childhood trauma in school settings: A framework for evidence-based practice. *School Mental Health, 11,* 40–53. https://doi.org/10.1007/s12310-018-9256-5

Chen, E., Brody, G. H., & Miller, G. E. (2022). What are the health consequences of upward mobility? *Annual Review of Psychology, 73*(1), 599–628. https://doi.org/10.1146/annurev-psych-033020-122814

Chen, E., Yu, T., Siliezar, R., Drage, J. N., Dezil, J., Miller, G. E., & Brody, G. H. (2020). Evidence for skin-deep resilience using a co-twin control design: Effects on low-grade inflammation in a longitudinal study of youth. *Brain, Behavior, and Immunity, 88,* 661–667. https://doi.org/10.1016/j.bbi.2020.04.070

Conradt, E., Hawes, K., Guerin, D., Armstrong, D. A., Marsit, C. J., Tronick, E., & Lester, B. M. (2016). The contributions of maternal sensitivity and maternal depressive symptoms to epigenetic processes and neuroendocrine functioning. *Child Development, 87*(1), 73–85. https://doi.org/10.1111/cdev.12483

Conradt, E., Lester, B. M., Appleton, A. A., Armstrong, D. A., & Marsit, C. J. (2013). The roles of DNA methylation of NR3C1 and 11β-HSD2 and exposure to maternal mood disorder in utero on newborn neurobehavior. *Epigenetics, 8*(12), 1321–1329. https://doi.org/10.4161/epi.26634

Council for Accreditation of Counseling and Related Educational Programs. (2024). *2024 CACREP standards.* https://www.cacrep.org/wp-content/uploads/2023/06/2024-Standards-Combined-Version-6.27.23.pdf

Cuevas, A. G., Ong, A. D., Carvalho, K., Ho, T., Chan, S. W., Allen, J. D., Chen, R., Rodgers, J., Biba, U., & Williams, D. R. (2020). Discrimination and systemic inflammation: A critical review and synthesis. *Brain, Behavior and Immunity, 89,* 465–479. https://doi.org/10.1016/j.bbi.2020.07.017

Dantzer, R., O'Connor, J. C., Lawson, M. A., & Kelley, K. W. (2011). Inflammation-associated depression: From serotonin to kynurenine. *Psychoneuroendocrinology, 36*(3), 426–436. https://doi.org/10.1016/j.psyneuen.2010.09.012

Del Giudice, M., Ellis, B. J., & Shirtcliff, E. A. (2011). The adaptive calibration model of stress reactivity. *Neuroscience and Biobehavioral Reviews, 35*(7), 1562–1592. https://doi.org/10.1016/j.neubiorev.2010.11.007

Douthit, K. (2015a). Bringing the laboratory into the office: How epigenetics can inform counseling practice. *Counseling Today, 58*(3), 18–23.

Douthit, K. (2015b). Psychoneuroimmunology: Tapping the potential of counseling to heal the mind-body. *Counseling Today, 57*(9), 12–15.

Dowlati, Y., Herrmann, N., Swardfager, W., Liu, H., Sham, L., Reim, E. K., & Lanctôt, K. L. (2010). A meta-analysis of cytokines in major depression. *Biological Psychiatry, 67*(5), 446–457. https://doi.org/10.1016/j.biopsych.2009.09.033

Furman, D., Campisi, J., Verdin, E., Carrera-Bastos, P., Targ, S., Franceschi, C., Ferrucci, L., Gilroy, D. W., Fasano, A., Miller, G. W., Miller, A. H., Mantovani, A., Weyand, C. M., Barzilai, N., Goronzy, J. J., Rando, T. A., Effros, R. B., Lucia, A., Kleinstreuer, N., & Slavich, G. M. (2019). Chronic inflammation in the etiology of disease across the life span. Nature Medicine, 25(12), 1822–1832. https://doi.org/10.1038/s41591-019-0675-0

Glassman, A. H., & Miller, G. E. (2007). Where there is depression, there is inflammation . . . sometimes! *Biological Psychiatry, 62*(4), 280–281. https://doi.org/10.1016/j.biopsych.2007.05.032

Hänsel, A., Hong, S., Cámara, R. J., & Von Kaenel, R. (2010). Inflammation as a psychophysiological biomarker in chronic psychosocial stress. *Neuroscience and Biobehavioral Reviews, 35*(1), 115–121. https://doi.org/10.1016/j.neubiorev.2009.12.012

Hong, H., Kim, B. S., & Im, H.-I. (2016). Pathophysiological role of neuroinflammation in neurodegenerative diseases and psychiatric disorders. *International Neurology Journal, 20*(1 Suppl.), S2–S7. https://doi.org/10.5213/inj.1632604.302

James, S. A. (1994). John Henryism and the health of African-Americans. Culture, Medicine, and Psychiatry, 18(2), 163–182. https://doi.org/10.1007/BF01379448

Janeway, C. A., Travers, P., Walport, M., & Shlomchik, M. (2001). *Immunobiology: The immune system in health and disease* (5th ed.). Garland Science.

Jones, P. A., & Takai, D. (2001). The role of DNA methylation in mammalian epigenetics. *Science, 293*(5532), 1068–1070. https://doi.org/10.1126/science.1063852

Juster, R. P., McEwen, B. S., & Lupien, S. J. (2010). Allostatic load biomarkers of chronic stress and impact on health and cognition. *Neuroscience and Biobehavioral Reviews, 35*(1), 2–16. https://doi.org/10.1016/j.neubiorev.2009.10.002

Kaliman, P., Álvarez-López, M. J., Cosín-Tomás, M., Rosenkranz, M. A., Lutz, A., & Davidson, R. J. (2014). Rapid changes in histone deacetylases and inflammatory gene expression in expert meditators. *Psychoneuroendocrinology, 40*, 96–107. https://doi.org/10.1016/j.psyneuen.2013.11.004

Kaufman, S. H. E. (2019). Immunology's coming of age. *Frontiers in Immunology, 10*, Article 684. https://doi.org/10.3389/fimmu.2019.00684

Kendall-Tackett, K. (2009). Psychological trauma and physical health: A psychoneuroimmunology approach to etiology of negative health effects and possible interventions. *Psychological Trauma: Theory, Research, Practice, and Policy, 1*(1), 35–48. https://doi.org/10.1037/a0015128

Kiecolt-Glaser, J. K., Loving, T. J., Stowell, J. R., Malarky, W. B., Lemeshow, S., Dickinson, S. L., & Glaser, R. (2005). Hostile marital interactions, proinflammatory cytokine production, and wound healing. *Archives of General Psychiatry, 62*(12), 1377–1384. https://doi.org/10.1001/archpsyc.62.12.1377

Knopik, V. S., Maccani, M. A., Francazio, S., & McGeary, J. E. (2012). The epigenetics of maternal cigarette smoking during pregnancy and effects on child development. *Developmental Psychopathology, 24*(4), 1377–1390. https://doi.org/10.1017/S0954579412000776

Koss, K. J., & Gunnar, M. R. (2018). Annual research review: Early adversity, the hypothalamic-pituitary-adrenocortical axis, and child psychopathology. *Journal of Child Psychology and Psychiatry, 59*(4), 327–346. https://doi.org/10.1111/jcpp.12784

Lester, B. M., Conradt, E., & Marsit, C. (2016). Introduction to the special section on epigenetics. *Child Development, 87*(1), 29–37. https://doi.org/10.1111/cdev.12489

Liu, X., Lin, X., Heath, M. A., Zhou, Q., Ding, W., & Qin, S. (2018). Longitudinal linkages between parenting stress and oppositional defiant disorder (ODD) symptoms among Chinese children with ODD. *Journal of Family Psychology, 32*(8), 1078–1086. https://doi.org/10.1037/fam0000466

Magan, D., & Yadav, R. K. (2022). Psychoneuroimmunology of meditation. *Annals of Neurosciences, 29*(2–3), 170–176. https://doi.org/10.1177/09727531221109117

Majd, M., Saunders, E. F. H., & Engeland, C. G. (2020). Inflammation and the dimensions of depression: A review. *Frontiers in Neuroendocrinology, 56*, Article 100800. https://doi.org/10.1016/j.yfrne.2019.100800

Marmot, M. G., Smith, G. D., Stansfeld, S., Patel, C., North, F., Head, J., White, I., Brunner, E., Feeney, A., & Davey Smith, G. (1991). Health inequalities among British civil servants: The Whitehall II study. *The Lancet, 337*(8754), 1387–1393. https://doi.org/10.1016/0140-6736(91)93068-K

Marmot, M., & Wilkerson, R. (Eds.). (2005). *Social determinants of health*. Oxford University Press.

Marshall, G. D. (2011). The adverse effects of psychological stress on immunoregulatory balance: Applications to human inflammatory diseases. *Immunology and Allergy Clinics of North America, 31*(1), 133–140. https://doi.org/10.1016/j.iac.2010.09.013

McClelland, S., Korosi, A., Cope, J., Ivy, A., & Baram, T. Z. (2011). Emerging roles of epigenetic mechanisms in the enduring effects of neonatal stress and experience on learning and memory. *Neurobiology of Learning and Memory, 96*(1), 79–88. https://doi.org/10.1016/j.nlm.2011.02.008

McEwen, B. S. (2003). Mood disorders and allostatic load. *Biological Psychiatry, 54*(3), 200–207. https://doi.org/10.1016/S0006-3223(03)00177-X

McEwen, B. S. (2007). Biology and physiology of stress and adaptation: Central role of the brain. *Physiological Reviews, 87*(3), 873–904. https://doi.org/10.1152/physrev.00041.2006

McEwen, B. S. (2016). In pursuit of resilience: Stress, epigenetics, and brain plasticity. *Annals of the New York Academy of Sciences, 1373*(1), 56–64. https://doi.org/10.1111/nyas.13020

McGowan, P. O., & Szyf, M. (2010). The epigenetics of social adversity in early life: Implications for mental health outcomes. *Neurobiology of Disease, 39*(1), 66–72. https://doi.org/10.1016/j.nbd.2009.12.026

Miao, Z., Wang, Y., & Sun, Z. (2020). The relationship between stress, mental disorders, and epigenetic regulation of BDNF. *International Journal of Molecular Sciences, 21*(4), 1375. https://doi.org/10.3390/ijms21041375

Mill, J., & Petronis, A. (2008). Pre- and peri-natal environmental risks for attention-deficit hyperactivity disorder (ADHD): The potential role of epigenetic processes in mediating susceptibility. *Journal of Child Psychology and Psychiatry, 49*(10), 1020–1030. https://doi.org/10.1111/j.1469-7610.2008.01909.x

Myers, J. E., Sweeney, T. J., & Witmer, J. M. (2000). The wheel of wellness counseling for wellness: A holistic model for treatment planning. *Journal of Counseling & Development, 78*(3), 251–266. https://doi.org/10.1002/j.1556-6676.2000.tb01906.x

Nestler, E. J. (2014). Epigenetic mechanisms of depression. *JAMA Psychiatry, 71*(4), 454–456. https://doi.org/10.1001/jamapsychiatry.2013.4291

Oberlander, T. F., Weinberg, J., Papsdorf, M., Grunau, R., Misri, S., & Devlin, A. M. (2008). Prenatal exposure to maternal depression, neonatal methylation of human glucocorticoid receptor gene (NR3C1) and infant cortisol stress responses. *Epigenetics, 3*(2), 97–106. https://doi.org/10.4161/epi.3.2.6034

Palma-Gudiel, H., Prather, A. A., Lin, J., Oxendine, J. D., Guintivano, J., Xia, K., Rubinow, D. R., Wolkowitz, O., Epel, E. S., & Zannas, A. S. (2021). HPA axis regulation and epigenetic programming of immune-related genes in chronically stressed and non-stressed mid-life women. *Brain, Behavior, and Immunity, 92*, 49–56. https://doi.org/10.1016/j.bbi.2020.11.027

Pieterse, A. L., Todd, N. R., Neville, H. A., & Carter, R. T. (2012). Perceived racism and mental health among Black American adults: A meta-analytic review. *Journal of Counseling Psychology, 59*(1), 1–9. https://doi.org/10.1037/a0026208

Ponniah, K., & Hollon, S. D. (2009). Empirically supported psychological treatments for adult acute stress disorder and posttraumatic stress disorder: A review. *Depression and Anxiety, 26*(12), 1086–1109. https://doi.org/10.1002/da.20635

Radtke, K. M., Schauer, M., Gunter, H. M., Ruf-Leuschner, M., Sill, J., Meyer, A., & Elbert, T. (2015). Epigenetic modifications of the glucocorticoid receptor gene are associated with the vulnerability to psychopathology in childhood maltreatment. *Translational Psychiatry, 5*(5), Article e571. https://doi.org/10.1038/tp.2015.63

Ridley, C. R. (2005). *Overcoming unintentional racism in counseling and therapy: A practitioner's guide to intentional intervention* (2nd ed.). Sage.

Russotti, J., Warmingham, J. M., Handley, E. D., Rogosch, F. A., & Cicchetti, D. (2020). Characterizing competence among a high-risk sample of emerging adults: Prospective predictions and biological considerations. *Development and Psychopathology, 32*(5), 1937–1953.

Sapolsky, R. M. (2004). *Why zebras don't get ulcers*. Holt.

Shariq, A. S., Brietzke, E., Rosenblat, J. D., Barendra, V., Pan, Z., & McIntyre, R. S. (2018). Targeting cytokines in reduction of depressive symptoms: A comprehensive review. *Progress in Neuropsychopharmacology & Biological Psychiatry, 83*, 86–91. https://doi.org/10.1016/j.pnpbp.2018.01.003

Sue, D. W., & Sue, D. (2016). *Counseling the culturally diverse: Theory and practice* (7th ed.). Wiley.

Summers, N. (2016). *Fundamentals of case management practice: Skills for the human services* (5th ed.). Cengage.

Surtees, P. G., Wainwright, N. W. J., Bockholdt, S. M., Luben, R. N., Warcham, N. J., & Khaw, K. T. (2008). Major depression, C-reactive protein, and incident ischemic heart disease in health men and women. *Psychosomatic Medicine, 70*(8), 850–855. https://doi.org/10.1097/PSY.0b013e318183acd5

Swartz, H. A. (1999). Interpersonal psychotherapy. In M. E. Hersen & A. S. Bellack (Eds.), *Handbook of comparative interventions for adult disorders* (pp. 139–155). Wiley.

Syed, S. A., & Nemeroff, C. B. (2017). Early life stress, mood, and anxiety disorders. *Chronic Stress, 1*, 1–16. https://doi.org/10.1177/2470547017694461

Szyf, M. (2011). DNA methylation, the early-life social environment, and behavior disorders. *Neurodevelopmental Disorders, 3*(3), 238–249. https://doi.org/10.1007/s11689-011-9079-2

Tarullo, A. R., & Gunnar, M. R. (2006). Child maltreatment and the developing HPA axis. *Hormones and Behavior, 50*(4), 632–639. https://doi.org/10.1016/j.yhbeh.2006.06.010c

van Goozen, S. H. M., Fairchild, G., Snoek, H., & Harold, G. T. (2007). The evidence for a neurobiological model of childhood antisocial behavior. *Psychology Bulletin, 133*(1), 146–182. https://doi.org/10.1037/0033-2909.133.1.149

Walker, E. F., & Diforio, D. (1997). Schizophrenia: A neural diathesis-stress model. *Psychological Review, 104*(4), 667–685. https://doi.org/10.1037/0033-295X.104.4.667

Williams, D. R., Lawrence, J., & Davis, B. (2019). Racism and health: Evidence and needed research. *Annual Review of Public Health, 40*, 105–125. https://doi.org/10.1146/annurev-publhealth-040218-043750

Wilson, C. J., Finch, C. E., & Cohen, H. J. (2002). Cytokines and cognition: The case for a head-to-toe inflammatory paradigm. *Journal of the American Geriatrics Society, 50*(12), 2041–2056. https://doi.org/10.1046/j.1532-5415.2002.50619.x

Zuckerman, M. (1999). *Vulnerability to psychopathology: A biosocial model*. American Psychological Association.

Chapter 4

Neurophysiology of Traumatic Stress

Laura K. Jones

Psychological trauma is pervasive in the United States. According to findings from the World Health Organization World Mental Health surveys, an estimated 70% of individuals throughout the world have experienced at least one incident of trauma during their lifetime (Kessler et al., 2017). Research has further indicated that most people experience not just one but multiple experiences of trauma over the course of their life (Kessler et al., 2017). Since the emergence of COVID-19, the incidence of trauma exposure has likely increased significantly (Boyraz & Legros, 2020; Horesh & Brown, 2020). It is inevitable that counselors will work with a trauma survivor at some point in their career.

Trauma survivors are a unique population of clients who warrant a unique skill set (Briere & Scott, 2013). Unprepared counselors can unknowingly retraumatize clients, predominantly by pushing them into processing traumatic memories before they are emotionally or physiologically prepared (Wells et al., 2003). It is essential for counselors working with trauma survivors to take a trauma-informed approach that considers these unique experiences and is informed by an understanding of how the brain and body respond to traumatic experiences. This chapter examines not only the neurophysiological responses to stress and traumatic stress but the implications of this knowledge for how counselors interact with clients after traumatic events. These concepts are illustrated through the story of Julian.

2024 CACREP Standards

This chapter addresses 2024 Council for Accreditation of Counseling and Related Educational Programs (CACREP) Standards pertinent to the Foundational Counseling Curriculum (Section 3) areas of Lifespan Development (Standard C) and Counseling Practice and Relationships (Standard E):

- Systemic, cultural, and environmental factors that affect lifespan development, functioning, behavior, resilience, and overall wellness (Standard C.11.)
- The influence of mental and physical health conditions on coping, resilience, and overall wellness for individuals and families across the life span (Standard C.12.)
- Effects of crises, disasters, stress, grief, and trauma across the life span (Standard C.13.)
- Evidence-based counseling strategies and techniques for prevention and intervention (Standard E.15.)
- Crisis intervention, trauma-informed, community-based, and disaster mental health strategies (Standard E.20.)

This chapter also addresses the following Entry-Level Specialized Practice Area (Section 5) standard for Marriage, Couple, and Family Counseling (Standard F):

- Impact of interpersonal violence on marriages, couples, and families (Standard F.3.)

Clinical Case Study: Julian

Julian is a 22-year-old cisgender American female of Eastern European heritage (second generation). She is a law student at a prestigious university in the southeastern United States and identifies as heterosexual. Julian was referred from the university counseling center after experiencing a panic attack after one of her classes. She reported that she has never experienced a panic attack before this situation but said that over the past 6 months she has experienced nightmares and difficulty sleeping, she has felt agitated and jumpy, and her grades have been dropping. She also reported that she has generally felt down and has increased her use of alcohol, suggesting that she drinks around two glasses of wine a night to relax and not be so "amped up" so she can sleep. She reported that on the day of the panic attack, she had just been in a role-play scenario acting as a defense attorney for a hypothetical male client accused of rape. Julian stated that she thought she was just nervous during the trial. Julian reported very shallow, rapid breathing, racing heartbeat, profuse sweating, tense shoulders, and dry mouth. She made it through the scenario. After class, she sat alone in the library studying for a test and started thinking about the survivor's story in the scenario. At that point, she began to feel as though she could not breathe, and swallowing became challenging. Her heart was pounding, and she thought she was having a heart attack. One of her best friends passed by her and asked whether she was OK. She nodded yes, but the friend perceived that something was wrong. Julian indicated that the friend put a hand on her back and sat down with her. She said that she instantly started feeling soothed as she felt her friend's hand and heard her voice.

Nature of Stress

To fully conceptualize the effects of traumatic stress on the body, a better understanding of the nature of stress in general is needed. The phrase "I'm just really stressed right now" is ubiquitous in counseling. Clients seem more comfortable discussing stress than many other complaints. It is socially acceptable, if not expected, to be stressed in this Western culture. However, the concept of stress is often misunderstood. Stress can be characterized as any pressure that is put on the system (i.e., body and brain) from a psychological, physical, or environmental source—or, as Selye (1975) conceptualized it, stress is any demand on the body for change.

Stress is often considered to be a detrimental experience, but that is not always the case. In small, contained doses, stress can be beneficial to people's attention, motivation, processing, and overall health, including immune system functioning. It helps people's bodies and minds to adapt and warns them that changes are needed to their internal or external environments. Stress that has beneficial effects on the system is called *eustress* (Selye, 1975). However, the body can be pushed only so hard and for so long. Eventually, a person exceeds their ability to readily adapt. At that point, stress can have deleterious effects. Even if a person experiences low-intensity stress, it can be harmful if it is chronic. Stress that leads to adverse responses is called *distress* (Selye, 1975). Distress can also be associated with too little stress. In such cases, an individual may feel frustration, a lack of motivation, boredom, and apathy. Between these two extremes of distress exists an optimal level of arousal and associated bodily functioning that corresponds to optimal or peak performance. Thus, one's goal should not be to get rid of stress entirely.

The body's response to stress is an adaptive process. This ability to exhibit resiliency and adaptively respond to life stressors is termed *allostasis*. Most of you have likely heard of *homeostasis*, or a balance of bodily states. However, homeostasis is a static experience, and bodily functioning is rarely static. Allostasis is essentially either homeostasis in action or the body's attempt to reach homeostasis. It is a response by the whole organism in a changing and often challenging environment not only to present stressors but also in anticipation of future stressors (Raglan & Schulkin, 2014). *Allostatic load* refers to the wear and tear on the body in the face of chronic or extreme stress, those circumstances in which one tends to feel overwhelmed and more vulnerable to a range of physiological and psychological difficulties (Raglan & Schulkin, 2014). The point at which the system begins to break down and growth and maintenance are no longer possible is termed *allostatic overload*. States of high allostatic load and overload can negatively affect every system of the body and have been associated with adverse mental and physical health outcomes (Bobba-Alves et al., 2022; Carbone, 2021; Guidi et al., 2021). Epigenetic changes, or changes in how our DNA is read and transcribed, can also result from the allostatic overload brought on by chronic or traumatic stress (B. S. McEwen, 2020; C. A. McEwen, 2022). It has been proposed that these effects and the associated epigenetic changes that can occur as a result may have the potential to carry over to future generations by way of both the sperm and the egg (Sharma, 2019; Yehuda & Lehrner, 2018), such as with intergenerational trauma.

Thus, the goal is to learn to cope with stress in a healthy way or, to put it in terms of the brain and body, to physiologically adapt and regulate one's autonomic response in the face of changing psychological, physical, or environmental pressures. Although we most often discuss the negative outcomes of stress (as does this chapter), burgeoning research has begun to explore the neurophysiology of resilience and posttraumatic growth, including the underlying epigenetic changes that cause someone to be highly resilient to stress, such as blunted inflammatory (immune system) responses (Mehta et al., 2020). Understanding this adaptive nature of the body in response to stress is essential in understanding the body's response to traumatic stress and the clinical implications of working with survivors. Let's now take a closer look at the effects that traumatic events can have on one's body, life, and relationships.

> **Reflective Questions**
>
> Which symptoms of PTSD may Julian have been experiencing?
>
> What aspects of these symptoms may play an adaptive or protective role for Julian?

Posttraumatic Stress

Although the terms *trauma* and *posttraumatic stress* are often used interchangeably, they represent different concepts. A trauma (i.e., traumatic event) is defined in the *Diagnostic and Statistical Manual of Mental Disorders, Fifth Edition, Text Revision* (*DSM-5-TR;* American Psychiatric Association [APA], 2022) as the experience of actual or threatened death, serious injury, or sexual violation. According to the *DSM-5-TR,* a person can experience these three situations in four ways: personal exposure, direct observation, learning of an event occurring to a family member or friend, or repeated exposure to aversive details of the situation (except through media, television, or movie exposure, unless work related). Traumatic events are often characterized as existing along a spectrum, from single-incident trauma (e.g., a car accident) to complex trauma, or as chronic, ongoing traumatic experiences that occur within a specific context (e.g., child maltreatment, partner violence; Courtois, 2004).

Traumatic and posttraumatic stress encompass not the actual event but the symptoms that arise as a result of the traumatic event. Whether behavioral, physical, emotional, or cognitive, these symptoms that counselors often see and hear result from the physiological responses of the body (e.g., neurological, autonomic, gastrointestinal, endocrine, immune) to that traumatic event. They are people's best efforts at processing (consciously or preconsciously) the event(s) and staying safe in the midst or aftermath of the trauma. In this way, the body's response to a traumatic event is inherently an adaptive process. The body is trying to keep the person safe and protect them from future potential harm. However, for some individuals this traumatic stress response lasts long after the threat for harm has dissipated and leads to a host of maladaptive social, emotional, cognitive, and physical outcomes, such as affective dysregulation, reexperiencing of the trauma, decreased self-monitoring, irritability, engagement in risky behaviors, and poor impulse control. Such unrelenting traumatic stress can also lead to poor physical health outcomes (Nichter et al., 2019; Pacella et al., 2013).

After exposure to a traumatic event, many adult survivors frequently experience a heightened state of vigilance (e.g., scanning the environment for perceived threats), an implicit or explicit desire to avoid reminders of the trauma, and affective responding (e.g., intense feelings of anxiety, fear related to the past trauma, or persistent negative mood). For many, such responses begin to dissipate over time. For others, however, the reexperiencing of the event becomes a debilitating state, precipitating multiple *DSM-5-TR* diagnoses—most notably, depression and those that fall under trauma- and stressor-related disorders. Most commonly reported among survivors are symptoms of posttraumatic stress disorder (PTSD; APA, 2022). Nearly 1 in 8 adult trauma survivors develops PTSD, a condition that encompasses interconnected neurological, physiological, psychological, and interpersonal consequences (APA, 2022; Breslau & Kessler, 2001).

The *DSM-5-TR* characterizes PTSD as a combination of symptoms that fall into one of the following four symptom clusters: (a) the intrusion of flashbacks, memories, nightmares, or reminders of the event; (b) avoidance of anything associated with the event; (c) negative alterations of mood and cognitions associated with anything related to the traumatic event, including an inability to remember aspects of the event, negative mood state and inability to experience positive emotions, and exaggerated beliefs about the self in relation to others; and (d) changes in autonomic arousal and reactivity, which can include hypervigilance, reckless behaviors, sleep disturbance, irritability and anger, and being easily startled.

> **Reflective Question**
>
> What aspects of these symptoms may play an adaptive or protective role for Julian?

Neurophysiology of Posttraumatic Stress

Underlying each of these symptoms lies a host of aberrations in brain and body functioning. When exposed to traumatic stress, nearly every part of the brain and much of the body is affected, as illustrated by the following scenario. (Please note that if you have experienced personal trauma, envisioning this scenario may be *triggering*. In other words, you may begin experiencing symptoms similar to those you experienced when previously exposed to trauma.)

> Imagine that you are walking late at night down a dark city street. There is only one streetlight in the distance, and it begins flickering and then goes out. You are alone. You feel your senses becoming more acute. You are hyperaware of everything around you. You may begin to breathe a bit faster and feel your heart rate start to increase. As you begin to pick up your pace, a sound catches your attention, and you look over to your left. When you look back, you notice someone standing directly in front of you with what appears to be a knife.

Did you notice anything about your body as you were reading through this scenario? Can you imagine what your body would be experiencing if you lived it? When you are awake and alert, information from the body (visceral sensations) and senses is constantly being brought into the brain, all but smell being processed by the thalamus and directed toward the appropriate subcortical or cortical areas used in optimally processing the information. Any distinct smells during the event move from the olfactory bulb directly to limbic regions of the brain, notably the amygdala and hippocampus. When you are exposed to a trauma or an acutely stressful experience, incoming sensory information is registered as threatening, and your body begins to sound its alarm system and adapt in such a way as to keep you safe. Your amygdala sends signals to the hypothalamus to release hormones that will initiate one's fight, flight, freeze, and/or fawn defenses. The *fawn response* occurs when someone responds to a threat by attempting to pacify it, such as when a person tries to please or placate an aggressor in a manner inconsistent with the person's actual thoughts and feelings. (Note that little to no empirical research exists at this time to explain the neurophysiology of the fawn response.) This response is a two-tiered system (Godoy et al., 2018; Gunnar & Quevedo, 2007). The first tier, known as the sympathetic-adrenal-medullary (SAM) axis, activates the sympathetic nervous system. The hypothalamus sends messages down to the adrenal glands and the adrenal medulla (i.e., inside of the adrenal glands) to release epinephrine and norepinephrine (as hormones). In tandem with the activation of the hypothalamus is the activation of the locus coeruleus in the brain stem. Among other functions, the locus coeruleus is the primary center for norepinephrine (as neurotransmitter) production. These catecholamines (epinephrine and norepinephrine) initiate energy metabolism and activate the sympathetic nervous system, causing many of the classic symptoms that you experienced when reading the preceding scenario. This leads to an accelerated heart rate, an increased pulse and blood pressure, an increased breathing rate, stimulated glucose release by the liver, dilated pupils, and inhibited gut and intestinal functioning. The body is sending all of its energy and resources to the parts of the body that will help to keep you alive, to fight your way out of the situation or flee from it as quickly as you can. You are not stopping to make careful decisions and to weigh your options, functions that would recruit higher cortical areas of your brain. That would waste time and potentially threaten your survival. In this moment, you are functioning from the more primitive and instinctual areas of your brain.

The hypothalamic-pituitary-adrenal (HPA) axis is the second tier of the fight-or-flight response. As discussed in Chapter 1, the HPA axis is responsible for regulating the release of corticosteroids—namely, the "stress hormone" cortisol from the adrenal cortex (i.e., outer region; see Figure 4.1). Although it has other functions, cortisol primarily helps the body to regulate the amount of glucose that is circulating through the bloodstream at any given time. Under prolonged or chronic stress, it can also impede functioning of the immune, endocrine (e.g., sex and thyroid hormones), circulatory, and gastrointestinal systems and lead to atrophy of the hippocampus. During this process, endogenous opioids can also be released into the system

(Foster et al., 2017; Morey et al., 2015; Sherin & Nemeroff, 2011; Whitworth et al., 2005). These opioids act as an analgesic, limiting sensitivity to emotional and physical pain (van der Kolk et al., 1989). SAM activation produces short-term effects, with rapid responses, whereas the HPA axis produces both short- and long-term changes (Godoy et al., 2018).

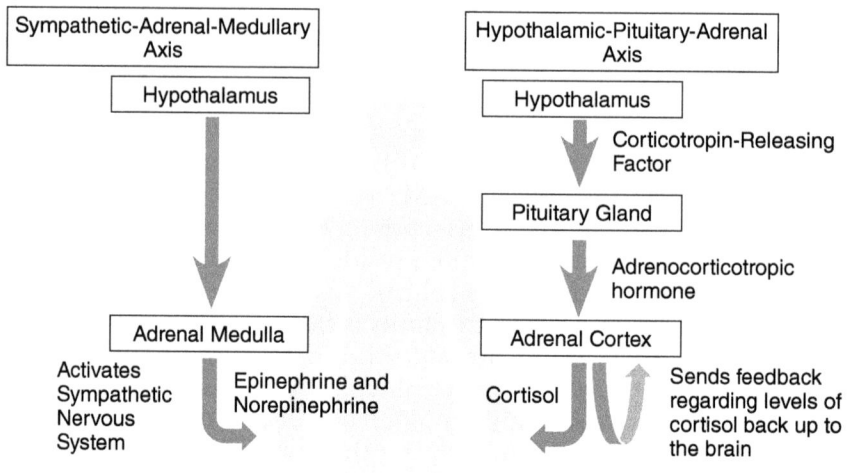

FIGURE 4.1

Hypothalamic-Pituitary-Adrenal (HPA) and Sympathetic-Adrenal-Medullary (SAM) Axes

Note. Derivative work of human body silhouette, Mikael Häggström, 2009, made transparent by Frédéric Michel, via Wikimedia Commons. Public domain.

Once the threat is no longer pervasive and the secretion of cortisol has reached an optimal threshold, the adrenal glands send chemical signals back up to the hypothalamus to discontinue production of corticotropin-releasing factor. Thus, the supply of cortisol in the system diminishes (Sriram et al., 2012). This is part of the allostatic process. The body is constantly responding to changes and perceived threats and has built-in mechanisms to reregulate itself. However, in the face of marked trauma or chronic unrelenting stress, this negative feedback loop becomes dysregulated.

This process of chronic threat responding—namely, the overactivation of the amygdala and resulting sympathetic nervous system and HPA responses—also disrupts the functioning of the frontal and prefrontal cortices as well as the hippocampus. Within the prefrontal cortex is a region, known as the *orbitofrontal cortex*, that has marked effects on interpersonal functioning. Not only does this area of the cortex regulate autonomic responses, but it is explicitly involved in the regulation of emotion and attuned interpersonal communication involving eye contact, response flexibility, and social cognition (Cavada & Schultz, 2000). The dorsolateral prefrontal cortex also shows decreased activation. This area plays a role in controlling unwanted memories and gives a sense of time to experiences, thereby giving an individual a sense that the experience will come to an end (Anderson et al., 2004). The decreased activation in these areas impairs the ability of the prefrontal cortex to optimally regulate the limbic and autonomic nervous systems (i.e., top-down control). This, in turn, can contribute to a host of maladaptive social, emotional, cognitive, and physical outcomes, such as affective dysregulation, reexperiencing of the trauma, decreased self-monitoring, irritability, engagement in risky behaviors, and poor impulse control. The intense fear experienced in trauma can lead to decreased activation in the region of the frontal cortex associated with language production, known as Broca's area (Shin et al., 1999). This may in turn affect survivors' ability to put words to experiences, particularly when reexperiencing a traumatic event.

Chronic traumatic stress responses can also disrupt declarative long-term memory consolidation in the hippocampus, leading to impaired integration of factual information with the emotional memory of the traumatic event (Hayes et al., 2011). Impeded functioning of the thalamus in response to trauma may also have an impact on memory of the event (van der Kolk, 2014). The thalamus helps to integrate one's sensory experiences with declarative or autobiographical components of memory. Because the functioning of the thalamus is impaired, one is left with a memory of the trauma that is trapped in one's body and senses. Rather than giving a fluid narrative of the event, survivors may describe sensory experiences such as sights, sounds, and bodily sensations, all accompanied by intense emotional reactions (van der Kolk, 2014).

The functioning of the anterior cingulate cortex and insula is also disrupted in traumatic stress. The rostral or forward part of the anterior cingulate shows a decrease in activation, which suggests a failure to appropriately weigh distracting emotional information and inhibit the functioning of the amygdala (Offringa et al., 2013). Conversely, the dorsal (top) anterior cingulate cortex increases in activation. Such functioning is related to the exaggerated fear response that is classic in PTSD (Shin et al., 2011). Furthermore, increased activation of the anterior insula in PTSD may be associated with a heightened awareness of internal bodily arousal during reexperiencing and hyperarousal symptoms (Hopper et al., 2007). Conversely, in states of hypoarousal, such as with emotional numbing and dissociative experiences, inverse patterns in the insula are seen, whereby the person has decreased activation and impaired awareness of bodily states (Hopper et al., 2007).

Although the SAM and HPA axes, along with the allied effects of epinephrine, norepinephrine, and cortisol, have taken center stage in research detailing the physiological underpinnings of traumatic responses, researchers have more recently begun to investigate and recognize the pronounced role of gamma-aminobutyric acid (GABA) and oxytocin in PTSD symptomatology and recovery (Giovanna et al., 2020; Rosso et al., 2022). GABA, the body's primary inhibitory neurotransmitter, plays a role in stress responses and the regulation of anxiety. Decreased levels of GABA have been demonstrated in PTSD, and lower GABA in cortical areas may play a role in difficulties around arousal regulation and symptom severity in PTSD (Rosso et al., 2022).

Similar to corticotropin-releasing factor in the HPA axis, oxytocin is also produced in the paraventricular nucleus of the hypothalamus, which extends neuronal projections directly to the amygdala, hippocampus, and brain stem (Campbell, 2010). Oxytocin is essential to the development and fostering of social and intimate bonds and has been found to play an integral role in deciphering the emotions of others (Hurlemann et al., 2010) and assessing interpersonal trustworthiness (Zak et al., 2004). In addition, oxytocin has been found to increase prefrontal cortex activity and has been hypothesized to play a role in different stages of trauma processing (Engel et al., 2019). In part, decreased amygdala activity may improve emotion regulation and decrease avoidance behavior (Olff et al., 2010). Given the palliative effects of oxytocin on physiological stress responses, authors have proposed the use of pharmacological oxytocin to augment and enhance the efficacy of cognitive behavior therapy and exposure therapy with survivors (Giovanna et al., 2020). However, endogenous concentrations of oxytocin (oxytocin occurring naturally in the body as opposed to being administered) do not appear to be different between individuals who have experienced trauma and those who have not, regardless of PTSD diagnosis (Engel et al., 2019).

Porges (2011) posited that the autonomic nervous system responds to trauma following a "phylogenetic hierarchy" (p. 155), whereby an individual typically first responds from the most evolutionarily recent cortical components. When this fails, more primitive structural defense systems are engaged. The autonomic nervous system is composed of three branches: (a) the sympathetic nervous system, (b) the dorsal vagal branch of the parasympathetic nervous system, and (c) the ventral vagal branch of the parasympathetic nervous system. Each of these corresponds to a level of autonomic arousal—namely, hyperarousal, hypoarousal, and optimal arousal, respectively. When hyperaroused, individuals experience emotional dysregulation, hypervigilance, and a reliance on survival mechanisms. In a hypoaroused state, individuals often experience dissociation, emotional numbing, and immobility, sometimes considered the freeze response.

According to Porges's (2011) theory, the ventral vagal branch of the parasympathetic nervous system, also known as the *social engagement system*, represents a state of regulated arousal, generating feelings of safety and promoting behaviors that enhance social bonds (Fosha et al., 2009; Porges, 2003). These behaviors are thought to include eye gaze, facial expressions, voice tone, social gestures, and even the ability to extract a human voice from background noises by modulating the function of a set of muscles of the inner ear. Individuals with PTSD often experience a dysregulated autonomic response and fluctuate between hypo- and hyperaroused states, and they are not functioning from the social engagement system. Steuwe et al. (2012) found that, compared with individuals without PTSD, survivors of trauma with PTSD experience threat-mediated arousal in areas of the brain associated with the autonomic nervous system when exposed to direct eye contact. Survivors of trauma also have difficulty with *affective prosody*, or the ability to properly interpret emotional cues in the rhythm, pitch, stress, and intonation of language (Freeman et al., 2009). This may be due to the dysregulation of the autonomic nervous system and an inability to effectively modulate muscles of the inner ear that detect such variations (Porges, 2011). It is important to remember that despite an overabundance of support for Porges's theory among clinicians who work with trauma, researchers have called into question various aspects, specifically around the *ventral vagal complex* (Neuhuber & Berthoud, 2022; Taylor et al., 2022).

Sex Differences in Posttraumatic Stress

Notable differences exist in the prevalence, symptoms, and duration of posttraumatic stress between males and females. Males are more likely to experience traumatic events, yet females are more than twice as likely to develop PTSD, with PTSD symptomatology lasting as much as 4 times longer even when the extent of trauma exposure and type of trauma experienced are controlled (Blain et al., 2010; Kessler, 2000; Olff et al., 2007; Tolin & Foa, 2006). Males tend to experience more anger and to reexperience symptoms, whereas females experience greater degrees of emotional numbing, restricted affect, and avoidance responses as well as higher levels of psychological reactivity to traumatic stimuli (Litz et al., 2000; Orsillo et al., 2004; Spahic-Mihajlovic et al., 2005).

The role of sex hormones may be one factor contributing to such differences. Both testosterone and estrogen serve as a protective factor against developing PTSD and anxiety disorders and lead to enhanced fear regulation and extinction (Christiansen & Berke, 2020; Daskalakis et al., 2016; Ney et al., 2019; Seligowski et al., 2020). Relatedly, progesterone appears to both directly and indirectly exert memory-enhancing effects at the time of trauma and may lead to an increase in negative emotional memories (Ney et al., 2019). However, individuals with a menstrual cycle experience a constantly cycling level of such hormones. This has led to the suggestion that the phase of menstrual cycle at the time of a trauma can potentially influence how likely a person is to develop PTSD. For example, an individual might be more psychologically resilient if they experienced a trauma just before ovulation, when estrogen levels peak, than they would be if they experienced trauma when estrogen levels are lower and progesterone levels higher (such as in the luteal phase). Such research has begun to elucidate why individuals with a menstrual cycle may be more likely to experience posttraumatic stress responses than individuals who are testosterone predominant and experience more consistent levels of gonadal hormones and, thus, more consistent protective benefits of testosterone.

Complex Posttraumatic Stress

Complex trauma or ongoing experiences that entail exposure to multiple types of trauma, especially interpersonal traumas occurring during childhood, may lead to a type of posttraumatic response distinct from PTSD (Cloitre et al., 2011; Courtois, 2004). In addition to having symptoms of PTSD, individuals experiencing complex PTSD often demonstrate difficulties in the following five core areas: emotion regulation, relational capacities, attention and consciousness (e.g., dissociation), belief systems, and somatic distress (Cloitre et al., 2011). Furthermore, the younger the individual is at the time of the first trauma, the more likely that individual is to experience the symptoms of complex PTSD in addition to those of PTSD (van der Kolk et al., 2005).

Unique patterns of brain and neuroendocrine changes are seen in complex trauma survivors, particularly those who experience child maltreatment (van der Kolk et al., 2005). As with symptom expression, both earlier age at onset and longer duration of the complex trauma lead to more pronounced structural and functional changes in the brain (Andersen et al., 2008). There may also be sensitive periods for the influence of early, complex traumatic stress on brain functioning and development (Oh et al., 2022; Teicher et al., 2006). Changes stemming from early child maltreatment in particular include decreases in the volume of and synaptic density in the amygdala, hippocampus, corpus callosum, and prefrontal cortex (Andersen et al., 2008; Teicher et al., 2006). Exposure to early life trauma may also lower morning cortisol levels and blunt the diurnal rhythm of cortisol, a marker for HPA dysregulation (Bevans et al., 2008). Lower morning levels of cortisol are also linked to childhood aggression and difficulties with affect regulation (Cicchetti & Rogosch, 2007; Murray-Close et al., 2008). As additional research is conducted on complex PTSD, a better understanding of the unique structural and functional neurophysiological changes that occur in the face of prolonged and repeated traumatic stress will begin to emerge.

> **Reflective Questions**
>
> Consider the symptoms of PTSD identified earlier.
>
> What might be the neurophysiological origins of Julian's PTSD symptoms?
>
> How did these neurophysiological changes lead to maladaptive experiences?

Trauma-Focused Interventions

Multiple therapies have emerged and been refined, stemming from such knowledge of the brain. Cognitive behavior therapy interventions, such as prolonged exposure therapy (Foa et al., 2007), cognitive processing therapy (Resick & Schnicke, 1993), and trauma-focused cognitive behavior therapy (Ponniah & Hollon, 2009), are considered best practices in the field and have a solid research base. Additional therapies that honor the role of the body, nondeclarative memory, and *bottom-up* processing (i.e., initiating changes in the body and subcortical areas of the brain that then affect the cortex, as opposed to *top-down* interventions that start by altering cognitive understandings that aim to influence the functioning of more subcortical and sensory structures) in conceptualizations of trauma are beginning to gain even more solid empirical evidence of their effectiveness. These therapies include eye movement desensitization and reprocessing (Shapiro & Solomon, 1995), neurofeedback (Chapin & Russell-Chapin, 2014), and sensorimotor psychotherapy (Ogden & Minton, 2000). These approaches work to help clients regulate the response of their body and brain to emotionally laden traumatic memories and, in doing so, allow survivors to develop a sense of agency and regain ownership of the body (van der Kolk, 2014).

Researchers have effectively demonstrated that of the interventions currently investigated, none are successful in addressing "the full range of clinical problems observed in trauma survivors" (McFarlane & Yehuda, 2000, p. 941). Tragically, somewhere between 20% and 50% of clients who are trauma survivors drop out of psychotherapy (Bryant et al., 2007; Schottenbauer et al., 2008), with those higher in avoidance coping and numbing being more prone to discontinue counseling (Bryant et al., 2007). Not surprisingly, impairments in emotion regulation, a hallmark of PTSD, directly interfere with participation in psychotherapy (Freeman et al., 2009). Given these very real therapeutic challenges, what can counselors do to improve clinical outcomes with this population?

Briere and Scott (2013) suggested that regardless of theoretical underpinnings, effective trauma-informed therapy for PTSD can be broken down into several core principles that follow a phase-oriented approach to trauma treatment first cited by Herman (1992). The International Society for Traumatic Stress Studies in its best practices for complex PTSD similarly recommended a sequenced and phase-based approach to working with survivors (Cloitre et al., 2011). Given the difficulties cited about interpersonal trust and the posited importance of the ventral vagal social engagement system in autonomic arousal, the first consideration in trauma care is safety and security. The therapeutic relationship is recognized as the essential context in which healing from trauma can occur (Herman, 1992) and is strongly predictive of therapeutic outcomes (Roth & Fonagy, 2006). Providing and ensuring safety encompasses both physical and psychological safety. When clients do not perceive a sense of safety, disclosing and working through the trauma may be retraumatizing. Thus, counselors should take time to attend to the therapeutic relationship initially.

After the establishment of a trusting relationship, clients can benefit from knowledge about and skills to cope with their condition. Providing culturally, linguistically, and developmentally appropriate psychoeducation about what trauma does to the body and brain can empower clients. Psychoeducation can be augmented by visual brain models—either three dimensional or the hand model presented in Figure 1.4—pictures, or metaphors. As with all psychoeducation, it is important that clients personally relate to the information and make it applicable to their understanding and experience. Psychoeducation may also lead to self-compassion regarding survivors' present experience. Developing self-compassion has been associated with reducing PTSD symptoms such as negative self-perceptions (Seligowski et al., 2015).

Trauma survivors need to learn distress tolerance and affect regulation techniques before trauma is processed cognitively, because thinking and talking about the traumatic event may cause distress. This is where biofeedback, mindfulness, and grounding exercises may be used. Deep breathing and yoga can also be helpful, although they should be approached with caution. Certain yoga poses and the sensation of filling the lungs with air during deep-breathing exercises can be triggering for some clients. For example, the sensations of deep breathing may, either implicitly or explicitly, remind the survivor of running away during the trauma. Identifying, discriminating, and labeling emotions may also be beneficial. As described in Chapter 2, the cognitive labeling of physiologically based emotions (feelings) can help to calm autonomic responses (Roth & Fonagy, 2006). This building of distress tolerance helps clients to develop coping skills to manage what distress may arise as the difficult trauma-related memories and sensations are encountered and later processed.

Subsequent phases of trauma treatment typically target cognitive and emotional processing of the trauma (Briere & Scott, 2013). A central notion exists among many trauma theorists that the key to lasting treatment for PTSD resides in the successful integration of bifurcated memories of the traumatic event (the explicit and implicit experiences of the event), whereby the declarative memory of the traumatic event is reintegrated with the powerfully distressing implicit, emotional, and somatic traumatic memory (Foa & Rothbaum, 1998). This can be accomplished through any number of the aforementioned therapies, each with its own unique way of processing the trauma. Many of these therapies involve telling or writing out the trauma story. However, some approaches, such as eye movement desensitization and reprocessing, do not require an overt retelling of the story at all. In addition, Briere and Scott (2013) suggested that improving identity and relational functioning may also be a necessary treatment component for individuals who have experienced complex or attachment traumas early in life.

My Brain-Based Approach to the Case of Julian

On first hearing Julian's story, I (a counselor) wondered whether Julian had experienced past trauma, especially interpersonal trauma given the nature of the trial that triggered her panic attack. Despite this hypothesis, I did not jump into asking her about past trauma. Instead, I held onto this information and recognized that the primary task with Julian would be to develop a strong therapeutic relationship, relying heavily on person-centered techniques of empathy and reflective listening. Slowly and consistently forming this strong, nonjudgmental environment helped to keep Julian feeling safe and accepted within that relationship.

As an established base was formed with Julian, she began to disclose more detail around events 6 months before the panic attack in which she had experienced a sexual assault. I provided psychoeducation about how trauma can affect the body and the brain, as well as trust within interpersonal relationships. Discussing the adaptive nature of trauma responses and even the potential role of estrogen in this process is important. Next, I discussed the phases of general trauma therapy and the importance of self-regulation skills. Immediacy was used to help Julian identify moments when she started to become hyper- or hypoaroused. We discussed what arousal felt like in her body and started to label some of those sensations. I used a feelings chart for Julian to begin applying a wider range of words to her sensations. Exercises that would help her to maintain control over her arousal and emotions were located. In session, we practiced mindfulness, diaphragmatic breathing, and grounding exercises so that Julian could practice these outside of session as she became more comfortable. On those occasions when Julian began to disassociate or have flashbacks in session, I recognized that her Broca's area might not have been fully functioning and that she might have been having difficulty describing what she was experiencing in that moment.

This process of building a safe connection, understanding Julian's emotions, and finding effective means of regulating Julian's arousal and emotions was a long, slow process. I knew that until this was accomplished, working through Julian's traumatic memories would likely retraumatize her and have a negative impact on her ultimate recovery. When she felt safe and grounded enough to discuss her past trauma in detail, we discussed her optimal level of challenge and created a stress-fear exposure hierarchy. Together, we slowly worked through each level of that hierarchy, starting with the least challenging level and progressing to the most challenging. Julian was given full autonomy to determine when she was ready to move to the next level of the hierarchy. During this process, we eventually addressed the traumatic memories and associated triggers through imaginal exposure. Julian began to slowly retell her memories.

As she told her story, we regularly paused to reflect back on the material to establish a slow and steady pace to the story. Julian was taught to stop intermittently to discuss her emotions as she was telling her story. As she retold her narrative, Julian used her newly learned verbal feeling labels and her grounding skills to manage her level of arousal and stress. Throughout the course of her story, I closely watched her nonverbal expressions of emotion and provided feedback. I also intermittently reiterated that Julian was safe in the counseling room as opposed to reliving the trauma. The slow pace of the therapy allowed Julian to maintain an optimal level of arousal, strengthening the connections between her prefrontal cortex and limbic system and allowing for greater executive control over her subcortical limbic structures and a reintegration of the trauma memory. Moreover, by stopping to discuss and put words to Julian's emotions, we began to integrate her implicit and explicit memories of the events. This process occurred repeatedly until Julian was able to get through the entire story and maintain her level of arousal. This can be a very protracted process.

During the final stage of therapy, Julian began to adapt to life without her traumatic response, learning how to function differently in her social environment given her enhanced corticolimbic connections, self- and emotion regulation, and improved neural and memory integration. At this point, I began to work with Julian on developing her interpersonal functioning and communication so that she could enhance her connections with others. Julian wanted to become involved in a support group for survivors to establish role models for healthy interpersonal interactions and to build trust outside the therapeutic relationship. I emphasized Julian's strengths throughout our work, empowering her to reach out and help others.

Conclusion

During and after exposure to a traumatic event, the brain and body undergo a cascade of neurophysiological changes. In the face of danger or threat to life, these changes are adaptive and protective in nature. However, once the acute threat is no longer present, the persistence of such responses can become maladaptive and lead to unrelenting symptoms of mental and physical distress. Trauma survivors are a unique population of clients who require counselors to have specialized knowledge of how trauma affects the body and brain to offer safe and effective care to clients.

This chapter provided a basic introduction to the neurophysiology of stress and traumatic stress as well as how such knowledge has informed and is continuing to inform best-practice models. There is so much more that could be added to this chapter to provide a more in-depth understanding of trauma. Furthermore, as new research continues to emerge, therapeutic approaches to trauma will continue to strengthen and unfold, in the hope that counselors can deliver even more effective interventions for trauma survivors that increase therapeutic retention rates, lead to symptom reduction, and aid in neurophysiological regulation and optimal functioning. The contents of this chapter and book are intended to provide a foundation for you to continue your reading and understanding of the neurophysiology of trauma responses and posttraumatic stress.

Quiz

1. In the face of extreme or chronic stress, which of the following statements regarding cortisol is true?
    a. A negative feedback loop for the HPA axis is initiated, which impairs the body's ability to regulate levels of cortisol.
    b. Cortisol levels stay consistent.
    c. Cortisol is released from the cingulate cortex.
    d. A negative feedback loop for the HPA axis is disrupted, which impairs the body's ability to regulate levels of cortisol.

2. Which of the following structures is not implicated in impaired traumatic memories?
    a. Amygdala.
    b. Pineal gland.
    c. Thalamus.
    d. Hippocampus

References

American Psychiatric Association. (2022). *Diagnostic and statistical manual of mental disorders* (5th ed., text rev.). https://doi.org/10.1176/appi.books.9780890425787

Andersen, S. L., Tomada, A., Vincow, E. S., Valente, E., Polcari, A., & Teicher, M. H. (2008). Preliminary evidence for sensitive periods in the effect of childhood sexual abuse on regional brain development. *Journal of Neuropsychiatry and Clinical Neurosciences, 20*(3), 292–301.

Anderson, M. C., Ochsner, K. N., Kuhl, B., Cooper, J., Robertson, E., Gabrieli, S. W., Glover, G. H., & Gabrieli, J. D. (2004). Neural systems underlying the suppression of unwanted memories. *Science, 303*(5655), 232–235.

Bevans, K., Cerbone, A., & Overstreet, S. (2008). Relations between recurrent trauma exposure and recent life stress and salivary cortisol among children. *Development and Psychopathology, 20*(1), 257–272.

Blain, L. M., Galovski, T. E., & Robinson, T. (2010). Gender differences in recovery from posttraumatic stress disorder: A critical review. *Aggression and Violent Behavior, 15*(6), 463–474.

Bobba-Alves, N., Juster, R. P., & Picard, M. (2022). The energetic cost of allostasis and allostatic load. *Psychoneuroendocrinology, 146*, Article 105951.

Boyraz, G., & Legros, D. N. (2020). Coronavirus disease (COVID-19) and traumatic stress: Probable risk factors and correlates of posttraumatic stress disorder. *Journal of Loss and Trauma, 25*(6–7), 503–522.

Breslau, N., & Kessler, R. C. (2001). The stressor criterion in *DSM-IV* posttraumatic stress disorder: An empirical investigation. *Biological Psychiatry, 50*(9), 699–704.

Briere, J., & Scott, C. (2013). *Principles of trauma therapy: A guide to symptoms, evaluation and treatment.* Sage.

Bryant, R. A., Moulds, M. L., Mastrodomenico, J., Hopwood, S., Felmingham, K., & Nixon, R. D. V. (2007). Who drops out of treatment for post-traumatic stress disorder? *Clinical Psychologist, 11*(1), 13–15. https://doi.org/10.1080/13284200601178128

Campbell, A. (2010). Oxytocin and human social behavior. *Personality and Social Psychology Review, 14*(3), 281–295. https://doi.org/10.1177/1088868310363594

Carbone, J. T. (2021). Allostatic load and mental health: A latent class analysis of physiological dysregulation. *Stress, 24*(4), 394–403.

Cavada, C., & Schultz, W. (2000). The mysterious orbitofrontal cortex. Foreword. *Cerebral Cortex, 10*(3), 205.

Chapin, T., & Russell-Chapin, L. A. (2014). *Neurotherapy and neurofeedback: Brain-based treatment for psychological and behavioral problems.* Routledge.

Christiansen, D. M., & Berke, E. T. (2020). Gender-and sex-based contributors to sex differences in PTSD. *Current Psychiatry Reports, 22,* 1–9.

Cicchetti, D., & Rogosch, F. A. (2007). Personality, adrenal steroid hormones, and resilience in maltreated children: A multilevel perspective. *Development and Psychopathology, 19*(3), 787–809.

Cloitre, M., Courtois, C. A., Charuvastra, A., Carapezza, R., Stolbach, B. C., & Green, B. L. (2011). Treatment of complex PTSD: Results of the ISTSS expert clinician survey on best practices. *Journal of Traumatic Stress, 24*(6), 615–627.

Council for Accreditation of Counseling and Related Educational Programs. (2024). *2024 CACREP standards.* https://www.cacrep.org/wp-content/uploads/2023/06/2024-Standards-Combined-Version-6.27.23.pdf

Courtois, C. A. (2004). Complex trauma, complex reactions: Assessment and treatment. *Psychotherapy: Theory, Research, Practice, Training, 41*(4), 412–425.

Daskalakis, N. P., McGill, M. A., Lehrner, A., & Yehuda, R. (2016). Endocrine aspects of PTSD: Hypothalamic-pituitary-adrenal (HPA) axis and beyond. In C. R. Martin, V. R. Preedy, & V. B. Patel (Eds.), *Comprehensive guide to post-traumatic stress disorder* (pp. 245–260). Springer International.

Engel, S., Klusmann, H., Laufer, S., Pfeifer, A. C., Ditzen, B., van Zuiden, M., Knaevelsrud, C., & Schumacher, S. (2019). Trauma exposure, post-traumatic stress disorder and oxytocin: A meta-analytic investigation of endogenous concentrations and receptor genotype. *Neuroscience & Biobehavioral Reviews, 107,* 560–601.

Foa, E. B., Hembree, E. A., & Rothbaum, B. O. (2007). *Prolonged exposure therapy for PTSD: Emotional processing of traumatic experiences.* Oxford University Press.

Foa, E. B., & Rothbaum, B. O. (1998). *Treating the trauma of rape: Cognitive-behavioral therapy for PTSD.* Guilford Press.

Fosha, D., Siegel, D. J., & Solomon, M. F. (2009). *The healing power of emotion: Affective neuroscience, development, and clinical practice.* Norton.

Foster, J. A., Rinaman, L., & Cryan, J. F. (2017). Stress and the gut-brain axis: Regulation by the microbiome. *Neurobiology of Stress, 7,* 124–136.

Freeman, T. W., Hart, J., Kimbrell, T., & Ross, E. D. (2009). Comprehension of affective prosody in veterans with chronic posttraumatic stress disorder. *Journal of Neuropsychiatry and Clinical Neurosciences, 21*(1), 52–58. https://doi.org/10.1176/appi.neuropsych.21.1.52

Giovanna, G., Damiani, S., Fusar-Poli, L., Rocchetti, M., Brondino, N., de Cagna, F., Mori, A., & Politi, P. (2020). Intranasal oxytocin as a potential therapeutic strategy in post-traumatic stress disorder: A systematic review. *Psychoneuroendocrinology, 115,* Article 104605. https://doi.org/10.1016/j.psyneuen.2020.104605

Godoy, L. D., Rossignoli, M. T., Delfino-Pereira, P., Garcia-Cairasco, N., & de Lima Umeoka, E. H. (2018). A comprehensive overview on stress neurobiology: Basic concepts and clinical implications. *Frontiers in Behavioral Neuroscience, 12,* Article 127.

Guidi, J., Lucente, M., Sonino, N., & Fava, G. A. (2021). Allostatic load and its impact on health: A systematic review. *Psychotherapy and Psychosomatics, 90*(1), 11–27.

Gunnar, M., & Quevedo, K. (2007). The neurobiology of stress and development. *Annual Review of Psychology, 58,* 145–173.

Häggström, M. (2009). *Human body silhouette* [Digital image]. https://commons.wikimedia.org/wiki/File%3AHuman_body_silhouette.svg

Hayes, J. P., LaBar, K. S., McCarthy, G., Selgrade, E., Nasser, J., Dolcos, F., & Morey, R. A. (2011). Reduced hippocampal and amygdala activity predicts memory distortions for trauma reminders in combat-related PTSD. *Journal of Psychiatric Research, 45*(5), 660–669. https://doi.org/10.1016/j.jpsychires.2010.10.007

Herman, J. L. (1992). *Trauma and recovery: The aftermath of violence from domestic abuse to political terror.* Basic Books.

Hopper, J. W., Frewen, P. A., Van der Kolk, B. A., & Lanius, R. A. (2007). Neural correlates of reexperiencing, avoidance, and dissociation in PTSD: Symptom dimensions and emotion dysregulation in responses to script-driven trauma imagery. *Journal of Traumatic Stress, 20*(5), 713–725. https://doi.org/10.1002/jts.20284

Horesh, D., & Brown, A. D. (2020). Traumatic stress in the age of COVID-19: A call to close critical gaps and adapt to new realities. *Psychological Trauma: Theory, Research, Practice, and Policy, 12*(4), 331–335.

Hurlemann, R., Patin, A., Onur, O. A., Cohen, M. X., Baumgartner, T., Metzler, S., Dziobek, I., Gallinat, J., Wagner, M., Maier, W., & Kendrick, K. M. (2010). Oxytocin enhances amygdala-dependent, socially reinforced learning and emotional empathy in humans. *Journal of Neuroscience, 30*(14), 4999–5007. https://doi.org/10.1523/jneurosci.5538-09.2010

Kessler, R. C. (2000). Posttraumatic stress disorder: The burden to the individual and to society. *Journal of Clinical Psychiatry, 61,* 4–14.

Kessler, R. C., Aguilar-Gaxiola, S., Alonso, J., Benjet, C., Bromet, E. J., Cardoso, G., Degenhardt, L., de Girolamo, G., Dinolova, R. V., Ferry, F., Florescu, F., Gureje, O., Haro, J. P., Huang, Y., Karam, E. G., Kawakami, N., Lee, S., Lepine, J.-P., Levinson, D., . . . Koenen, K. C. (2017). Trauma and PTSD in the WHO world mental health surveys. *European Journal of Psychotraumatology, 8*(sup5), Article 1353383.

Litz, B. T., Orsillo, S. M., Kaloupek, D., & Weathers, F. (2000). Emotional processing in posttraumatic stress disorder. *Journal of Abnormal Psychology, 109*(1), 26–39.

McEwen, B. S. (2020). A life-course, epigenetic perspective on resilience in brain and body. In A. Chen (Ed.), *Stress resilience* (pp. 1–21). Academic Press.

McEwen, C. A. (2022). Connecting the biology of stress, allostatic load and epigenetics to social structures and processes. *Neurobiology of Stress, 17,* Article 100426.

McFarlane, A. C., & Yehuda, R. (2000). Clinical treatment of posttraumatic stress disorder: Conceptual challenges raised by recent research. *Australian and New Zealand Journal of Psychiatry, 34*(6), 940–953. https://doi.org/10.1046/j.1440-1614.2000.00829.x

Mehta, D., Miller, O., Bruenig, D., David, G., & Shakespeare-Finch, J. (2020). A systematic review of DNA methylation and gene expression studies in posttraumatic stress disorder, posttraumatic growth, and resilience. *Journal of Traumatic Stress, 33*(2), 171–180.

Morey, J. N., Boggero, I. A., Scott, A. B., & Segerstrom, S. C. (2015). Current directions in stress and human immune function. *Current Opinion in Psychology, 5*, 13–17.

Murray-Close, D., Han, G., Cicchetti, D., Crick, N. R., & Rogosch, F. A. (2008). Neuroendocrine regulation and physical and relational aggression: The moderating roles of child maltreatment and gender. *Developmental Psychology, 44*(4), 1160–1176.

Neuhuber, W. L., & Berthoud, H. R. (2022). Functional anatomy of the vagus system: How does the polyvagal theory comply? *Biological Psychology, 174*, Article 108425.

Ney, L. J., Gogos, A., Hsu, C. M. K., & Felmingham, K. L. (2019). An alternative theory for hormone effects on sex differences in PTSD: The role of heightened sex hormones during trauma. *Psychoneuroendocrinology, 109*, Article 104416.

Nichter, B., Norman, S., Haller, M., & Pietrzak, R. H. (2019). Physical health burden of PTSD, depression, and their comorbidity in the US veteran population: Morbidity, functioning, and disability. *Journal of Psychosomatic Research, 124*, Article 109744.

Offringa, R., Brohawn, K. H., Staples, L. K., Dubois, S. J., Hughes, K. C., Pfaff, D. L., VanElzakker, M. B., Davis, F. C., & Haber, S. N. (2013). Diminished rostral anterior cingulate cortex activation during trauma-unrelated emotional interference in PTSD. *Biology of Mood & Anxiety Disorders, 3*(1), Article 10.

Ogden, P., & Minton, K. (2000). Sensorimotor psychotherapy: One method for processing traumatic memory. *Traumatology, 6*(3), 149–173.

Oh, S. J., Nam, K. R., Lee, N., Kang, K. J., Lee, K. C., Lee, Y. J., & Choi, J. Y. (2022). Developmental complex trauma induces the dysfunction of the amygdala-mPFC circuit in the serotonergic and dopaminergic systems. *Biochemical and Biophysical Research Communications, 605*, 104–110.

Olff, M., Langeland, W., Draijer, N., & Gersons, B. P. R. (2007). Gender differences in posttraumatic stress disorder. *Psychological Bulletin, 133*(2), 183–204. https://doi.org/10.1037/0033-2909.133.2.183

Olff, M., Langeland, W., Witteveen, A., & Denys, D. (2010). A psychobiological rationale for oxytocin in the treatment of posttraumatic stress disorder. *CNS Spectrums, 15*(8), 522–530.

Orsillo, S. M., Batten, S. V., Plumb, J. C., Luterek, J. A., & Roessner, B. M. (2004). An experimental study of emotional responding in women with posttraumatic stress disorder related to interpersonal violence. *Journal of Traumatic Stress, 17*, 241–248. https://doi.org/10.1023/b:jots.0000029267.61240.94

Pacella, M. L., Hruska, B., & Delahanty, D. L. (2013). The physical health consequences of PTSD and PTSD symptoms: A meta-analytic review. *Journal of Anxiety Disorders, 27*(1), 33–46.

Ponniah, K., & Hollon, S. D. (2009). Empirically supported psychological treatments for adults with acute stress disorder and posttraumatic stress disorder: A review. *Depression and Anxiety, 26*(12), 1086–1109.

Porges, S. W. (2003). Social engagement and attachment. *Annals of the New York Academy of Sciences, 1008*(1), 31–47. https://doi.org/10.1196/annals.1301.004

Porges, S. W. (2011). *The polyvagal theory: Neurophysiological foundations of emotions, attachment, communication, and self-regulation.* Norton.

Raglan, G. B., & Schulkin, J. (2014). Introduction to allostasis and allostatic load. In M. Kent, M. Davis, & J. W. Reich (Eds.), *The resilience handbook: Approaches to stress and trauma* (pp. 44–52). Routledge/Taylor & Francis.

Resick, P. A., & Schnicke, M. (1993). *Cognitive processing therapy for rape victims: A treatment manual* (Vol. 4). Sage.

Rosso, I. M., Silveri, M. M., Olson, E. A., Eric Jensen, J., & Ren, B. (2022). Regional specificity and clinical correlates of cortical GABA alterations in posttraumatic stress disorder. *Neuropsychopharmacology, 47*(5), 1055–1062.

Roth, A., & Fonagy, P. (2006). *What works for whom: A critical review of psychotherapy research.* Guilford Press.

Schottenbauer, M. A., Glass, C. R., Arnkoff, D. B., Tendick, V., & Gray, S. H. (2008). Nonresponse and dropout rates in outcome studies on PTSD: Review and methodological considerations. *Psychiatry: Interpersonal and Biological Processes, 71*(2), 134–168. https://doi.org/10.1521/psyc.2008.71.2.134

Seligowski, A. V., Harnett, N. G., Merker, J. B., & Ressler, K. J. (2020). Nervous and endocrine system dysfunction in posttraumatic stress disorder: An overview and consideration of sex as a biological variable. *Biological Psychiatry: Cognitive Neuroscience and Neuroimaging, 5*(4), 381–391.

Seligowski, A. V., Miron, L. H., & Orcutt, H. K. (2015). Relations among self-compassion, PTSD symptoms, and psychological health in a trauma-exposed sample. *Mindfulness, 6,* 1033–1041. https://doi.org/10.1007/s12671-014-0351-x

Selye, H. (1975). Confusion and controversy in the stress field. *Journal of Human Stress, 1*(2), 37–44.

Shapiro, F., & Solomon, R. M. (1995). *Eye movement desensitization and reprocessing.* Wiley.

Sharma, U. (2019). Paternal contributions to offspring health: Role of sperm small RNAs in intergenerational transmission of epigenetic information. *Frontiers in Cell and Developmental Biology, 7,* Article 215.

Sherin, J. E., & Nemeroff, C. B. (2011). Post-traumatic stress disorder: The neurobiological impact of psychological trauma. *Dialogues in Clinical Neuroscience, 13*(3), 263–278.

Shin, L. M., Bush, G., Milad, M. R., Lasko, N. B., Brohawn, K. H., Hughes, K. C., Macklin, M. L., Gold, A. L., Karpf, R. D., Orr, S. P., Rauch, S. L., & Pitman, R. K. (2011). Exaggerated activation of dorsal anterior cingulate cortex during cognitive interference: A monozygotic twin study of posttraumatic stress disorder. *American Journal of Psychiatry, 168*(9), 979–985. https://doi.org/10.1176/appi.ajp.2011.09121812

Shin, L. M., McNally, R. J., Kosslyn, S. M., Thompson, W. L., Rauch, S. L., Alpert, N. M., Metzger, L. J., Lasko, N. B., Orr, S. P., & Pitman, R. K. (1999). Regional cerebral blood flow during script-driven imagery in childhood sexual abuse-related PTSD: A PET investigation. *American Journal of Psychiatry, 156*(4), 575–584.

Spahic-Mihajlovic, A., Crayton, J. W., & Neafsey, E. J. (2005). Selective numbing and hyperarousal in male and female Bosnian refugees with PTSD. *Journal of Anxiety Disorders, 19*(4), 383–402.

Sriram, K., Rodriguez-Fernandez, M., & Doyle, F. J., III. (2012). Modeling cortisol dynamics in the neuro-endocrine axis distinguishes normal, depression, and post-traumatic stress disorder (PTSD) in humans. *PLoS Computational Biology, 8*(2), Article e1002379.

Steuwe, C., Daniels, J., Frewen, P., Densmore, M., Pannasch, S., Beblo, T., Reiss, J., & Lanius, R. (2012). Effect of direct eye contact in PTSD related to interpersonal trauma: An fMRI study of activation of an innate alarm system. *Social Cognitive and Affective Neuroscience, 9*(1), 88–97. https://doi.org/10.1093/scan/nss105

Taylor, E. W., Wang, T., & Leite, C. A. (2022). An overview of the phylogeny of cardiorespiratory control in vertebrates with some reflections on the "polyvagal theory." *Biological Psychology, 172*, Article 108382.

Teicher, M. H., Tomoda, A., & Andersen, S. L. (2006). Neurobiological consequences of early stress and childhood maltreatment: Are results from human and animal studies comparable? *Annals of the New York Academy of Sciences, 1071*(1), 313–323.

Tolin, D. F., & Foa, E. B. (2006). Sex differences in trauma and posttraumatic stress disorder: A quantitative review of 25 years of research. *Psychological Bulletin, 132*(6), 959–992.

van der Kolk, B. A. (2014). *The body keeps the score: Brain, mind, and body in the healing of trauma*. Penguin.

van der Kolk, B. A., Greenberg, M. S., Orr, S. P., & Pitman, R. K. (1989). Endogenous opioids, stress induced analgesia, and posttraumatic stress disorder. *Psychopharmacology Bulletin, 25*(3), 417–421.

van der Kolk, B. A., Roth, S., Pelcovitz, D., Sunday, S., & Spinazzola, J. (2005). Disorders of extreme stress: The empirical foundation of a complex adaptation to trauma. *Journal of Traumatic Stress, 18*(5), 389–399.

Wells, M., Trad, A., & Alves, M. (2003). Training beginning supervisors working with new trauma therapists. *Journal of College Student Psychotherapy, 17*(3), 19–39. https://doi.org/10.1300/J035v17n03_03

Whitworth, J. A., Williamson, P. M., Mangos, G., & Kelly, J. J. (2005). Cardiovascular consequences of cortisol excess. *Vascular Health and Risk Management, 1*(4), 291–299.

Yehuda, R., & Lehrner, A. (2018). Intergenerational transmission of trauma effects: Putative role of epigenetic mechanisms. *World Psychiatry, 17*(3), 243–257.

Zak, P. J., Kurzban, R., & Matzner, W. T. (2004). The neurobiology of trust. *Annals of the New York Academy of Sciences, 1032*(1), 224–227. https://doi.org/10.1196/annals.1314.025

Chapter 5

Clinical Neuroscience of Substance Use Disorders

Sean B. Hall, Laura K. Jones, and Kiera Walker

In 1997, Alan Leshner, then director of the National Institute on Drug Abuse, introduced the brain disease model of addiction (BDMA). In his seminal paper, Leshner likened addiction to a metaphorical "switch" that, when flipped, triggers compulsive, uncontrollable drug use. He later extended this metaphor to suggest that repeated drug use "hijacks" the brain's natural motivational system to prioritize substance use above all else (Hall et al., 2015).

Leshner recognized that his metaphor oversimplified the true nature of addiction (Hall et al., 2015). He knew that although addiction might be rooted in one's biology, it is also influenced by both learning and social context. Nonetheless, Leshner recognized the importance of shifting public perception toward the idea that substance use should be viewed as a public health issue. He learned from his experience at the National Institute of Mental Health that recasting disorders like schizophrenia and depression as medical conditions reduced stigma and increased funding for research. He hoped to leverage this strategy for addiction (Hall et al., 2015). Over time, Leshner's concept took hold. His 1997 paper is now recognized as the beginning of a paradigm shift in addiction research. The BDMA remains the dominant research paradigm in addiction science.

Addiction can be defined as a mental disposition characterized by repeated periods of heightened motivation to participate in a conditioned behavior despite negative consequences (Kelly et al., 2022). Under the BDMA, researchers seek to understand what motivates people to seek and use drugs in spite of negative consequences. Much of this research has focused on disentangling the neural substrates of motivation and reward (Volkow et al., 2016).

In this chapter, substance use and addiction are viewed through the lens of conditioned behavior and motivation to continue the use of substances despite the negative consequences. This information can help counselors to enhance their case conceptualization and to link evidence-based interventions to empirically derived targets. Throughout the chapter, the term *drug* is used to refer to an external (exogenous) chemical substance that has addictive properties and includes alcohol in addition to other substances.

2024 CACREP Standards

This chapter addresses 2024 Council for Accreditation of Counseling and Related Educational Programs (CACREP) Standards pertinent to the Foundational Counseling Curriculum (Section 3) areas of Lifespan Development (Standard C), Counseling Practice and Relationships (Standard E), and Assessment and Diagnostic Processes (Standard G):

- Theories of learning (Standard C.3.)
- Theories and neurobiological etiology of addictions (Standard C.5.)
- Biological, neurological, and physiological factors that affect lifespan development, functioning, behavior, resilience, and overall wellness (Standard C.10.)
- Counseling strategies and techniques used to facilitate the client change process (Standard E.10.)
- Evidence-based counseling strategies and techniques for prevention and intervention (Standard E.15.)
- Procedures to identify substance use, addictions, and co-occurring conditions (Standard G.12.)

This chapter also addresses the following Entry-Level Specialized Practice Area (Section 5) standards for Addiction Counseling (Standard A):

- Neurological, behavioral, psychological, physical, and social effects of psychoactive substances and addictive disorders on the user and significant others (Standard A.1.)
- Risk and protective factors for substance use disorders (Standard A.2.)
- Pharmacological interventions used to address substance use withdrawal, craving, and relapse prevention (Standard A.7.)

Clinical Case Study: Imara

As Imara prepared to speak, she felt a surge of anxiety. Despite this, she believed that sharing her story would be valuable to others. She refused to be controlled by fear. "It's too late to back out now!" Imara thought. She took a deep breath and began to speak. At first, her voice sounded raspy and unsteady. Fortunately, the anxiety she fought to suppress gradually began to settle. She was comforted by the subtle expressions of support and gentle nods of agreement that fluttered through the audience.

Imara shared how she grew up in Miami. Her mother taught at a local school, and her father owned a small information technology company. Her family had emigrated from Haiti to the United States when she was 6 years old. For Imara, the transition was difficult. She struggled to make new friends and was sometimes bullied by other children. During high school, Imara was known to be an exceptionally talented student. However, on the inside, she worried deeply about being judged, criticized, or humiliated. On the one hand, Imara dreaded social events and frequently used her schoolwork as an excuse to avoid them. But on the other hand, she felt lonely. After graduation, she saw college as an opportunity to reinvent herself. She was determined to overcome the fear and worry that isolated her from others. Imara began using alcohol to reduce the feelings of dread that arose during social events. When she drank, the harsh criticism she directed toward herself would dampen. For just a short period, she could savor the feeling of being unencumbered by the crippling fear of rejection. Over time, something changed. The effects of alcohol shifted from helping her manage difficult emotions to amplifying them. Alcohol went from helping her to connect more easily with strangers to isolating her emotionally and physically from the people she cared about the most. One day Imara was struck by a startling revelation. She was alone now more than ever. Imara recalled that as family and friends began to express concern over her behavior, she withdrew. She unexpectedly found herself choosing alcohol over everyone and everything. Over time, her drinking increased. So too did her troubles at work and with the legal system. Her long-term goals were now secondary to the daily pursuit and consumption of alcohol. In truth, Imara no longer enjoyed drinking. She resented the need to drink in order to feel normal and reasonably healthy. Over time, Imara began recognizing how her life had changed since that first sip. She had become a different person: one who was obsessed with a substance and one who was willing to sacrifice her career, family, and freedom to this seemingly trivial thing. But why?

Imara encouraged the audience to explore this question for themselves. She also encouraged them to reflect upon why alcohol consumed their attention for so long. Why did it affect them so differently than it did others? Why was it so hard to resist the urge to drink when reminders of alcohol would unexpectedly arise? Imara challenged the audience to explore their thoughts, emotions, and responses. For her, sobriety came after she could identify who she was and prioritize what was most important to her.

Foundational Concepts in the Neurobiology of Addiction

To understand how Imara transitioned from recreational to compulsive drinking, a review of several concepts that are foundational to a neurobiological understanding of addiction is important. These concepts include homeostatic and allostatic regulation, epigenetics, tolerance and sensitization, the opponent process theory of acquired motivation (Koob, 2020; Solomon & Corbit, 1974), and the incentive-sensitization theory of addiction (Berridge & Robinson, 2016; Robinson & Berridge, 1993). As a review from Chapter 1, *homeostasis* is best conceived as a closed-loop feedback system that reacts to changing environmental demands. Under this system, an internal sensor monitors specific physiological states. A control center then compares these states against "ideal" conditions known as *set points*. If large differences exist between the current and ideal states, an effector mechanism is activated to make corrections (Billman, 2020).

In contrast, *allostatic* regulation is the body's capacity to preemptively recalibrate set points in various physiological systems to meet forthcoming environmental demands. Individuals in allostatic overload who engage in unhealthy behaviors are at an increased risk of systemic illnesses (health conditions affecting multiple physiological systems), such as cardiovascular, autoimmune, and psychiatric disorders (Finlay et al., 2022). *Epigenetics*, discussed in Chapter 2, is when external or internal conditions activate or silence gene expression by modifying the chromatin structure of DNA (Masterpasqua, 2009). Drugs exploit this mechanism to remodel the structure of neurons (Robison & Nestler, 2011). Such modifications are thought to affect key processes involved in neurotransmission, which in turn fuel the cognitive and behavioral phenotypes (e.g., rewards and cravings) of addiction.

> **Reflective Question**
>
> What might have increased the likelihood of Imara becoming addicted to alcohol?

Tolerance and Sensitization

Tolerance to the intended effect of a drug occurs when the potency of that drug gradually diminishes with repeated use. Thus, one must consume increasingly larger doses to maintain the desired outcome. Various forms of tolerance exist. For example, *pharmacokinetic tolerance* generally refers to the upregulation of metabolizing enzymes, which in turn lowers the drug concentration level. *Pharmacodynamic tolerance* refers to the desensitization of receptors in response to repeated drug use. *Behavioral tolerance* occurs when, after repeated pairings, environmental cues become conditioned to decrease one's response to a drug's desired effect. For example, subjects may be less sensitive to the intoxicating effects of alcohol when it is consumed either in a familiar mix (e.g., beer, cocktail) or in a familiar environmental context (Hancock & McKim, 2017).

In contrast, *sensitization* (or reverse tolerance) occurs when the potency of a drug gradually increases with subsequent use. Over time, relatively low doses of the drug will stimulate effects similar to those typically observed with higher doses. Evidence for the sensitizing effects of drugs originates from studies demonstrating gradual increases in locomotor activity in rodents following repeated administration of stimulants, opioids, nicotine, PCP (phencyclidine), ethanol, and MDMA (ecstasy). In humans, a great deal of research has focused on the sensitizing effects of drugs on the motivation to use (Robinson & Berridge, 1993). For example, the pharmacologic properties of drugs may alter neuronal gene expression, leading to structural alterations that sensitize individuals both to the rewarding properties of the drug and to associated stimuli (De Roo et al., 2008).

Opponent Process Theory of Acquired Motivation

The opponent process theory (Koob, 2020; Solomon & Corbit, 1974) assumes an oscillation between two opposing psychological states consisting of *reactions* (primary process) and *after reactions* (secondary process). If an individual is exposed to a rewarding stimulus, a pleasurable emotional reaction rises sharply before achieving peak intensity. If the rewarding stimulus is maintained, peak reward will begin to decline and level off. However, if the rewarding stimulus is withdrawn, the pleasurable emotional reaction terminates and the individual experiences a negative emotional state. This after reaction also increases sharply before reaching its peak and ultimately leveling off. The opponent process theory is one framework for conceptualizing the transition from recreational to compulsive use, whereby the motivation to use may shift from the desire to experience pleasure (positive reinforcement) to the need to avoid the negative interoceptive and emotional states associated with cessation (Koob & Volkow, 2010). The opponent process theory has been used to explain a variety of substance use addiction processes, including for opioids (Koob, 2020).

Incentive-Sensitization Theory of Addiction

The incentive-sensitization theory of addiction outlines the putative mechanisms underlying the increased motivation to use drugs despite negative consequences (Berridge & Robinson, 2016; Robinson & Berridge, 1993). Under this framework, reward is composed of three elements: liking, wanting, and predictive learning.

The *liking* component is subdivided into the experience of pleasure and the objective neural or behavioral reactions to rewarding stimuli (Berridge et al., 2009). *Wanting* refers to incentive salience, which is regarded as an "attention-grabbing" form of motivation that mobilizes one to approach and consume rewards in the presence of drug-related cues. The *predictive learning* component refers to the associations between drug-related cues and expected rewards. The subjective experience of pleasure (liking) is controlled by tiny hedonic hotspots located most notably in the nucleus accumbens, ventral pallidum, and prefrontal cortex (PFC). These regions interact to paint a "pleasure gloss" onto sensory information, which amplifies the subjective experience of pleasure—a process mediated, in part, by endogenous opioid and cannabinoid systems (Berridge & Kringelbach, 2015).

Excessive wanting emerges when repeated drug use elicits permanent neuroadaptations in the mesolimbic dopamine (DA) system, which increases the incentive salience of drug-related cues (e.g., internal, environmental). Over time, these cues become intense motivational magnets that trigger bursts of reward-seeking activity. In other words, chronic use may sensitize some individuals to experience an excessive degree of wanting, which is amplified beyond the level experienced by people who do not use drugs and persists despite a reticence to consume the drug (Berridge, 2022).

The Addiction Cycle

Substance use disorders are presumed to fall along a continuum ranging from impulsive to compulsive behavior (Koob & Volkow, 2010). Impulsivity is the instinctive urge to perform a specific act. It is often measured by evaluating either an individual's preference for smaller-sooner versus larger-later rewards (i.e., delay discounting) or an individual's inability to inhibit or resist a prepotent response (go/no-go task; Kwako & Koob, 2017). In contrast, compulsive behavior refers to repetitive, stereotyped behaviors that seemingly increase along a progressive-ratio reinforcement schedule. Often these behaviors are performed to reduce negative emotions, such as those associated with recurrent intrusive or unwanted thoughts seen in obsessive-compulsive disorder (Schreiber et al., 2011). Koob and Volkow (2010) argued that whereas engagement in impulsive behavior is motivated by immediate pleasure or gratification, compulsive behavior is motivated by wanting to remove negative thoughts and emotions. They further reasoned that transitioning from recreational to compulsive use reflects a shift in one's motivation from positive to negative reinforcement. For example, the motivation to consume a drug of abuse may focus initially on experiencing positive hedonic arousal and positive outcome expectancies. Over time, these pleasurable sensations taper off and are replaced by negative hedonic arousal and negative outcome expectancies associated with cessation. To describe this process, Koob and Volkow outlined three stages of the addiction cycle, each associated with distinguishing neurobiological characteristics. These stages are binge/intoxication (characterized by positive reinforcement), withdrawal/negative affect (characterized by negative reinforcement), and preoccupation/anticipation (craving).

Binge/Intoxication

In the binge/intoxication stage of the addiction cycle, the liking and wanting aspects of reward are largely mediated by neural circuits distributed throughout the basal ganglia. The *basal ganglia* are a cluster of nuclei, located deep within the forebrain, that contain the striatum, globus pallidus, subthalamic nucleus, and substantia nigra. The striatum divides into the dorsal and ventral regions (Lipton et al., 2019). The dorsal striatum is composed of the caudate nucleus and putamen, whereas the ventral striatum is made up of the nucleus accumbens and olfactory tubercle. Broadly explained, dopaminergic neurons in the ganglia help to regulate both motivation (Felger & Miller, 2012) and habit learning (Koob et al., 2014). For addiction researchers, DA has been an active line of inquiry for increasing knowledge of reward and motivation (Schultz, 2002). The primary reward circuit, known as the *mesocorticolimbic DA system,* starts in the ventral tegmental area and projects to the nucleus accumbens, the olfactory tubercle, the amygdala, and regions near the ventral and dorsal striatum (caudate and putamen). Notably, the mesocorticolimbic DA system can be further subdivided into the mesolimbic pathway (projects from the ventral tegmental area to the nucleus accumbens) and mesocortical pathway (also originates in the ventral tegmental area but sends its projections to the dorsolateral PFC and the orbitofrontal cortices; Stahl, 2013). Dopaminergic projections in the mesolimbic pathway have been implicated in reward prediction (Nieh et al., 2013), whereas neurons in the mesocortical pathway have been linked to executive functions, such as working memory, behavioral flexibility, and decision-making (Floresco & Magyar, 2006). Last, the nigrostriatal DA pathway originates in the substantia nigra pars compacta and projects to the striatum. This region has been implicated in movement disorders such as Parkinson's disease (Benazzouz et al., 2014). These dopaminergic pathways are depicted in Figure 5.1.

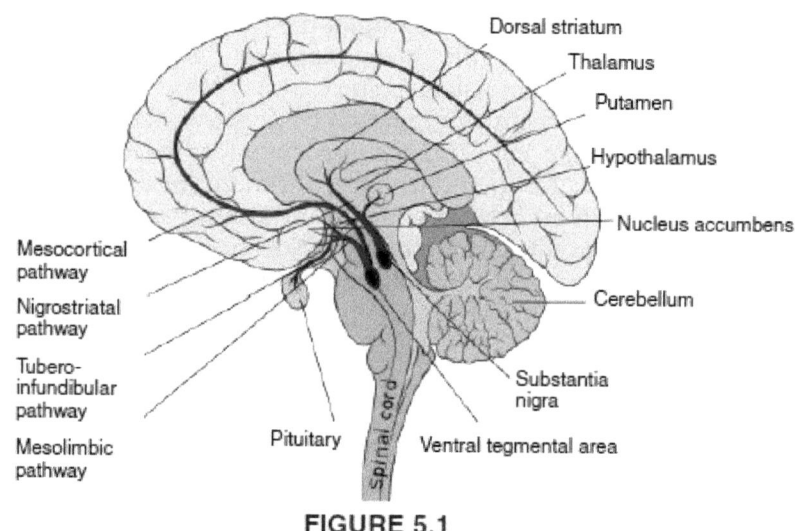

FIGURE 5.1

Major Dopaminergic Pathways

Note. From Patrick J. Lynch, 2015, via Wikimedia Commons. Used under Creative Commons Attribution 4.0 International License.

DA binds with at least five subtypes of G protein–coupled receptors, the D1- through D5-like receptors (known as D1R through D5R), that are distributed throughout various regions in the brain. DA cells in the ventral tegmental area are made up of GABAergic medium spiny neurons (MSNs), which express either D1R or D2R (Robison & Nestler, 2011). These MSNs may operate in a tonic or phasic firing state (Grace & Bunney, 1984). During tonic firing, dopaminergic cells release short (single-spike, 1–8 hertz [Hz]) spontaneous bursts of DA into the extracellular space (Volkow & Morales, 2015). In contrast, phasic firing refers to a series of high-frequency bursts (15 Hz; induced by glutamatergic stimulation), which flood the synaptic cleft with DA (Grace, 1991; Volkow & Morales, 2015). Unlike in tonic firing, which is designed to sustain a constant amount of "background" DA in the synapse, the accumulation of DA after phasic firing is rapidly reabsorbed by the presynaptic neuron (Grace, 1991). According to Schultz (2002), phasic firing in DA neurons is induced by a stimulus that holds motivational value as opposed to aversive or unpleasant stimuli.

Dopaminergic neurons in the mesolimbic pathway (ventral tegmental area projections toward the nucleus accumbens) are separable into direct and indirect pathways. The direct pathway begins in the striatum and projects to the substantia nigra pars reticulata and internal globus pallidus, whereas the indirect pathway extends to the external globus pallidus (Smith et al., 2013). Notably, MSNs in each pathway have distinct characteristics (Hikida et al., 2010) that may be integral to reward-based learning (Yawata et al., 2012). For example, excitatory MSNs in the direct pathway (high affinity for DA) express D2R, and inhibitory MSNs located along the indirect pathway (low affinity for DA) express D1R.

Both pathways synchronize to enhance reward-based learning, whereby MSNs in the direct pathway modulate persistent reward and MSNs in the indirect pathway modulate transient punishment (Volkow & Morales, 2015). It is hypothesized that the D1R/D2R ratio influences the wanting and craving aspects of drug addiction. Given that peak reward occurs when DA rapidly floods each pathway (which activates D2R and inhibits D1R), routes of administration more commonly associated with rapid uptake (e.g., inhalation, intravenous administration) are more rewarding and, therefore, more addictive than gradual ones.

To enhance reward prediction, glutamatergic tracts from the amygdala, hippocampus, and ventral PFC also innervate the aforementioned projections. When stimulated by enough DA to trigger phasic firing, these mechanisms help the individual form associations between the reward cues and the drug (Volkow & Morales, 2015). Afterward, exposure to reward cues can activate a conditioned response, whereby phasic firing is triggered in expectation of reward. This conditioned response represents the wanting or motivational aspect of addiction introduced in the incentive-sensitization theory. The associated stimulus is now granted incentive salience, motivating the individual to seek and acquire the drug. This framework is thought to underlie the maintenance of compulsive drug-seeking behavior despite a reduction in hedonic pleasure of the drug as a result of tolerance (Berridge et al., 2009).

Withdrawal/Negative Affect

During compulsive drug use, temporary cessation has been linked to a withdrawal syndrome characterized by symptoms both physical (specific to the drug of choice) and psychological (motivational deficit). Understanding the fundamental mechanisms that underlie these symptoms helps conceptualize the role of negative reinforcement in maintaining drug use. In the opponent process theory (Koob, 2020; Solomon & Corbit, 1974), excessive drug use stimulates the brain's natural reward system (primary process) and an anatomic region referred to by some addiction researchers as the "extended amygdala" initiates an antireward response (secondary process). The extended amygdala plays a prominent role in reward learning. It consists of highly related regions of the brain, including the central nucleus of the amygdala, bed nucleus of the stria terminalis, nucleus accumbens shell, and sublenticular substantia innominata (Alheid & Heimer, 1988).

The withdrawal/negative affect stage of the addiction cycle is characterized by the activation of unpleasant, antireward symptoms, which include irritability, dysphoric mood, alexithymia, and anhedonia. Over time, the intensity of the reward diminishes while the intensity of the antireward increases. As drug dependence progresses, the brain's reward system is weakened, and the motivational deficits associated with withdrawal begin to emerge. These processes are governed by changes within and between systems (e.g., central nervous system, peripheral nervous system, endocrine system). *Within-system changes* (e.g., within the mesolimbic DA system) are neuroadaptations that affect the same pathways implicated in reward. These changes are mediated by neuroplastic alterations among MSNs in the nucleus accumbens. Some examples are decreased long-term potentiation and increased cAMP-response element binding protein phosphorylation. Recall that these epigenetic mechanisms influence synaptic plasticity (e.g., decreases in D2Rs) and molecular signaling. Ultimately, within-system changes are associated with decreased sensitivity in the mesocorticolimbic system to DA stimulation, thereby reducing both the drug's liking and wanting properties (Koob et al., 2014; Koob & Le Moal, 2008).

For instance, naturally occurring reinforcers trigger the phasic activation of DA neurons immediately after the reward (e.g., food, sex) is consumed. After repeated pairings, a transition occurs, whereby increases in DA release are triggered only when the reward is expected but not while it is consumed (Schultz, 2002). The gradual transition from voluntary drug use to drug addiction can be affected by the type of drug, the frequency or pattern of use, and the age of onset (Volkow & Morales, 2015). In addition, drug use can also affect D2 autoreceptors, which play a role in the presynaptic regulation of extracellular DA. By affecting D2 autoreceptors, drugs may gradually suppress the amount of DA released into the extracellular space (Volkow & Morales, 2015). Some individuals are naturally predisposed to express more D2R on the surface of postsynaptic neurons, which may increase their vulnerability to drug use (Nader et al., 2008). Chronic use may reduce this D2R availability, leading to downregulation (i.e., decreased sensitivity to DA) of the postsynaptic receptors in the ventral and dorsal striatum (Volkow et al., 1993).

Withdrawal from chronic drug use has been linked to dysregulation in the hypothalamic-pituitary-adrenal axis, including elevated levels of adrenocorticotropic hormone, cortisol, and corticotrophin-releasing hormone in the extended amygdala (Koob & Le Moal,

2008). Withdrawal has also been linked to increases in the level of norepinephrine in the bed nucleus of the stria terminalis and neuropeptide Y in the extended amygdala. Together these adaptations may underlie compulsive drug use to stave off dysphoric mood states present during withdrawal. Even as the drug's liking and wanting properties subside, compulsive use may be motivated by the desire to reestablish baseline qualia and reduce the negative hedonic states associated with withdrawal (Redish et al., 2008).

Preoccupation/Anticipation

The preoccupation/anticipation stage of the addiction cycle outlines the neurobiological systems thought to underlie both craving and reinstatement. The appetitive urge to relapse (craving) can be activated when clients are exposed to a drug-, cue-, or stress-induced trigger (Shaham et al., 2003). We might wonder why some people find it so difficult to suppress this urge. Chronic drug use impairs those brain regions that are critical for exerting *top-down* control over one's behavior. Recall that the PFC plays a role in mediating executive function, whereby three core executive functions—that is, inhibition, working memory, and cognitive flexibility—form the basis of top-down mental operations from which other higher order cognitive processes can be coordinated (Diamond, 2013). For instance, such core executive functions may be recruited to engage in different aspects of problem-solving, reasoning, or planning (Diamond, 2013). Executive dysfunction may mediate the loss of inhibitory control and reward-related processing among drug users, resulting in an overvaluation of the drug and an undervaluation of other rewards.

When exposed to drug cues, the phasic firing of DA neurons in the nucleus accumbens and dorsal striatum is regulated by glutamatergic inputs from the PFC, hippocampus, amygdala, and thalamic nuclei. Such hyperactivation is thought to enhance reward prediction (Volkow & Morales, 2015). For example, when exposed to cocaine, chronic users show increased activation in the ventral anterior cingulate cortex (emotional learning and assigning emotional valence to stimuli; Devinsky et al., 1995) and in the orbitofrontal cortices (craving, response inhibition, learned associations, assignment of value; Stalnaker et al., 2015). In contrast, chronic drug users who inhibit their craving show decreased activity in the ventral PFC (Nelson & Guyer, 2011), nucleus accumbens, and a component of Broca's area (Volkow et al., 2010). Such findings provide further evidence that regions in the PFC may help inhibit prepotent drug cravings when exposed to reminder cues. These findings may indicate the presence of an inhibitory (STOP) signal, primarily mediated by the ventrolateral PFC and the orbitofrontal cortices, and a reward-seeking (GO) signal, primarily mediated by the anterior cingulate cortex and the dorsolateral PFC.

In brain imaging studies, researchers have found increased insular cortex activation during intense cravings for cocaine (Garavan et al., 2000), alcohol (Myrick et al., 2004), heroin (Sell et al., 1999), and nicotine (McClernon et al., 2005). The insular cortex interprets interoceptive signals (e.g., muscle tension, gastrointestinal discomfort, pain, heart rate) to "mark" the hedonic potential of specific outcomes, allowing individuals to discriminate between outcomes with more or less value (Naqvi et al., 2014). These hedonic markers may partly explain why people continue to use drugs despite the negative consequences. Abnormal functioning in the insula during addiction may place individuals at greater risk for increasing the value of smaller-sooner rewards and for difficulty identifying potentially risky situations (Claus et al., 2011; Naqvi et al., 2014). The insula thus may have an important role in addiction because of its capacity for integrating external stimuli, interoceptive signals, motivational states, and projections to networks involved in executive function and motor movement (DeVito et al., 2013).

Limitations of the BDMA

Proponents and critics agree that the BDMA does not offer a satisfying explanation of addiction (Heilig et al., 2021). However, experts differ as to whether it offers a stronger explanation than alternative theories. Some scholars believe the BDMA will continue to progress (Volkow et al., 2019), while others argue that a better theory is needed (Verschure & Wiers, 2022). A central argument in this debate surrounds whether addiction should be conceptualized as a disease. Formally, the label "disease" is used when one or more disorders cause some disposition to undergo a *clinically abnormal biological process*. A clinically abnormal process meets all three of the following criteria: (a) It is not part of the individual's natural life plan (e.g., aging or pregnancy); (b) it is associated with increased risk of pain, suffering, death, or impairment in social, academic, or occupational function; and (c) it exceeds a threshold marking the boundary between subclinical and clinical impairments in clinical function (Ceusters & Smith, 2010). Typically, this threshold is operationalized using an empirically defined cutoff measured via diagnostic laboratory tests. Although scientists disagree on the exact causal mechanisms underlying psychiatric disorders, they generally agree that psychiatric disorders are more than simple brain diseases and likely result from complex interactions among biological, psychological, and social factors across the life span (Volkow et al., 2019). To date, researchers have been

> **Reflective Question**
>
> Why might Imara continue to use alcohol despite the negative consequences?

unable to identify a specific, unitary malfunction in brain physiology to explain the onset of addictive behavior (Larsen & Hastings, 2018).

Another element of this debate involves how addiction controls one's goal-directed behavior. Under the BDMA, researchers assume dysfunction in the brain's motivation system channels all goal-directed behavior to acquire and consume the drug. This view presumes that drugs usurp free will. A consequence of this perspective is that the disease, not the individual, is responsible for using, despite harmful consequences. Others argue that addiction certainly influences goal-directed behavior but never takes complete control, and the individual does not fully surrender free will. The first perspective implies that the brain becomes the target of intervention. The second perspective implies that the intervention target is expanded beyond the brain to include social contexts and behavioral components (Rotgers, 2022).

Implications for Clinical Intervention

Counseling Approaches

The neural processes targeted for addiction treatment are currently delineated as top-down or bottom-up. *Top-down processes* are interventions that endeavor to revise behaviors and cognitions, interposed through augmented prefrontal cortical function and amended executive control. In contrast, *bottom-up processes* are interventions that target projections beneath the cortex (Potenza et al., 2011). Behavioral therapies are typically utilized in top-down processes and pharmacological treatments in bottom-up processes. The three phases of addiction consist of detoxification, initial recovery, and relapse prevention; pharmacotherapies and different behavioral interventions may be beneficial and necessary at each phase to meet the goals of treatment (Potenza et al., 2011).

Brief motivational models (Burke et al., 2003), contingency management (Dutra et al., 2008), and cognitive behavioral models (Dutra et al., 2008; Magill & Ray, 2009; Tolin, 2010) have shown strong empirical promise in treating addiction. Motivational interviewing is a principle-driven approach designed to assist clients in determining and, if necessary, changing behaviors contributing to their present distress (Dabbo & Dabbo, 2010). Using microskills, using empathy, effectively responding to resistance, and revealing discrepancies in a client's current behavior compared to their desired goals, the counselor helps the client to find congruence.

Contingency management is an empirically supported intervention with an operant conditioning theoretical foundation that helps clients initiate and maintain drug abstinence (Bigelow & Silverman, 1999). Two standard models are voucher-based reinforcement therapy

(Higgins et al., 1991) and the fishbowl technique (Petry et al., 2005). Voucher-based reinforcement therapy rewards clients if their urine or breath samples test negative for drug use; the clients are then given a voucher and are able to exchange that voucher for services and products conducive to living drug free (Prendergast et al., 2006). The fishbowl technique is so termed because clients who maintain drug abstinence reach into a vessel containing pieces of paper with positive reinforcement messages such as "good job" or gift cards that may range from $1 to $100 (Prendergast et al., 2006). Both models reward for sustained drug abstinence by increasing the value of the vouchers or providing bonus opportunities to draw from the fishbowl. Both have shown efficacious results (Prendergast et al., 2006).

Cognitive behavior therapy (CBT) is derived from the Marlatt and Donovan (2005) relapse prevention model that aims to bring awareness of and develop skills to cope with affective, cognitive, and contextual triggers that may contribute to substance use (Magill & Ray, 2009). CBT strategies include training individuals on how to decline drugs, participate in endeavors unrelated related to drug use, recognize triggers that may lead to drug use, and determine the functionality of drug use in their everyday life (Magill & Ray, 2009). CBT is one of the most comprehensively studied interventions in the field, and these treatments are currently utilized in many clinical environments (Magill & Ray, 2009).

The future of behavioral therapy is expected to include cognitive strategies intended to help strengthen brain function; this encompasses treatments such as mindfulness-based therapies that seem helpful in inhibiting negative mood states such as depression (Brewer et al., 2010), which can be onset with drug abstinence (Potenza et al., 2011). Mindfulness helps facilitate self-regulation of attention by orienting the individual to their present experience and teaching them to accept their current reality instead of avoiding it (Brewer et al., 2010).

Medication-Assisted Treatment

Canonical approaches for pharmacological treatments are aimed at the drug "reward" system. In medication-assisted treatment (MAT), medications are used to reduce the positive reinforcement of hedonic pleasure from drug use (agonist) or create unpleasant effects when using the drug (antagonist). For example, naltrexone (e.g., Narcan) is an agonist medication that nulls the effects of alcohol by binding to endorphin receptors and blocking the hedonic effects. Disulfiram (e.g., Antabuse) is an antagonist medication that creates unpleasant aversive effects in the user when consuming alcohol. Medications have also been utilized to target negative reinforcement of drug use, such as irritability or drug cravings, that occurs when abstaining from drugs (Potenza et al., 2011).

Enormous effort has been put into investigating behavioral therapies and pharmacotherapies for addiction treatment; however, no single treatment is completely effective (Vocci et al., 2005). Thus, combinatorial strategies are being utilized to help enhance treatment. Methods to determine combinatorial treatment include (a) choosing combination therapies that address the shortcomings of individual therapy, (b) utilizing treatments that emphasize similar processes in various ways, and (c) choosing treatments that have a synergistic effect when used together but no individual effect alone (Potenza et al., 2011). Currently, researchers seeking to enhance treatment for addictions are focusing on viewing individual differences as endophenotypes (i.e., stable phenotypes that have a genetic connection; Potenza et al., 2011). Endophenotypes are more compelling, as prescribers can readily link them to a biological mechanism (Fineberg et al., 2010).

A Brain-Based Approach to the Case of Imara

Given Imara's history and struggles with alcohol use, a combination of counseling approaches and MAT was used to address both the psychological and physiological aspects of addiction. The counselor emphasized the importance of combining these approaches to enhance Imara's overall treatment outcomes.

In the initial phase of counseling, motivational interviewing assisted Imara to explore her ambivalence toward alcohol use. The counselor used reflective listening and microskills to understand Imara's perspective and guide her in recognizing the discrepancies between her current behavior and her desired goals. Motivational interviewing helped Imara find intrinsic motivation to change by exploring the reasons behind her alcohol use and the impact it had on her life.

Given the severity of Imara's alcohol use and its negative impact on various aspects of her life, the counselor worked with a psychopharmacologist who provided MAT as a complementary intervention. Naltrexone, an agonist medication, was selected to reduce the positive reinforcement Imara experienced from alcohol. By blocking endorphin receptors, it diminished the pleasurable effects of alcohol, making it less rewarding. The counselor worked collaboratively with a prescribing health care professional to monitor Imara's response to the medication and adjust the dosage as needed.

Once Imara decided to reduce her alcohol use with the goal of eventual abstinence, contingency management was implemented to reinforce abstinence. The counselor discussed and implemented a voucher-based reinforcement therapy to provide positive reinforcement for sustained drug abstinence. This strengthened Imara's commitment to recovery. As Imara cut down and eventually eliminated drug use, she experienced negative mood states associated with withdrawal.

Mindfulness assisted Imara to accept (rather than avoid) her negative experiences in the present moment, reducing the likelihood of relapse triggered by negative emotions. As treatment progressed, the counselor integrated CBT strategies to help Imara develop awareness and coping skills for triggers that were contributing to her substance use. Imara identified and challenged negative thought patterns, developed effective refusal skills, and explored alternative coping mechanisms.

By combining counseling approaches and MAT, the counselor provided Imara with a comprehensive and individualized intervention plan to support her in achieving and maintaining recovery from alcohol addiction. The integration of these interventions addressed both the behavioral and pharmacological aspects of addiction, increasing the likelihood of success in Imara's recovery journey.

Quiz

1. Dopaminergic neurons in the mesolimbic pathway are largely associated with:
    a. Reward prediction.
    b. Executive function.
    c. Movement disorders.
    d. Decision-making.
2. Which of the following is *not* an evidence-based counseling intervention for addiction?
    a. Motivational interviewing.
    b. Cognitive behavior therapy.
    c. Contingency management.
    d. Interpersonal and social rhythm therapy.
3. Which of the following elements make up the addiction cycle?
    a. Detoxification, initial recovery, and relapse prevention.
    b. Liking, needing, and opponent processes.
    c. Binge/intoxication, withdrawal/negative affect, and preoccupation/anticipation.
    d. Positive reinforcement, negative reinforcement, and decision-making.

References

Alheid, G. F., & Heimer, L. (1988). New perspectives in basal forebrain organization of special relevance for neuropsychiatric disorders: The striatopallidal, amygdaloid, and corticopetal components of substantia innominata. *Neuroscience, 27*(1), 1–39. https://doi.org/10.1016/0306-4522(88)90217-5

Benazzouz, A., Mamad, O., Abedi, P., Bouali-Benazzouz, R., & Chetrit, J. (2014). Involvement of dopamine loss in extrastriatal basal ganglia nuclei in the pathophysiology of Parkinson's disease. *Frontiers in Aging Neuroscience, 6*, Article 87. https://www.frontiersin.org/articles/10.3389/fnagi.2014.00087

Berridge, K. C. (2022). Is addiction a brain disease? The incentive-sensitization view. In N. Heather, M. Field, A. C. Moss, & S. Satel (Eds.), *Evaluating the brain disease model of addiction* (pp. 74–86). Routledge. https://doi.org/10.4324/9781003032762-8

Berridge, K. C., & Kringelbach, M. L. (2015). Pleasure systems in the brain. *Neuron, 86*(3), 646–664. https://doi.org/10.1016/j.neuron.2015.02.018

Berridge, K. C., & Robinson, T. E. (2016). Liking, wanting, and the incentive-sensitization theory of addiction. *American Psychologist, 71*(8), 670–679. https://psycnet.apa.org/doi/10.1037/amp0000059

Berridge, K. C., Robinson, T. E., & Aldridge, J. W. (2009). Dissecting components of reward: 'Liking', 'wanting', and learning. *Neurosciences, 9*(1), 65–73. https://doi.org/10.1016/j.coph.2008.12.014

Bigelow, G. E., & Silverman, K. (1999). Theoretical and empirical foundations of contingency management treatments for drug abuse. In S. T. Higgins & K. Silverman (Eds.), *Motivating behavior change among illicit-drug abusers: Research on contingency management interventions* (pp. 15–31). American Psychological Association. https://doi.org/10.1037/10321-001

Billman, G. E. (2020). Homeostasis: The underappreciated and far too often ignored central organizing principle of physiology. *Frontiers in Physiology, 11*, Article 200. https://www.frontiersin.org/articles/10.3389/fphys.2020.00200

Brewer, J. A., Bowen, S., Smith, J. T., Marlatt, G. A., & Potenza, M. N. (2010). Mindfulness-based treatments for co-occurring depression and substance use disorders: What can we learn from the brain? *Addiction, 105*(10), 1698–1706. https://doi.org/10.1111/j.1360-0443.2009.02890.x

Burke, B. L., Arkowitz, H., & Menchola, M. (2003). The efficacy of motivational interviewing: A meta-analysis of controlled clinical trials. *Journal of Consulting and Clinical Psychology, 71*(5), 843–861. https://doi.org/10.1037/0022-006X.71.5.843

Ceusters, W., & Smith, B. (2010). Foundations for a realist ontology of mental disease. *Journal of Biomedical Semantics, 1*(1), Article 10. https://doi.org/10.1186/2041-1480-1-10

Claus, E. D., Kiehl, K. A., & Hutchison, K. E. (2011). Neural and behavioral mechanisms of impulsive choice in alcohol use disorder. *Alcohol: Clinical and Experimental Research, 35*(7), 1209–1219. https://doi.org/10.1111/j.1530-0277.2011.01455.x

Council for Accreditation of Counseling and Related Educational Programs. (2024). *2024 CACREP standards.* https://www.cacrep.org/wp-content/uploads/2023/06/2024-Standards-Combined-Version-6.27.23.pdf

Dabbo, S., & Dabbo, N. (2010). Motivational interviewing: Preparing people for change. *University of Toronto Medical Journal, 87*(2), 101–102. https://doi.org/10.5015/utmj.v87i2.1178

De Roo, M., Klauser, P., Garcia, P. M., Poglia, L., & Muller, D. (2008). Spine dynamics and synapse remodeling during LTP and memory processes. In W. S. Sossin, J. C. Lacaille, V. F. Castellucci, & S. Belleville (Eds.), *Progress in brain research* (pp. 199–207). Elsevier. https://doi.org/10.1016/S0079-6123(07)00011-8

Devinsky, O., Morrell, M. J., & Vogt, B. A. (1995). Contributions of anterior cingulate cortex to behaviour. *Brain, 118*(1), 279–306. https://doi.org/10.1093/brain/118.1.279

DeVito, E. E., Meda, S. A., Jiantonio, R., Potenza, M. N., Krystal, J. H., & Pearlson, G. D. (2013). Neural correlates of impulsivity in healthy males and females with family histories of alcoholism. *Neuropsychopharmacology, 38*(10), 1854–1863. https://doi.org/10.1038/npp.2013.92

Diamond, A. (2013). Executive functions. *Annual Review of Psychology, 64*(1), 135–168. https://doi.org/10.1146/annurev-psych-113011-143750

Dutra, L., Stathopoulou, G., Basden, S. L., Leyro, T. M., Powers, M. B., & Otto, M. W. (2008). A meta-analytic review of psychosocial interventions for substance use disorders. *American Journal of Psychiatry, 165*(2), 179–187. https://doi.org/10.1176/appi.ajp.2007.06111851

Felger, J. C., & Miller, A. H. (2012). Cytokine effects on the basal ganglia and dopamine function: The subcortical source of inflammatory malaise. *Frontiers in Neuroendocrinology, 33*(3), 315–327. https://doi.org/10.1016/j.yfrne.2012.09.003

Fineberg, N. A., Potenza, M. N., Chamberlain, S. R., Berlin, H. A., Menzies, L., Bechara, A., Sahakian, B. J., Robbins, T. W., Bullmore, E. T., & Hollander, E. (2010). Probing compulsive and impulsive behaviors, from animal models to endophenotypes: A narrative review. *Neuropsychopharmacology, 35*(3), 591–604. https://doi.org/10.1038/npp.2009.185

Finlay, S., Rudd, D., McDermott, B., & Sarnyai, Z. (2022). Allostatic load and systemic comorbidities in psychiatric disorders. *Psychoneuroendocrinology, 140*, Article 105726. https://doi.org/10.1016/j.psyneuen.2022.105726

Floresco, S. B., & Magyar, O. (2006). Mesocortical dopamine modulation of executive functions: Beyond working memory. *Psychopharmacology, 188*(4), 567–585. https://doi.org/10.1007/s00213-006-0404-5

Garavan, H., Pankiewicz, J., Bloom, A., Cho, J.-K., Sperry, L., Ross, T. J., Salmeron, B. J., Risinger, R., Kelley, D., & Stein, E. A. (2000). Cue-induced cocaine craving: Neuroanatomical specificity for drug users and drug stimuli. *American Journal of Psychiatry, 157*(11), 1789–1798. https://doi.org/10.1176/appi.ajp.157.11.1789

Grace, A. A. (1991). Phasic versus tonic dopamine release and the modulation of dopamine system responsivity: A hypothesis for the etiology of schizophrenia. *Neuroscience, 41*(1), 1–24. https://doi.org/10.1016/0306-4522(91)90196-U

Grace, A. A., & Bunney, B. S. (1984). The control of firing pattern in nigral dopamine neurons: Single spike firing. *The Journal of Neuroscience, 4*(11), 2866–2876. https://doi.org/10.1523/JNEUROSCI.04-11-02866.1984

Hall, W., Carter, A., & Forlini, C. (2015). The brain disease model of addiction: Is it supported by the evidence and has it delivered on its promises? *The Lancet Psychiatry*, *2*(1), 105–110. https://doi.org/10.1016/S2215-0366(14)00126-6

Hancock, S. D., & McKim, W. A. (2017). *Drugs and behavior: An introduction to behavioral pharmacology* (8th ed.). Pearson.

Heilig, M., MacKillop, J., Martinez, D., Rehm, J., Leggio, L., & Vanderschuren, L. J. M. J. (2021). Addiction as a brain disease revised: Why it still matters, and the need for consilience. *Neuropsychopharmacology*, *46*(10), 1715–1723. https://doi.org/10.1038/s41386-020-00950-y

Higgins, S. T., Delaney, D. D., Budney, A. J., Bickel, W. K., Hughes, J. R., Foerg, F., & Fenwick, J. W. (1991). A behavioral approach to achieving initial cocaine abstinence. *American Journal of Psychiatry*, *148*(9), 1218–1224. https://doi.org/10.1176/ajp.148.9.1218

Hikida, T., Kimura, K., Wada, N., Funabiki, K., & Nakanishi, S. (2010). Distinct roles of synaptic transmission in direct and indirect striatal pathways to reward and aversive behavior. *Neuron*, *66*(6), 896–907. https://doi.org/10.1016/j.neuron.2010.05.011

Kelly, R. M., Hastings, J., & West, R. (2022). How an addiction ontology can unify competing conceptualizations of addiction. In N. Heather, M. Field, A. C. Moss, & S. Satel (Eds.), *Evaluating the brain disease model of addiction* (pp. 484–496). Routledge. https://doi.org/10.4324/9781003032762-46

Koob, G. F. (2020). Neurobiology of opioid addiction: Opponent process, hyperkatifeia, and negative reinforcement. *Biological Psychiatry, 87*(1), 44–53.

Koob, G. F., Arends, M. A., & Le Moal, M. (2014). *Drugs, addiction, and the brain*. Academic Press.

Koob, G. F., & Le Moal, M. (2008). Addiction and the brain antireward system. *Annual Review of Psychology*, *59*(1), 29–53. https://doi.org/10.1146/annurev.psych.59.103006.093548

Koob, G. F., & Volkow, N. D. (2010). Neurocircuitry of addiction. *Neuropsychopharmacology*, *35*(1), 217–238. https://doi.org/10.1038/npp.2009.110

Kwako, L. E., & Koob, G. F. (2017). Neuroclinical framework for the role of stress in addiction. *Chronic Stress*, *1*. https://doi.org/10.1177/2470547017698140

Larsen, R. R., & Hastings, J. (2018). From affective science to psychiatric disorder: Ontology as a semantic bridge. *Frontiers in Psychiatry*, *9*, Article 487. https://www.frontiersin.org/articles/10.3389/fpsyt.2018.00487

Leshner, A. I. (1997). Addiction is a brain disease, and it matters. *Science*, *278*(5335), 45–47. https://doi.org/10.1126/science.278.5335.45

Lipton, D. M., Gonzales, B. J., & Citri, A. (2019). Dorsal striatal circuits for habits, compulsions and addictions. *Frontiers in Systems Neuroscience*, *13*, Article 28. https://www.frontiersin.org/articles/10.3389/fnsys.2019.00028

Lynch, P. J. (2015). Dopaminergic pathways [Digital image]. https://commons.wikimedia.org/wiki/File:Dopaminergic_pathways.svg

Magill, M., & Ray, L. A. (2009). Cognitive-behavioral treatment with adult alcohol and illicit drug users: A meta-analysis of randomized controlled trials. *Journal of Studies on Alcohol and Drugs, 70*(4), 516–527. https://doi.org/10.15288/jsad.2009.70.516

Marlatt, G. A., & Donovan, D. M. (2005). *Relapse prevention: Maintenance strategies in the treatment of addictive behaviors* (2nd ed.). Guilford Press.

Masterpasqua, F. (2009). Psychology and epigenetics. *Review of General Psychology, 13*(3), 194–201. https://doi.org/10.1037/a0016301

McClernon, F. J., Hiott, F. B., Huettel, S. A., & Rose, J. E. (2005). Abstinence-induced changes in self-report craving correlate with event-related FMRI responses to smoking cues. *Neuropsychopharmacology, 30*(10), 1940–1947. https://doi.org/10.1038/sj.npp.1300780

Myrick, H., Anton, R. F., Li, X., Henderson, S., Drobes, D., Voronin, K., & George, M. S. (2004). Differential brain activity in alcoholics and social drinkers to alcohol cues: Relationship to craving. *Neuropsychopharmacology, 29*(2), 393–402. https://doi.org/10.1038/sj.npp.1300295

Nader, M. A., Czoty, P. W., Gould, R. W., & Riddick, N. V. (2008). Positron emission tomography imaging studies of dopamine receptors in primate models of addiction. *Philosophical Transactions of the Royal Society B: Biological Sciences, 363*(1507), 3223–3232. https://doi.org/10.1098/rstb.2008.0092

Naqvi, N. H., Gaznick, N., Tranel, D., & Bechara, A. (2014). The insula: A critical neural substrate for craving and drug seeking under conflict and risk. *Annals of the New York Academy of Sciences, 1316*(1), 53–70. https://doi.org/10.1111/nyas.12415

Nelson, E. E., & Guyer, A. E. (2011). The development of the ventral prefrontal cortex and social flexibility. *Developmental Cognitive Neuroscience, 1*(3), 233–245. https://doi.org/10.1016/j.dcn.2011.01.002

Nieh, E. H., Kim, S.-Y., Namburi, P., & Tye, K. M. (2013). Optogenetic dissection of neural circuits underlying emotional valence and motivated behaviors. *Optogenetics and Pharmacogenetics in Neuronal Function and Dysfunction, 1511*, 73–92. https://doi.org/10.1016/j.brainres.2012.11.001

Petry, N. M., Peirce, J. M., Stitzer, M. L., Blaine, J., Roll, J. M., Cohen, A., Obert, J., Killeen, T., Saladin, M. E., Cowell, M., Kirby, K. C., Sterling, R., Royer-Malvestuto, C., Hamilton, J., Booth, R. E., Macdonald, M., Liebert, M., Rader, L., Burns, R., ... Li, R. (2005). Effect of prize-based incentives on outcomes in stimulant abusers in outpatient psychosocial treatment programs: A national drug abuse treatment clinical trials network study. *Archives of General Psychiatry, 62*(10), 1148–1156. https://doi.org/10.1001/archpsyc.62.10.1148

Potenza, M. N., Sofuoglu, M., Carroll, K. M., & Rounsaville, B. J. (2011). Neuroscience of behavioral and pharmacological treatments for addictions. *Neuron, 69*(4), 695–712. https://doi.org/10.1016/j.neuron.2011.02.009

Prendergast, M., Podus, D., Finney, J., Greenwell, L., & Roll, J. (2006). Contingency management for treatment of substance use disorders: A meta-analysis. *Addiction, 101*(11), 1546–1560. https://doi.org/10.1111/j.1360-0443.2006.01581.x

Redish, A. D., Jensen, S., & Johnson, A. (2008). A unified framework for addiction: Vulnerabilities in the decision process. *Behavioral and Brain Sciences, 31*(4), 415–437. https://doi.org/10.1017/S0140525X0800472X

Robinson, T. E., & Berridge, K. C. (1993). The neural basis of drug craving: An incentive-sensitization theory of addiction. *Brain Research Reviews, 18*(3), 247–291. https://doi.org/10.1016/0165-0173(93)90013-P

Robison, A. J., & Nestler, E. J. (2011). Transcriptional and epigenetic mechanisms of addiction. *Nature Reviews Neuroscience, 12*(11), 623–637. https://doi.org/10.1038/nrn3111

Rotgers, F. (2022). My brain disease made me do it: Bioethical implications of the brain disease model of addiction. In N. Heather, M. Field, A. C. Moss, & S. Satel (Eds.), *Evaluating the brain disease model of addiction* (pp. 166–175). Routledge. https://doi.org/10.4324/9781003032762-19

Schreiber, L., Odlaug, B. L., & Grant, J. E. (2011). Impulse control disorders: Updated review of clinical characteristics and pharmacological management. *Frontiers in Psychiatry, 2*, Article 1. https://doi.org/10.3389/fpsyt.2011.00001

Schultz, W. (2002). Getting formal with dopamine and reward. *Neuron, 36*(2), 241–263. https://doi.org/10.1016/S0896-6273(02)00967-4

Sell, L. A., Morris, J., Bearn, J., Frackowiak, R. S. J., Friston, K. J., & Dolan, R. J. (1999). Activation of reward circuitry in human opiate addicts. *European Journal of Neuroscience, 11*(3), 1042–1048. https://doi.org/10.1046/j.1460-9568.1999.00522.x

Shaham, Y., Shalev, U., Lu, L., de Wit, H., & Stewart, J. (2003). The reinstatement model of drug relapse: History, methodology and major findings. *Psychopharmacology, 168*(1), 3–20. https://doi.org/10.1007/s00213-002-1224-x

Smith, R. J., Lobo, M. K., Spencer, S., & Kalivas, P. W. (2013). Cocaine-induced adaptations in D1 and D2 accumbens projection neurons (a dichotomy not necessarily synonymous with direct and indirect pathways). *Addiction, 23*(4), 546–552. https://doi.org/10.1016/j.conb.2013.01.026

Solomon, R. L., & Corbit, J. D. (1974). An opponent-process theory of motivation: I. Temporal dynamics of affect. *Psychological Review, 81*(2), 119–145.

Stahl, S. M. (2013). *Stahl's essential psychopharmacology: Neuroscientific basis and practical applications.* Cambridge University Press.

Stalnaker, T. A., Cooch, N. K., & Schoenbaum, G. (2015). What the orbitofrontal cortex does not do. *Nature Neuroscience, 18*(5), 620–627. https://doi.org/10.1038/nn.3982

Tolin, D. F. (2010). Is cognitive-behavioral therapy more effective than other therapies? A meta-analytic review. *Clinical Psychology Review, 30*(6), 710–720. https://doi.org/10.1016/j.cpr.2010.05.003

Verschure, P. F. M. J., & Wiers, R. W. (2022). Addiction biases choice in the mind, brain, and behavior systems: Beyond the brain disease model. In N. Heather, M. Field, A. C. Moss, & S. Satel (Eds.), *Evaluating the brain disease model of addiction* (pp. 384–404). Routledge. https://doi.org/10.4324/9781003032762-39

Vocci, F. J., Acri, J., & Elkashef, A. (2005). Medication development for addictive disorders: The state of the science. *American Journal of Psychiatry, 162*(8), 1432–1440. https://doi.org/10.1176/appi.ajp.162.8.1432

Volkow, N. D., Fowler, J. S., Wang, G.-J., Hitzemann, R., Logan, J., Schlyer, D. J., Dewey, S. L., & Wolf, A. P. (1993). Decreased dopamine D2 receptor availability is associated with reduced frontal metabolism in cocaine abusers. *Synapse, 14*(2), 169–177. https://doi.org/10.1002/syn.890140210

Volkow, N. D., Fowler, J. S., Wang, G.-J., Telang, F., Logan, J., Jayne, M., Ma, Y., Pradhan, K., Wong, C., & Swanson, J. M. (2010). Cognitive control of drug craving inhibits brain reward regions in cocaine abusers. *NeuroImage, 49*(3), 2536–2543. https://doi.org/10.1016/j.neuroimage.2009.10.088

Volkow, N. D., Koob, G. F., & McLellan, A. T. (2016). Neurobiologic advances from the brain disease model of addiction. *New England Journal of Medicine, 374*(4), 363–371. https://doi.org/10.1056/NEJMra1511480

Volkow, N. D., Michaelides, M., & Baler, R. (2019). The neuroscience of drug reward and addiction. *Physiological Reviews, 99*(4), 2115–2140. https://doi.org/10.1152/physrev.00014.2018

Volkow, N. D., & Morales, M. (2015). The brain on drugs: From reward to addiction. *Cell, 162*(4), 712–725. https://doi.org/10.1016/j.cell.2015.07.046

Yawata, S., Yamaguchi, T., Danjo, T., Hikida, T., & Nakanishi, S. (2012). Pathway-specific control of reward learning and its flexibility via selective dopamine receptors in the nucleus accumbens. *Proceedings of the National Academy of Sciences, USA, 109*(31), 12764–12769. https://doi.org/10.1073/pnas.1210797109

Chapter 6

Psychopharmacology Basics

Nancy E. Sherman and Thomas A. Field

Psychopharmacological medications can play an important role in wellness and recovery for some individuals and conditions, and they have become a staple of mental health treatment for many. As mental health concerns rose during the COVID-19 pandemic, more people began using psychotropic medication. In a Household Pulse Survey on mental health, the Centers for Disease Control and Prevention (2022) found that 23% of adults reported taking psychiatric medication for mental health, compared with only 16% of adults in 2019.

Although these medications cannot cure mental disorders, they may play an essential role in alleviating symptoms. To best care for clients, counselors should know how psychotropic medications work in the brain, their typical side effects, when to refer a client for a medication evaluation, and which mental disorders might best be treated using an additional or alternative approach. Counselors must also be aware of medical conditions and nonpsychotropic medicines that may cause psychological symptoms. This chapter is intended to present basic pharmacological information; not everything you need to know about psychotropic medication is within its scope.

With the advent of Medicare independent provider status and increasing participation as members of integrated behavioral health care and interdisciplinary treatment teams, professional counselors are increasingly expected to consult with medical professionals regarding their clients. Therefore, knowledge of psychopharmacology and medication management is becoming a more important competence for professional counselors. Professional counselors may well be clients' primary source of information about their medication because primary care providers prescribe more psychotropic drugs than do psychiatrists. Furthermore, counselors typically see clients weekly, whereas prescribers of psychotropic medications may see their clients only once every 4 to 6 weeks. Therefore, counselors must have some basic knowledge of psychotropic drug classification, pharmacokinetics, and neurobiological action. Counselors should also possess knowledge of psychiatric medication use with children and older adults. With this information, counselors can better identify symptoms that warrant psychiatric referral; monitor client symptoms as well as treatment and side effects; and, if indicated and within the limits of their scope of practice, help prepare clients to meet with their medical professional prescriber regarding psychotropic medication.

The case of Charlie illustrates essential elements of psychopharmacology and the use of psychotropic (psychiatric) medication. Key terms such as *pharmacokinetics* and *neurotransmission* are defined and explained. We provide information on the classification, contraindications, and abuse potential of psychotropic drugs and explain how these drugs work in the brain and body. We highlight special considerations for using psychiatric medication with children and older adults and present a brain-based approach to medication referral, monitoring, and psychoeducation.

2024 CACREP Standards

This chapter addresses 2024 Council for Accreditation of Counseling and Related Educational Programs (CACREP) Standards pertinent to the Foundational Counseling Curriculum (Section 3) areas of Lifespan Development (Standard C) and Counseling Practice and Relationships (Standard E):

- Biological, neurological, and physiological factors that affect lifespan development, functioning, behavior, resilience, and overall wellness (Standard C.10.)
- Classification, effects, and indications of commonly prescribed psychopharmacological medications (Standard E.18.)

Clinical Case Study: Charlie

Charlie is a 5-year-old boy who lives with his mother and stepfather. Charlie is one of four children, two of whom are stepsiblings. His parents work, and the children often stay with their maternal grandmother after school. Charlie's grandmother spoke to me (Nancy Sherman) about her concern that he had been prescribed risperidone (Risperdal) by his family physician. She had looked the drug up on the internet and learned it was for people with schizophrenia (or "crazy people," as she described them). The doctor told her that risperidone had been around for a long time and should work well for Charlie's problems. Although she was concerned about the drug, she said that Charlie's behavior had markedly improved and that he was sleeping through the night for the first time in years.

Charlie had been hyperactive since he was a baby. From age 4, Charlie had been treated for attention-deficit/hyperactivity disorder (ADHD) with various medications, first with stimulant medication and later with others. Nothing seemed to help until he began taking the risperidone. In addition to the hyperactivity, Charlie had difficulty learning social skills and playing with other children. He was obsessed with dinosaurs and engaged in some repetitive behavior. Charlie had been treated only by his pediatrician/family practice doctor and had received no counseling or psychiatric evaluation. Charlie's family had concerns about the long-term effects of the medication and believed the physician was not monitoring it.

Neurobiology of Psychotropic Drugs

As described in Chapter 1, brain communication takes place between neurons. Neurons communicate with each other through neurotransmitters. When two neurons are in close proximity, signaling between the neurons opens the presynaptic terminal buttons of the presynaptic neuron. When a neuron is excited, an electrical current travels through an axon to a terminal button. The presynaptic neuron must then release neurotransmitters to the postsynaptic neuron across a space called a *synapse*. Each particular type of neurotransmitter binds to a specific receptor. Receptors are named according to the kind of neurotransmitter they prefer to tie. For example, serotonin binds to serotonin receptors. Neurotransmitters sent into the synaptic cleft either bind to receptors, are destroyed chemically by enzymes, or are reabsorbed into the presynaptic neuron via reuptake.

Reflective Question

What other possible interventions might be helpful to Charlie and his parents?

We are not done yet! Neurotransmission is the first step in a long process that may eventually generate an electrical charge in the neuron called the *action potential*. To use a metaphor, neurotransmitters "open or close the door" for ions to enter the cell, and it is ions that generate the electrical charge. Once this occurs, the postsynaptic neuron fires an electrical impulse from the dendrites to the axon terminal buttons, sending the electrical charge down the neuron's axon. If another neuron is in proximity, this causes signaling, and the process repeats with the (formerly postsynaptic, now presynaptic) cell opening its axon terminal buttons and sending neurotransmitters into the synaptic cleft to bind with the next postsynaptic neuron.

Dynamics in Neurotransmission

Neurotransmission is complex and not always predictable. Each neuron has thousands of receptors, with many receptor types. Neurotransmitters can bind to multiple types of receptors and have different effects, depending on the receptor to which they bind. There are two main classifications of receptor types: (a) ligand-activated ionotropic receptors and (b) metabotropic receptors. *Ionotropic* receptors allow for direct entry of ions into the neuron once the neurotransmitter binds to the receptor. In contrast, *metabotropic* receptors do not directly open ion channels but initiate an indirect second signaling pathway that may open or close the ion channel. Although all common neurotransmitters bind with metabotropic receptors (indirect effect), only some bind with ionotropic receptors (direct effect). Thus, only some neurotransmitters have a direct (i.e., stronger or more pronounced) effect on opening ion channels.

Once a neurotransmitter is bound to a receptor, ion channels in the receiving neuron's dendrites may open or close, depending on the neurotransmitter. Some receptors allow positively charged ions (e.g., sodium) to enter, whereas others allow negatively charged ions (e.g., chloride) to enter. While effects can vary, most positively charged ions will increase the likelihood of eventually creating an action potential within the neuron. Most negatively charged ions will reduce the possibility of an action potential, thus halting the ongoing neurotransmission between neurons in the network. Some neurotransmitters are associated with opening channels that allow positively charged ions into the neuron, thus increasing the likelihood of action potential in what is called an *excitatory* effect. As shown in Table 6.1, acetylcholine, dopamine, glutamate, and norepinephrine are all known to have an excitatory effect. Other neurotransmitters are known to open channels to negatively charged ions, thus reducing the likelihood of action potential in what is called an *inhibitory* effect. As shown in Table 6.1, gamma-aminobutyric acid (GABA) and serotonin have inhibitory effects.

If repeated across the neural network, excitatory and inhibitory effects are associated with changes to physiological functioning. For example, norepinephrine (or noradrenaline) is related to the activation of the autonomic nervous system, resulting in increased use of energy resources such as with an increased heart rate. High levels of

TABLE 6.1
Neurotransmitters and Their Binding Effects

Neurotransmitter	Chemical Abbreviation	Binds With Ionotropic Receptors	Binds With Metabotropic Receptors	Effect
Acetylcholine	ACh	Yes	Yes	Excitatory
Dopamine	DA	No	Yes	Excitatory
Gamma-aminobutyric acid	GABA	Yes	Yes	Inhibitory
Glutamate	Glu	Yes	Yes	Excitatory
Norepinephrine	NE	No	Yes	Excitatory
Serotonin	5-HT	Yes	Yes	Inhibitory

such activation are often associated with anxiety and fear. In contrast, GABA is associated with reducing activation of the nervous system, resulting in preserving energy resources that accompany sleep and low energy expenditure. GABA is associated with anxiety and fear control and reduction.

Effects of Psychotropic Drug on Neurotransmission

Psychotropic drugs can have excitatory and inhibitory properties. Some psychotropic drugs enhance the release of excitatory neurotransmitters (i.e., open channels for positively charged ions), whereas other psychotropic drugs strengthen the release of inhibitory neurotransmitters (i.e., open channels for negatively charged ions).

In addition, psychotropic drugs can act in several different ways. One way is by mimicking the effects of naturally occurring neurotransmitters. Drugs can also block neurotransmission and alter neurotransmitters' shared storage, release, and removal. One important mechanism by which psychotropics act is to stop the reuptake of a neurotransmitter released from the presynaptic terminal. Selective serotonin reuptake inhibitors (SSRIs) work by blocking serotonin reuptake and leaving more of the neurotransmitter in the synapse. Increasing serotonin levels in the synapse is believed to improve mood (Preston et al., 2021). Drugs that bind and enhance the function of receptors are known as *agonists*, and those that bind and thereby block normal function are known as *antagonists*.

Pharmacokinetics

Pharmacokinetics can be defined as the study of what happens to a drug when it enters the body. The four pharmacokinetic processes are absorption, distribution, metabolism, and excretion. Every drug has a unique kinetic profile composed of these four factors (Preston et al., 2021).

Absorption is how a drug enters the bloodstream. Most psychotropic drugs are orally administered, so most medications are absorbed through the stomach and small intestine. A full or empty stomach, drug solubility, and blood flow affect the absorption rate. Some psychotropics are injected intravenously (directly into the bloodstream) or intramuscularly (into the muscle). The route of administration affects the absorption of the drug. For example, intravenously administered drugs are absorbed faster than orally or intramuscularly administered ones. Other ways of administration for psychotropic medications include sublingual (under the tongue) and transdermal (through the skin, usually by an adhesive patch).

Distribution refers to the circulation of medication throughout the body via the bloodstream after absorption. The amount of blood supply to an area determines the rate of distribution. For example, the heart and brain have an increased blood supply, and the medication will act more quickly in these areas. The drug also distributes to the locations of metabolism and excretion. Factors influencing medicine distribution include the chemical consistency of the medication, the amount administered, other drugs taken, local blood flow, and membrane permeability.

Metabolism is the process of how the body changes drugs to eliminate them. As a drug enters the body, it is treated as a foreign substance and is chemically altered by metabolism, producing by-products called *metabolites*. Some metabolites have the desired therapeutic effect of reducing the individual's symptoms. Other metabolites affect different body tissues and result in undesirable side effects. When liver function is impaired, there is a risk of excessive drug levels. Toxicity or poisoning can occur. If the liver metabolizes the drug too quickly, the person may not get enough of the drug to experience a therapeutic level that reduces symptoms.

All drugs are eventually removed from the body during a process called *excretion*. Most drugs are excreted through urine after processing by the kidneys. Without adequate kidney function, toxic levels of a drug can accumulate in the bloodstream. Drugs are metabolized and eliminated at varying rates. The *half-life* of a drug is the amount of time it takes for the plasma concentration of the blood to be reduced by one half. When a person takes medication regularly, there is an ongoing process of drug absorption from each dose and, simultaneously, a continuous process of drug removal by metabolism and elimination. A *steady state* occurs when the amount of drug taken is equivalent to the amount eliminated. Drugs usually take between five and six half-lives to reach a steady state. Drugs with a shorter half-life reach a steady state quickly, and those with a longer half-life are slower to get to a steady state (Kramer, 2003). Drug side effects can be less severe or even eliminated after reaching a steady state. For example,

SSRIs have reduced gastrointestinal side effects after reaching a steady state (Kramer, 2003).

> **Reflective Question**
>
> Gathering information about medication response is an essential step in the counseling process.
> What information about Charlie would you want to know?

An individual's response to a drug is affected by many factors that impact the psychokinetic processes. These factors include age, genetics, the expectation of the drug's effect, and interactions with other medicines. Because of these factors, determining drug dosage and route of administration can be a complex issue. In complex cases, it is essential to consider referring clients for medication evaluation to a specialized psychopharmacologist, such as a psychiatrist or psychiatric nurse practitioner, who has extensive knowledge of psychotropic medications.

Use of Psychiatric Drugs With Children and Older Adults

Children and older adults are two groups of particular concern when prescribing psychiatric drugs. According to Preston et al. (2021), medical professionals have only recently recognized significant neurobiological development and metabolism differences for child and older adult populations. The pharmacokinetics of drugs in children differ from those in adults because of age, body composition, liver and kidney function, and maturation of enzymatic systems (Preston et al., 2021). Medication dosing for children must consider such factors, making prescribing a complex art for many psychopharmacologists. For example, a 12-year-old who weighs 200 pounds would likely require a different therapeutic dose than a 12-year-old who weighs 100 pounds. The therapeutic amount may also differ between a 12-year-old and a 17-year-old who each weigh 200 pounds.

In older adults, the pharmacokinetic action of psychotropic drugs differs from that in younger adults. Age-related changes cause drugs to stay active longer in the body, prolonging their effect and potentially increasing side effects in older adults. Water in the body decreases, and the amount of fat increases. This combination can result in higher concentrations of water-soluble drugs because there is less water to dilute them and more fat to store medications (Ruscin & Linnebur, 2022). The liver does not metabolize drugs as efficiently in older adulthood, and the kidneys are less able to excrete drugs into urine, resulting in less elimination.

Psychotropic Drug Classification

Psychotropic drugs are placed in different categories and classified according to their primary effect on the brain and central nervous system. Table 6.2 displays neurotransmitters targeted by psychopharmacological drugs. The main classifications for psychiatric medications are antidepressants, anxiolytics, mood stabilizers, antipsychotics, and stimulants. In this section, we explore contraindications for use and abuse potential in each category. A drug has a *contraindication* when its use can have serious adverse effects, mainly if the drug interacts with other medication a person is taking. *Abuse potential* is the likelihood that a person might abuse a drug because of the hedonic effect the drug elicits. The only psychotropic drug classes with abuse potential are the anti-anxiety and stimulant medications. Although psychotropic drugs reduce symptoms and improve quality of life, they are not a cure for any mental disorder.

TABLE 6.2

Neurotransmitters Targeted by Psychopharmacological Drugs

Drug Class	Trade Name	Dopamine	GABA	Norepinephrine	Serotonin
Anxiolytic (benzodiazepine)	Ativan, Klonopin, Valium, Xanax		Enhance		
Antidepressant (SSRI)	Celexa, Lexapro, Luvox, Paxil, Prozac, Zoloft				Enhance
Antidepressant (SNRI)	Cymbalta, Pristiq, Strattera			Enhance	Enhance
Antidepressant (NDRI)	Wellbutrin	Enhance		Enhance	
Antidepressant (SMS)	Trintellix			Enhance	
Antipsychotic (first generation)	Haldol, Thorazine	Reduce			
Antipsychotic (second generation)	Abilify, Clozaril, Geodon, Invega, Risperdal, Seroquel, Zyprexa	Reduce			
Stimulant	Adderall, Concerta, Focalin, Ritalin, Vyvanse	Enhance			

Note. Enhance = the axon releases more of the neurotransmitter or the dendrite receives more of the neurotransmitter by inhibiting reuptake to the axon. Reduce = the axon releases less of the neurotransmitter or the dendrite receives less of the neurotransmitter by blocking the receiving channels. Mood stabilizers are not included in this table. GABA = gamma-aminobutyric acid; SSRI = selective serotonin reuptake inhibitor; SNRI = selective norepinephrine reuptake inhibitor; NDRI = norepinephrine and dopamine reuptake inhibitor; SMS = serotonin modulator and stimulator.

Antidepressants

Among adults, antidepressants are the most commonly prescribed psychotropic drug class, and anti-anxiety medication is the second most prescribed (Kantor et al., 2015). Antidepressants and stimulants are the medications most commonly used by children and adolescents (Sultan et al., 2018). Antidepressants block the reuptake process so that excitatory and inhibitory neurotransmitters such as serotonin and norepinephrine (depending on the medication) stay in the synaptic cleft longer; thus, greater receptor binding is expected. These medications treat depression, anxiety, and obsessive-compulsive disorder.

Antidepressants are relatively safe medications with low potential for abuse. The most commonly prescribed antidepressants are SSRIs and selective norepinephrine reuptake inhibitors (SNRIs). The side effects of SSRIs and SNRIs are less pronounced than those of older agents such as monoamine oxidase inhibitors and tricyclics.

All antidepressant medications seem to have equivalent effectiveness in relieving depression symptoms, although some antidepressants work better than others for a particular person. Antidepressants have a relatively long half-life, with agents such as fluoxetine (e.g., Prozac) taking 6 weeks to reach a therapeutic level (steady state). When a client believes an antidepressant is not relieving symptoms after taking it as prescribed for at least 2 months, counselors should encourage and coach the client to talk to their psychopharmacologist. Because of the risk of a severe adverse reaction called *serotonin syndrome*, a person should take only one antidepressant at a time unless titrating from one medication to another.

Anxiolytics

Treatment for anxiety-related disorders includes prescribing anxiolytics in addition to antidepressants. Unlike antidepressants, many anxiolytics enhance only inhibitory neurotransmitters, such as GABA, and not excitatory neurotransmitters. Anxiolytics have a direct effect of releasing more GABA into the synaptic cleft, rather than solely blocking the reuptake of the same amount of neurotransmitter released into the cleft (a weaker effect). This effect, combined with a much shorter half-life, means that anxiolytics produce a stronger inhibitory response than antidepressants and can more quickly reduce anxiety and fear responses. The most commonly used anxiolytics are benzodiazepines, which reduce acute anxiety symptoms such as panic attacks, extreme worry, and fear.

Benzodiazepines increase the neurotransmission of GABA, producing a sedating, sleep-inducing, muscle-relaxing, and anti-anxiety effect. Their short half-life and hedonic properties mean that benzodiazepines (e.g., Klonopin, Xanax, Valium) have a high potential for misuse and should be prescribed for only a short term or as needed. Tolerance can occur, and sudden withdrawal can be severe (Sadock & Sadock, 2021). For this reason, benzodiazepines are prescribed with caution to individuals with active substance use disorders. Benzodiazepines are contraindicated during pregnancy because their use during the first trimester may cause congenital disabilities and other problems in the infant (Sadock & Sadock, 2021). Benzodiazepines are cautiously prescribed to older adults, who may become more somnolent and confused when using them (Ruscin & Linnebur, 2022). This cognitive impairment can result in falls and fractures. Billioti de Gage et al. (2014) found that benzodiazepine use was also associated with an increased risk of Alzheimer's disease in a large sample of older adults.

Beta blockers also may be used to manage the physical effects of anxiety, such as rapid heartbeat, trembling, and sweating, by blocking the release of epinephrine and norepinephrine. Other anxiolytics include buspirone (Buspar), which is often used to augment (i.e., used concurrently with) antidepressant treatment. Barbiturates are an older group of anti-anxiety agents that are less commonly used today because of their side effect profile.

Mood Stabilizers

Mood stabilizers treat mania in bipolar disorder and mood swings associated with other mental illnesses. Lithium is the most popular mood stabilizer. Only recently has research clarified our understanding of how it works in the brain. Mood stabilizers have strong inhibitory effects and, thus, reduce overactivation in neuronal functioning associated with both mania states and seizures. Anticonvulsant medications developed to treat seizure disorders, such as valproic acid (Depakote), also have mood-stabilizing effects. A meta-analysis of brain imaging data from people with bipolar disorder obtained from 11 international research groups found that individuals with bipolar disorder not taking lithium had reduced cerebral and hippocampal volumes compared with a comparison group of people without bipolar disorder (Hallahan et al., 2011). People with bipolar disorder who were taking lithium had significantly increased hippocampal and amygdala volume, which suggests that the use of lithium can make up for structural brain alterations in people with bipolar disorder (Sani et al., 2018).

Mood stabilizer medications can have severe side effects and must be monitored closely for the development of problems. For example, failure to remain adequately hydrated when taking lithium can result in toxicity. Thus, people taking lithium should avoid diuretics, such as caffeine and alcohol, and drink plenty of water. Most mood stabilizers are contraindicated during pregnancy (Sadock & Sadock, 2021), and pregnant women with bipolar disorder typically have to stop these medications, which results in an increased risk of manic episodes.

Antipsychotics

Antipsychotic medications are major tranquilizers that are effective in reducing symptoms of psychosis, such as hallucinations and delusions, that are associated with schizophrenia and acute mania. Antipsychotics work by blocking dopamine receptors (agonists). Newer second-generation atypical antipsychotics also block serotonin receptors. As a result, antipsychotics have a strong inhibitory effect and are used to reduce neural overactivity associated with psychosis and mania and to treat extreme aggression found in autism.

As potent inhibitory medications, antipsychotics are also associated with side effects that include reduced energy, lethargy, and drowsiness at higher dosages. Dopamine is a neurotransmitter associated with movement, and older first-generation antipsychotics can have the problematic side effect of initiating involuntary muscle movements known as *extrapyramidal* symptoms. If these persist, they can become a chronic condition known as *tardive dyskinesia*. Newer second-generation atypical antipsychotics are less likely to cause extrapyramidal side effects. However, because they also block serotonin (associated with appetite and metabolism regulation), they are more likely to cause metabolic side effects resulting in weight gain, increased cholesterol, and increased blood sugar that can progress to Type 2 diabetes (Sadock & Sadock, 2021). Because of these problematic side effects, lack of adherence to antipsychotic treatment can be a significant treatment barrier. Antipsychotics are not known for producing hedonic effects or for having habit-forming qualities (i.e., tolerance); thus, the misuse potential of antipsychotic drugs is small. Aripiprazole (Abilify) has been approved by the U.S. Food and Drug Administration (FDA) for adjunctive treatment in major depression and Tourette's syndrome.

Stimulants

Stimulants are most commonly prescribed to treat ADHD symptoms. Stimulants have excitatory effects by increasing dopamine transmission in the synapse through enhanced presynaptic release and reuptake inhibition. As a result, stimulants increase alertness, attention, and energy and elevate blood pressure, respiration, and heart rate (Sadock & Sadock, 2021). Most stimulants are amphetamine (Adderall) or methylphenidate (Ritalin) products. Stimulants also treat narcolepsy and chronic depression that has not responded to other treatments.

Stimulants have a high potential for misuse because of several factors. First, they have hedonic properties when taken at a high dose. Second, stimulants improve performance for most people, and high school and college students may purchase or share stimulants to enhance attention and concentration during academic study sessions or examinations. Third, individuals with eating disorders can misuse stimulants for their side effects of appetite suppression. Although more empirical research is needed, a longitudinal study found no evidence of problematic long-term consequences such as stunted growth (Harstad et al., 2014). As for all psychotropic medications, pregnant individuals should discuss the use of stimulants with their prescriber.

Technically, stimulants are part of a larger class of medications called cognitive enhancers. Medications are often prescribed to older adults with similar properties to stimulants used for ADHD. Cognitive enhancers such as donepezil (Aricept) have excitatory properties by boosting the release and reuptake inhibition of dopamine, norepinephrine, and acetylcholine, especially in the hippocampus. These medications have been found to reduce the rate of hippocampal atrophy and progression of mild cognitive impairment and forms of neurocognitive disorders such as Alzheimer's disease (Dubois et al., 2015).

In recent years, the greater frequency of stimulant usage has resulted in shortages, with many clients struggling to fill their prescriptions. Shortages can result in unintentional medication "holidays," and clients may seek guidance from counselors regarding how best to manage their symptoms of conditions such as ADHD without medication. The interventions for which there is the most empirical support are behavioral and cognitive behavioral therapies (Young et al., 2020). Neurofeedback has also been FDA-approved as a treatment for ADHD, although it is expensive and the cost is rarely covered by health insurance. These interventions often enhance client motivation through the use of contingency-based rewards for desired behavior and attentional focus (Young et al., 2020).

> **Reflective Questions**
>
> Risperidone is an atypical antipsychotic medication. Why might Charlie be taking this medication? What are the benefits and risks?

Medication Referrals and Consultation

Counselor referral to a prescriber of psychotropic medication is indicated by several factors (Preston et al., 2021). Clients experiencing an acute change in functioning need a general medical evaluation and may require a subsequent medication referral. Referrals are also indicated for acute conditions such as delirium, seizures, and neurobiological symptoms of depression (e.g., changes in sleep, appetite, concentration, anhedonia, hypoactivity, loss of energy). When a client is experiencing symptoms of schizophrenia or mania, referral to a psychopharmacologist as part of the treatment plan is considered a best practice.

Counselors should immediately refer clients using lamotrigine (Lamictal) to their psychopharmacologist when they are experiencing severe and potentially life-threatening side effects, such as Stevens-Johnson syndrome (Sadock & Sadock, 2021). Similarly, clients withdrawing from substance use disorders may require specialized medical treatment to avoid medical complications associated with withdrawal. Clients with medical issues should be carefully monitored, and the psychopharmacologist should be notified if their condition progressively declines, such as with neurocognitive disorders.

Consultation With Psychopharmacologists

When providing counseling services, consultation with psychopharmacologists can be crucial to ensuring the client is receiving integrated services in a complementary fashion that enhances the overall effectiveness of treatment. For example, clients taking stimulants for ADHD may be better able to pay attention during therapy sessions, and their medication adherence may optimize counseling sessions. Clients who believe their antidepressant has stopped working may have less energy to attend to the tasks of therapy (DeRubeis et al., 2008). Consultation often requires the client to sign a release so the practitioner can share protected health information with another provider. In the case of Charlie, the counselor should reach out to the primary care physician to better understand the provider's diagnosis and medication management plan. Although Charlie does not appear to be experiencing schizophrenia or mania, he is currently taking an antipsychotic, considered a medication best prescribed by a specialist (i.e., a psychopharmacologist). One might wonder whether Charlie needs to be seen by a psychopharmacologist to ensure he receives the best care in a complex case. This discussion with the primary care physician could be crucial.

Counselors can prepare for consultation meetings by preparing a list of symptoms the client is experiencing, using terminology from the mental status examination. Counselors should also know the medications taken by the client and any client-reported side effects. Counselors should also avoid making medication recommendations because this is outside their scope of practice and may cast doubt on other information the counselor has shared.

> **Reflective Question**
>
> Practice role-playing a consultation with the primary care physician about Charlie's case. How might you bring up concerns about the need for more specialized care?

Consultations With Clients

The client may sometimes ask the counselor for information about their medication. The counselor's role in this situation is to reinforce the importance of the client's relationship with the psychopharmacologist and prepare the client to bring up concerns regarding the medication with their medical provider in their next appointment. Clients may feel anxious about this encounter and hesitate to bring up vital information about medication-related concerns. Clients can feel more prepared by writing down questions and role-playing the conversation, and the counselor could attend the next medication management appointment. The counselor could also discuss their ancillary role in monitoring the client's symptoms and side effects. Although medication-related questions should be deferred to the client's psychopharmacologist, the counselor can provide information about how to adhere to a medication regimen. As with other neuroscientific information, counselors should clearly and concisely present the information in a way that "distills without diluting" it (Field & Ghoston, 2020, p. 227). For example, it is more important for a client to understand that an antidepressant must be taken as prescribed to reach a steady state than it is to understand the half-life and pharmacokinetics of antidepressants.

> **Reflective Question**
>
> Read through the case of Charlie again.
> Practice role-playing on preparing Charlie's family for their next appointment with the primary care physician.

Conclusion

Although counselors are not medical professionals, they need to have a working knowledge of the uses and misuses of psychotropic drugs to provide the best treatment possible. Counselors develop a therapeutic relationship that is often impossible with a medical professional. In this relationship, clients can learn how their medication works, the potential for side effects and problem use, and how to take their medication as prescribed. Counselors can provide proper monitoring and prepare clients to discuss their medicine(s) with their psychopharmacologist. When the client is a child, the counselor should include parents when asking questions about medication response.

Our Brain-Based Approach to the Case of Charlie

Charlie's parents lacked information about their son's condition. After consulting with the primary care provider, I (Nancy Sherman) received information that Charlie had been comprehensively assessed and diagnosed with autism. He was taking risperidone for acute aggressive episodes at home. I used this information to discuss Charlie's diagnosis and all available treatment options with his mother and grandmother. After this conversation, they understood the importance of an integrative approach and the relevance of risperidone in treating his aggressive symptoms.

With Charlie's permission, my neurocounseling process was to incorporate applied behavior analysis with neurofeedback to provide positive reinforcement for developing frustration tolerance and functional communication through classical and operant conditioning. I referred him to a practitioner specializing in applied behavior analysis treatment. The treatment plan included medication management for risperidone, with a goal for the family to eventually wean Charlie off the drug under the direction of the primary care provider. A necessary adjunctive intervention was to provide parent education and space for the family to process their adjustment to and feelings of loss at their son's autism diagnosis. Because they had the opportunity to review all treatment options, Charlie and his family could make an informed choice about what approach worked best for them.

> **Reflective Question**
>
> What might Charlie's counselor have missed had she not consulted with the primary care provider?

Quiz

1. When the amount of a drug available to the body is the same as the amount being eliminated, then _____ has been achieved.
   a. Half-life.
   b. Pharmacokinetic balance.
   c. Steady state.
   d. Partial state.
2. Which of the following is not a pharmacokinetic process?
   a. Excretion.
   b. Metabolism.
   c. Absorption.
   d. Accumulation.
3. When a serotonin neurotransmitter binds to a receptor, it has:
   a. An excitatory effect.
   b. An inhibitory effect.
   c. Excitatory and inhibitory effects.
   d. No effect, because serotonin neurotransmitters do not bind to receptors.

References

Billioti de Gage, S., Moride, Y., Ducruet, T., Kurth, T., Verdoux, H., Tournier, M., Pariente, A., & Bégaud, B. (2014). Benzodiazepine use and risk of Alzheimer's disease: Case-control study. *BMJ, 349*(7975), Article g5205. https://doi.org/10.1136/bmj.g5205

Centers for Disease Control and Prevention. (2022, July 20). *Mental health care: Household Pulse Survey.* https://www.cdc.gov/nchs/covid19/pulse/mental-health-care.htm

Council for Accreditation of Counseling and Related Educational Programs. (2024). *2024 CACREP standards.* https://www.cacrep.org/wp-content/uploads/2023/06/2024-Standards-Combined-Version-6.27.23.pdf

DeRubeis, R. J., Siegle, G. J., & Hollon, S. D. (2008). Cognitive therapy versus medication for depression: Treatment outcomes and neural mechanisms. *Nature Reviews Neuroscience, 9*(10), 788–796. https://doi.org/10.1038/nrn2345

Dubois, B., Chupin, M., Hampel, H., Lista, S., Cavedo, E., Croisile, B., Louis Tisserand, G., Touchon, J., Bonafe, A., Ousset, P. J., Ait Ameur, A., Rouaud, O., Ricolfi, F., Vighetto, A., Pasquier, F., Delmaire, C., Ceccaldi, M., Girard, N., Dufouil, C., . . . Hippocampus Study Group. (2015). Donepezil decreases annual rate of hippocampal atrophy in suspected prodromal Alzheimer's disease. *Alzheimer's & Dementia, 11*(9), 1041–1049. https://doi.org/10.1016/j.jalz.2014.10.003

Field, T. A., & Ghoston, M. G. (2020). *Neuroscience-informed counseling with children and adolescents.* American Counseling Association.

Hallahan, B., Newell, J., Soares, J. C., Brambilla, P., Strawkowski, S. M., Fleck, D. E., Kieseppa, T., Altshuler, L. L., Fornito, A., Malhi, G. S., McIntosh, A. M., Yurgelun-Todd, D. A., Labar, K. S., Sharma, V., MacQueen, G. M., Murray, R. M., & McDonald, C. (2011). Structural magnetic resonance imaging in bipolar disorder: An international collaborative mega-analysis of individual adult patient data. *Biological Psychiatry, 69*(4), 326–335. https://doi.org/10.1016/j.biopsych.2010.08.029

Harstad, E. B., Weaver, A. L., Katusic, S. K., Colligan, R. C., Kumar, S., Chan, E., & Barberesi, W. J. (2014). ADHD, stimulant treatment, and growth: A longitudinal study. *Pediatrics, 134*(4), e935–e944. https://doi.org/10.1542/peds.2014-0428

Kantor, E. D., Rehm, C. D., Haas, J. S., Chan, A. T., & Giovannucci, E. L. (2015). Trends in prescription drug use among adults in the United States from 1999–2012. *JAMA, 314*(17), 1818–1830. https://doi.org/10.1001/jama.2015.13766

Kramer, T. A. M. (2003). Side effects and therapeutic effects. *Medscape General Medicine, 5*(1), 28.

Preston, J. D., O'Neal, J. H., & Talaga, M. C. (2021). *Handbook of clinical psychopharmacology for therapists* (9th ed.). New Harbinger.

Ruscin, J. M., & Linnebur, S. A. (2022, September). *Pharmacokinetics in older adults.* Merck Manual Professional Version. https://www.merckmanuals.com/professional/geriatrics/drug-therapy-in-older-adults/pharmacokinetics-in-older-adults

Sadock, B. J., & Sadock, V. A. (2021). *Synopsis of psychiatry* (12th ed.). Lippincott Williams & Wilkins.

Sani, G., Simonetti, A., Janiri, D., Banaj, N., Ambrosi, E., De Rossi, P., Ciullo, V., Arciniegas, D. B., Piras, F., & Spalletta, G. (2018). Association between duration of lithium exposure and hippocampus/amygdala volumes in Type I bipolar disorder. *Journal of Affective Disorders, 232,* 341–348. https://doi.org/10.1016/j.jad.2018.02.042

Sultan, R. S., Correll, C. U., Schonbaum, M., King, M., Walkup, J. T., & Olfson, M. (2018). National patterns of commonly prescribed psychotropic medications to young people. *Journal of Child and Adolescent Psychopharmacology, 28*(3), 158–165. https://doi.org/ 10.1089/cap.2017.0077

Young, Z., Moghaddam, N., & Tickle, A. (2020). The efficacy of cognitive behavioral therapy for adults with ADHD: A systematic review and meta-analysis of randomized controlled trials. *Journal of Attention Disorders, 24*(6), 875–888. https://doi.org/10.1177/1087054716664413

Part II
COUNSELING ASSESSMENTS, RELATIONSHIPS, AND INTERVENTIONS

The second section of the text explicates how neuroscience both explains and informs clinical assessment, counseling relationships, and therapeutic interventions. These chapters build on the foundational knowledge presented in the first section by describing how to apply that foundational knowledge from intake to case closure and across a variety of modalities. We first introduce a neurocounseling approach to clinical assessment, along with assessing wellness and optimized performance, before focusing on the neuroscience of attention and its implications for empathy and microskills, leveraging neuroeducation to enhance client outcomes, adopting a neuroscience-informed approach to integrating counseling theories, and taking a neuroscience-informed approach to group and career-focused counseling.

■ ■ ■

Chapter 7

Neurocounseling Assessment

Lori A. Russell-Chapin

This chapter presents a comprehensive neurocounseling assessment model that is needed for any efficacious counseling treatment outcome. It emphasizes the importance of including a thorough assessment of every new counseling client. In neurocounseling, the main goal of treatment is achieving emotional, behavioral, and physiological self-regulation. The clinical evaluation goes even more in depth because medical history, head injuries, pregnancy, and birth complications are always essential to understanding the complexity and perhaps even origins of presenting symptoms and brain dysregulation. Both qualitative and quantitative evaluations should be used. The battery of tests must include a thorough psychosocial and medical history and self-report checklists and inventories that follow a wellness profile for conceptualization of a successful treatment plan. If warranted, electroencephalograms (EEGs) and continuous performance tests add another dimension to the neurocounseling assessment. This chapter includes suggestions and examples of possible tests to be included in the overall assessment. On the basis of those baseline scores and interviews, a successful conceptualization and treatment plan can be developed. The ensuing plan will then include goals evolving from both the assessment and the partnership between the client and counselor.

2024 CACREP Standards

This chapter addresses 2024 Council for Accreditation of Counseling and Related Educational Programs (CACREP) Standards pertinent to the Foundational Counseling Curriculum (Section 3) area of Assessment and Diagnostic Processes (Standard G):

- Ethical and legal considerations for selecting, administering, and interpreting assessments (Standard G.6.)
- Use of culturally sustaining and developmentally appropriate assessments for diagnostic and intervention planning purposes (Standard G.7.)
- Use of structured interviewing, symptom checklists, and personality and psychological testing (Standard G.10.)
- Diagnostic processes, including differential diagnosis and the use of current diagnostic classification systems (Standard G.11.)
- Procedures for assessing clients' experience of trauma (Standard G.14.)
- Procedures to identify client characteristics, protective factors, risk factors, and warning signs of mental health and behavioral disorders (Standard G.16.)

Clinical Case Study: Carrie

A family member referred Carrie because she believed Carrie was slowly killing herself with anorexia. Carrie is a 30-year-old White woman who had completed two different residential treatment programs for her eating disorder. Carrie also believed she was in trouble and was feeling hopeless and desperate. Her very low body weight of 75 pounds was causing multiple high-risk emotional and physiological problems. I asked for a release of information with her primary care physician to discuss a team approach for addressing Carrie's low weight and associated symptoms. Carrie's physician was already monitoring her low weight and had regular monthly medical checkups with Carrie. I was in charge of addressing Carrie's cognitive distortions and behavioral decisions related to her low body weight and helping her with associated mental health symptoms.
The initial interview consisted of a thorough psychosocial and medical history. The neurocounseling interview required additional questions such as information about her birth and delivery; head traumas; surgeries; and use of technology, including laptops, smartphones, and tablets.
During our initial interview, Carrie appeared extremely bright. Yet despite entering graduate school and continuing to hold down a part-time job in the helping professions, her overall demeanor was timid. Carrie's voice was almost inaudible, and eye contact was minimal. Her physical

appearance was startling and scary to me. I knew that Carrie was very sick and that we had a very long neurocounseling journey ahead of us.

She was compliant and completed the battery of tests that were required. Carrie's baseline pretreatment scores along with her posttreatment assessment scores are discussed throughout this chapter, in the order in which each assessment was presented. The inventories or types of inventories included in Carrie's assessment battery were as follows:

- Psychosocial-medical history interview
- Neurological risk assessment
- Neurocounseling observational notes
- Insomnia checklist
- Depression checklist
- Anxiety checklist
- Body Perception Questionnaire–Short Form
- Trauma checklist
- Millon Clinical Multiaxial Inventory–III (MCMI-III)
- Learning difficulties assessment
- Test of Variables of Attention (T.O.V.A.)
- Five-channel EEG
- Full 19-channel EEG

My first task with Carrie was to let her know that I knew she had a great deal of expertise and knowledge about her body and even food. She shyly smiled. This treatment had to be different because Carrie was beginning to give up on life. In my mind, building rapport and a therapeutic alliance was essential. Although our plan was to conduct neuro- and biofeedback, traditional talk counseling was a crucial part of the process, and my first session focused on building trust and rapport. Carrie shared that she was consuming approximately 700 calories a day and running more than 8 miles daily. I asked her if she thought that was the amount of exercise that a healthy person needs daily. She said "yes." It was then that I learned that Carrie had never had a menstrual cycle. She had struggled with an eating disorder since she was 12 years old. After our first session, I also realized that Carrie knew very little about healthy living. Thus, our counseling goals targeted improvements in healthy living and wellness in all aspects of daily life.

> **Reflective Questions**
>
> What are your major concerns in working with Carrie?
> How might she be at high risk?

Assessment

Assessment has many meanings, and numerous different types of assessments are available. An *assessment* is sometimes defined as a judgment of the quality, worth, importance, or value of something (American Psychological Association, n.d.). In neurocounseling, that definition makes sense because the counselor must gather information on enough data points to make an efficacious decision about counseling treatment, goals, and outcomes.

A clinical assessment should cover all major aspects of the client's life, and it also should include possible generational history as well. The epigenetics involved in a client's history could hold an essential key to the puzzle and presenting concern. If a client does not have that generational history and relatives are still living, this part of the assessment may even be fun for the client to complete. Recounting generational history back to the counselor is often therapeutic in itself.

The clinical neurocounseling assessment must contain both qualitative and quantitative portions from many different sources (Chapin, 2021). The qualitative and subjective assessments, such as structured interviews, psychosocial-medical history, and other targeted self-report inventories, often build necessary rapport and therapeutic alliance. The client needs to believe that the counselor truly understands their problems. The quantitative and objective measures, such as symptom checklists and EEGs, often offer the counselor and client a unique perspective from a normative and comparison point of view. Assisting a client in observing how a particular score deviates from an average provides a way to look at the client's problem from a different vantage point. The major advantage of having multiple sources of qualitative and quantitative information is in developing patterns from these assessments. If results from all tests and interviews are similar, then it is often easier and provides a stronger case for diagnosis, conclusions, and treatment strategies (Chapin & Russell-Chapin, 2014). If the results show a wide range of discrepancies, then additional specific assessments may be required.

> **Reflective Question**
>
> What are the advantages of using both qualitative and quantitative measurements in Carrie's situation?

Psychosocial-Medical History Interview

Many counselors include the client's psychosocial history as part of the initial intake by using a checklist or standardized form. The client can fill out the intake form ahead of time and bring it to the first session. Although this process may save time, the beginning stages of rapport building are lost. According to Demos (2005), a basic psychosocial medical history must include questions about personal family history, school history and performance, psychological history, and medical history. Information about lifestyle from a wellness perspective is important too. Inquiries about nutrition, sleep hygiene, exercise, work satisfaction and work environment, spirituality and religion, drug use, and any other possible resources and liabilities are musts to include in the psychosocial and medical history.

During the psychosocial-medical history interview, many details of Carrie's life unfolded. A large family of origin may have allowed Carrie to get lost in the family dynamics. A fire in the family home resulted in the loss of many of her treasured family heirlooms. Two head injuries may have affected Carrie's situation. Early traumas, such as bowel infections as a preschooler, influenced Carrie's overall development. Carrie had a bowel resection, and her physician completed an ileostomy surgery to facilitate weight gain. As part of her medical care, Carrie attended regular appointments with her physician to review the success of the ileostomy intervention.

Neurological Risk Assessment

If clients are not good historians, having them complete the Neurological Risk Assessment Checklist is a useful beginning because it gently nudges them into remembering certain potential sources of dysregulation. The assessment (see Figure 7.1) is a paper-and-pencil checklist that offers 15 different categories of possible neurological risks ranging from genetic predispositions to environmental toxins. If the client does not remember something or know an answer, then it may be wise to bring in another family member. Congruent information is the key, and the main goal for the counselor is to focus on any aspects of the interview that may explain possible sources of brain dysregulation, such as head injuries, birthing complications, emotional or physical trauma, substance abuse, prenatal exposure to toxins, or high fevers. The total number of items checked off indicates the potential neurological risks and brain dysregulation. Even one checked area may suggest brain dysregulation (Russell-Chapin, 2021). This assessment also assists in developing treatment goals and priorities. Carrie listed eight of the 15 sources of neurological risks, suggesting many possible reasons for her dysregulation ranging from surgical anesthesia to psychosocial stressors.

Neurological Dysregulation Risk Assessment

Name: __Carrie__   Age: __30__   Date: __10/2021__

Presenting Symptom(s): ___Anxiety, anorexia___

Complete the following survey to identify the potential source of neurological dysregulation and indicate whether or not it may be a risk factor for you or your child. Please be as honest as possible when answering questions.

■ Yes ❑ No	**Genetic Influences:**	Grandparents, parents, or siblings with mental health or learning disorders (including attention-deficit/hyperactivity disorder), posttraumatic stress disorder, depression, generalized anxiety disorder, substance abuse, personality or other severe psychological disorders (bipolar or schizophrenia).
❑ Yes ■ No	**Prenatal Exposure:**	Maternal distress, psychotropic medication use, alcohol or substance abuse, nicotine use, or possible exposure to environmental toxins including genetically modified foods, pesticides, petrochemicals, xenestrogens in plastics, heavy metals (lead/mercury), and fluoride, bromine, and chlorine in water.
❑ Yes ■ No	**Birth Complications:**	Forceps or vacuum delivery, oxygen loss, head injury, premature birth, difficult or prolonged labor, obstructed umbilical cord, or fetal distress.
■ Yes ❑ No	**Disease and High Fever:**	Sustained fever above 104 degrees due to bacterial infection, influenza, strep, meningitis, encephalitis, Reye's Syndrome, PANDAS, or other infections or disease processes.
■ Yes ❑ No	**Current Diagnosis:**	Of mental health, physical health, alcohol abuse, or learning disorder.
❑ Yes ■ No	**Poor Diet and Inadequate Exercise:**	Diet high in processed food; preservatives; simple carbohydrates (sugar and flour); genetically modified foods; foods treated with herbicides; pesticides, and hormones; low daily water intake, high caffeine intake; and lack of adequate physical exercise (20 minutes, 7 times a week).
❑ Yes ■ No	**Emotionally Suppressive Psychosocial Environment:**	Being raised or currently living in poverty; domestic violence; physical, emotional, or sexual abuse; alcoholic or mentally unstable family environment; emotional trauma; neglect; institutionalization; and inadequate maternal emotional availability or attachment.
■ Yes ❑ No	**Mild to Severe Head Injury:**	Experienced one or more blows to the head from a sports injury, fall, or auto accident (with or without loss of consciousness), or episodes of open head injury, coma, or stroke.
■ Yes ❑ No	**Prolonged Life Distress:**	Most commonly due to worry about money, work, economy, family responsibilities, relationships, personal safety, and/or health causing sustained periods of anxiety, irritability, anger, fatigue, lack of interest, low motivation or energy, nervousness, and/or physical aches and pains.
❑ Yes ■ No	**Stress-Related Disease:**	Includes heart disease, kidney disease, hypertension, obesity, diabetes, stroke, hormonal, and/or immunological disorders.
■ Yes ❑ No	**Prolonged Medication Use, Substance Use, or Other Addictions:**	Including legal or illegal drug use, substance abuse, or addiction (alcohol, drugs, nicotine, caffeine, medication, gambling, sex, spending, etc.) and overuse of screen technologies (cell phones, video games, television, computers, Internet, etc.).
❑ Yes ■ No	**Seizure Disorders:**	Caused by birth complications, stroke, head trauma, infection, high fever, oxygen deprivation, and/or genetic disorders and includes epilepsy, pseudoseizures, or epileptiform seizures.
❑ Yes ■ No	**Chronic Pain:**	Related to accidents, injury, or a disease process. Including back pain, headache and migraine pain, neck pain, facial pain, and fibromyalgia.
■ Yes ❑ No	**Surgical Anesthesia, Chemotherapy and/or Aging:**	Can cause mild cognitive impairment, insomnia and depression and be related to emotional trauma; loss and grief; chronic illness; physical decline; reduced mobility; pphysical, social, and emotional isolation; and decreased financial security.
■ Yes ❑ No	**Excessive I-Technology., Video Game, TV, and/or Screen Time:**	Use of 2 hours or more a day, beyond work and/or school, can result in brain activity similar to ADHD, epilepsy, absence seizures, congnitive decline, demential and alcohol or marijuana abuse.

FIGURE 7.1

Copyright © 2023 by Ted Chapin, PhD. All rights reserved. Adapted with permission.

Neurocounseling Observational Notes

The counselor needs to use knowledge of neurocounseling and keen observational and listening skills and take notes on many direct, and often indirect, behaviors of and anecdotes from the client. For example, if the client is often late or gets lost coming to the office, the counselor should jot that down in the case notes. If the client states they have difficulty with relationships and cannot seem to finish tasks, the counselor should write that down. These observations and statements are clues to the client's brain dysregulation. Every detail can lead the counselor to additional information about the functioning of the client's brain. Carrie spoke very quietly and was often difficult to hear. She rarely looked directly at people, especially at me as her counselor. She said she hoarded food in her room. Carrie's handshake was limp but warm. Her breathing was shallow. Her clothes were clean and baggy. She never smiled during the first interview.

Screening Inventories

The results of the intake history inform what objective test to administer to the client. Even though the client's history may not suggest a background of depression or anxiety, having the client fill out those self-reports often elicits additional information. When dealing with symptoms of learning difficulties and attention, tests of continuous performance provide good objective information, as do paper-and-pencil checklists. When selecting a checklist, ensure that it is both reliable (measures the same information every time and is scored consistently over multiple administrations) and valid (correlates with the presenting problem being assessed). The following assessments are typical in a neurocounseling evaluation.

Insomnia

Many counseling clients experience sleep concerns. Gaining a clear picture of how troubling their sleep disruptions are is important to the client's overall health. Several inventories exist to assess sleep problems. The Insomnia Severity Index (Morin et al., 2011) is a short and concise inventory of seven items that takes approximately 5 minutes to complete. Clients indicate the severity of their sleep disruption during the past 2 weeks. Clients rate the severity of insomnia symptoms such as difficulties falling asleep, nighttime awakening, and early morning awakening. Carrie's self-reported score indicated that she was struggling with clinical insomnia of moderate severity. She indicated feeling worried and distressed about her sleep problems, which were causing interference in her daily life.

Depression

Many depression screening tools are available. A commonly used assessment is the Beck Depression Inventory, Second Edition (BDI-II; Beck et al., 1996). The BDI-II is a relatively short test that takes around 15 minutes to complete. Clients self-report the degree to which they have been experiencing depressive symptoms for the past 2 weeks on a 4-point scale. Total scores indicate the severity of depressive symptoms as low to none, mild, moderate, or severe. Carrie's overall score indicated that she reported experiencing depressive symptoms to a severe degree.

Anxiety

As with depression, many anxiety inventories are on the market. A commonly used anxiety screening is the Burns Anxiety Inventory (Burns, 1993). This tool can usually be completed in 15 minutes because it has only 33 items. The client is asked to rate each statement according to how much a particular feeling has bothered them in the past several days. The inventory surveys mostly physiological manifestations of anxiety, such as heart racing, sweaty palms, and detachment from one's bodily sensations. As with the BDI-II, clients self-report the degree to which they experience anxiety on a 4-point scale. Total scores indicate low to absent, mild, moderate, or severe levels of anxiety. Carrie's overall score was very high, indicating that she was experiencing not only severe but extreme levels of anxiety. Some of Carrie's responses on question items guided treatment planning, such as her acknowledgment of feeling tense and stressed, with racing thoughts and fear of criticism by others and of being alone.

Body Perception

The Body Perception Questionnaire–Short Form (Porges, 2015) is a 46-item Likert scale with two subscales: Body Awareness and Autonomic Nervous System Reactivity. Although self-reported by the client, the ratings offer essential biofeedback information about perceptions and awareness of the body and its natural responses and a possible picture of the state of the client's fight, flight, or freeze reactions. On a scale ranging from 1 (*never*) to 5 (*always*), Carrie rated six items as *usually* and the rest as *occasionally*. These scores suggested that Carrie was not very aware of her body and its needs. This may offer some insight into her ability to detach mind from body.

Trauma

The counselor again has plenty of options when selecting a trauma checklist that fits the client's needs. Many inventories are available for children and adults. One such trauma checklist that is freely available is the PTSD Checklist for *DSM-5* (PCL-5; Weathers et al., 2013).

This self-report has 17 test items and takes only approximately 15 minutes to complete. The client is asked to rate how bothered an experience makes them feel on a scale ranging from 1 (*not at all*) to 5 (*extremely*). An example of a statement is "feeling jumpy or easily startled." The counselor adds up the total score, with higher scores indicating greater severity of trauma symptoms. Carrie's high score on the PCL-5 suggested that posttraumatic stress disorder was a possibility.

Other Screening Inventories and Assessments

Another assessment that may be useful is the Bender Visual-Motor Gestalt Test, Second Edition (Bender-Gestalt II; Brannigan & Decker, 2003). This test assists in determining the impact of traumatic brain injuries. The main emphasis of the Bender-Gestalt II is a psychological assessment of the client's visual-motor functioning, visual perception, neurological impairments, and emotional disturbances. Carrie was not administered this test.

Personality Inventories

Many standardized, valid, and reliable personality inventories are available. The Millon Clinical Multiaxial Inventory–IV (MCMI-IV; Millon et al., 2015) is often used in neurocounseling assessments. This test gives information about personality dynamics and presenting symptoms. The information gained from the MCMI-IV assists in treatment and goal setting for the client.

The MCMI-IV is a 195-item paper-and-pencil or computerized test that takes approximately 25 to 30 minutes to complete. The client rates the items as true or false and is instructed about the importance of responding truthfully to each item. The test analysis includes three validity components of disclosure, desirability, and debasement. These validity scales offer vital information to the counselor about the client's response style. Self-report inventories are prone to response bias, and validity scales help the counselor to understand whether the client is under- or overdisclosing. Clients can *underdisclose* to minimize problems in an attempt to appear desirable ("faking good"). Clients can *overdisclose* to exaggerate symptoms as part of a cry for help ("faking bad"). Understanding a client's response style helps the counselor understand the validity and accuracy of the client's self-reported symptoms.

After the test results were customized to Carrie, four scores were in the personality disorder range: schizoid, avoidant, depressive, and dependent. Carrie's severe presenting symptoms were high anxiety and dysthymia. Under severe clinical syndromes, a very high score on major depression also appeared. Her validity scales showed a profile of a person who was overly trusting and would self-disclose too much and too fast. Her profile indicated that she had very little positive regard for herself and was extremely self-critical.

Learning Difficulties Assessments

Numerous clients have learning and developmental difficulties. A multitude of self-report paper-and-pencil inventories and checklists address attentional problems. Note that some client problems can be related to overall aptitude, intelligence, or both. This type of testing was not needed for Carrie because she was actively holding down a part-time job with positive results and entering graduate school.

Attention Deficit

One example of a learning difficulties assessment is the Amen Clinic ADD Type Questionnaire (Amen, 2001). This instrument takes approximately 20 minutes to complete and has 71 items that assess six subtypes of attention-deficit/hyperactivity disorder (ADHD): inattention, hyperactive, overfocused, temporal lobe, limbic, and ring of fire. This inventory has two columns. The client completes the first column, and the client's significant other completes the second column. Both parties rate the degree to which they believe the client is experiencing certain symptoms. This allows the counselor to compare and analyze responses. The severity of symptoms such as distractibility is measured on a 5-point scale. The test provides interpretation of both overall scores and self-reported scores on subscales.

Carrie did not ask any other member of her family to fill out this inventory. Her self-reported scores were calculated, and she had significant problems in the overfocused and limbic categories of the questionnaire. These results might suggest that Carrie had great difficulty with obsessive-compulsive traits.

Continuous Performance Tests

Computerized continuous performance tests offer quantitative information about a person's attention. The T.O.V.A. is an auditory and visual nonlanguage test in which the person uses a microswitch to indicate their response (Greenberg & Waldmant, 1993). Clinicians are looking for both visual and auditory errors of omission and commission. In other words, the visual test scores might indicate whether the client clicked the microswitch too often when the X was presented, making errors of commission, or did not click when the X was presented, making errors of omission. In the auditory version, the client might click too often or too little when the stimulus sound is presented.

The results of Carrie's T.O.V.A. auditory test did not fall within normal limits. Her scores suggested possible attention problems as well as ADHD; however, her self-report did not suggest ADHD. To rule out further concerns, a hearing test from an audiologist might be recommended. Her visual scores on the T.O.V.A. were at the low end of normal.

Quantitative EEG

Still another type of assessment for trauma is the administration of a quantitative EEG (qEEG). Either a five-channel or a full 19-channel qEEG can be administered, depending on the client's need. EEG technology captures electrical energy at different frequencies, known as *brain waves*. Brain waves are often labeled by categories that correlate to possible behaviors. There are more categories than the typical five of delta, theta, alpha, beta, and gamma, but these five are the basics. Delta waves are often viewed at 0–3 hertz, or 0–3 cycles per second. These waves are slow, low waves and are associated with sleep and sometimes with trauma-related problems. Theta waves are typically 4–7 hertz and are associated with drowsiness and meditation. They are also slower waves. Alpha waves are categorized as 8–12 hertz and are needed for idling and transitioning from one brain wave state to another. Beta waves range from 13 to 30 hertz. They are considered busy waves. Low beta waves of 12–15 hertz assist in focused attention and problem-solving. When a person is too high in beta waves, there may be anxiety. Gamma waves are at least 30 hertz and are often found in bursts of insight.

Five-Channel qEEG

Because of Carrie's severe and chronic condition, she was given a five-channel qEEG. This shorter and quicker version of the EEG measured brain waves at five brain locations. As shown in Figure 7.2, these locations were Cz at the top of the head; F/3 and F/4 at the left and right front of the head, respectively; Fz at the middle front of the scalp; and O1 at the back of the left side of the head. The five-channel EEG often provides enough valuable data to proceed with treatment. It also saves time and money. This type of EEG is often called the Quick Q (Swingle, 2008) and often only takes 20 minutes to administer. The five-channel qEEG provides the overall representative state of the client's brain, and trauma may also be seen at the back of the head in the occipital lobes.

Counseling Assessments, Relationships, and Interventions

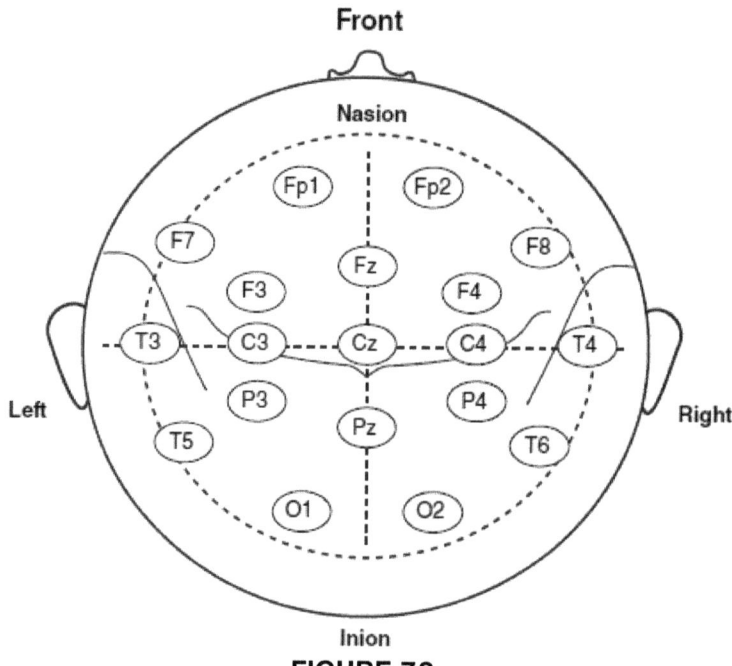

FIGURE 7.2

Head Map of Functions

Note. Copyright 2007 by John Anderson. All rights reserved. Reprinted with permission. See key below and on the next page for area definitions.

FIGURE 7.2 KEY

Nasion = The intersection of the frontal bone and two nasal bones of the human skull.
Inion = The most prominent projection of the occipital bone at the posteroinferior (lower rear) part of the human skull.
Fp1 = Attention, concentration, planning, verbal episodic retrieval, visual working memory. Orchestrates network planning, decision-making, and task completion. Increased or excess theta causes attention-deficit/hyperactivity disorder (ADHD). Increased or excess beta causes rigidity of focus.
Fp2 = Emotional attention, judgment, sense of self, self/impulse control, face/object processing, emotional inhibition, verbal episodic memory. Increased or excess theta causes impulsivity, poor social awareness, and anxiety. Increased or excess beta causes emotional overcontrol, decreased nuance.
Fz = Motor planning of both lower extremities (BLE) and midline, running, walking, kicking, response and emotional inhibition, grooming. Increased or excess theta causes ADHD. Increased or excess beta causes attention and motivational problems, obsessivecompulsive disorder (OCD).
F3 = Motor planning, right upper extremities (RUE), right fine motor coordination, visual episodic retrieval, mood elevation, object processing, emotional interpretation, positive mood. Increased or excess theta causes depression and decreased executive function. Increased or excess beta causes OCD.
F4 = Motor planning, left upper extremities (LUE), left fine motor coordination, verbal episodic and semantic retrieval, regulation of attention or impulse. Increased or excess theta causes lack of tact, disorganized dialogue, poor use of analogy and irony. Increased or excess beta causes hypervigilance.
F7 = Verbal expression, speech fluency, cognitive mood regulation, visual and auditory working memory, attentional gate, Broca's area. Increased or excess theta causes speech and work finding problems. Increased or excess beta causes increased input control.

FIGURE 7.2 KEY *(Continued)*

F8 = Emotional expression, drawing, endogenous mood regulation, face recognition, emotional processing, visual/spatial working memory, sustained attention. Increased or excess theta causes lack of prosody. Increased or excess beta causes oversensitivity to speech intonation in others.

C3 = Sensorimotor integration RUE, alerting response, right-hand writing, short-term memory. Increased or excess theta causes poor handwriting. Increased or excess beta causes motor hyperactivity.

Cz = Sensorimotor integration of BLE, ambulation, basal ganglia, thalamic efferents, substantia nigra. Increased or excess theta causes ADHD. Increased or excess beta causes attention and motivational problems.

C4 = Sensorimotor integration LUE, calming response, left-hand writing, short-term memory. Increased or excess beta causes hypervigilance.

T3 = Logical, verbal memory formation and storage, phonologic processing, hearing. Increased or excess theta causes memory and language problems. Increased or excess beta causes irritability.

T4 = Emotional and autobiographical memory formation and storage, hearing, personality, musical ability, organization. Increased or excess theta causes anger, sadness, aggression, tone of voice interpretation problems.

T5 = Logical, verbal understanding, word recognition, auditory processing, short-term memory, inner voice, meaning construction. Increased or excess theta causes anger, sadness, aggression, tone of voice interpretation problems.

P3 = Right-side perception and cognitive processing, spatial relations, multimodal sensations, calculations, praxis, verbal reasoning. Increased or excess theta causes problems with memory, organization, digit span, calculations. Increased or excess beta causes increased thinking or worrying.

Pz = Perception midline, spatial relation, praxis, route finding, attention shifting/integration. Increased or excess beta causes problems with perseveration, sensory vigilance.

P4 = Left-side perception and cognitive processing, spatial relations, multimodal interactions, praxis, nonverbal reasoning, visual-spatial sketchpad, vigilance, victim mentality. Increased or excess theta causes increased self-concern and rationalization. Increased or excess beta causes emotional rumination.

T6 = Emotional understanding, facial and symbol recognition, auditory processing, long-term memory. Increased or excess theta causes poor memory for faces and melodies.

O1 = Right visual processing, pattern recognition, color, movement, black and white and edge perception.

O2 = Left visual processing, pattern recognition, color, movement, black and white and edge perception.

Reflective Questions

Consider that listening is obviously central in providing a safe relationship.

How would you personally define listening?

How might different theories listen differently—cognitive behavior, humanistic, narrative, multicultural, feminist, and so forth?

Research has shown that the listener inevitably influences the way in which clients construct and think about their issues ("There is no immaculate listening"). What type of listener are you?

Carrie's EEG showed many healthy and normal responses, which was surprising and confusing because of the severe conditions and symptoms that were presented. Sleep disturbances were illuminated at O1 with low theta and beta ratios. Dysregulations were found in the prefrontal lobes at F3 and F4, probing for depression and lethargy. The Fz site showed low gamma and beta ratios, suggesting excessive passivity. Because of the severity of Carrie's symptoms and relatively healthy five-channel qEEG, the 19-channel qEEG was recommended.

19-Channel qEEG

When clients come to the counselor with chronic issues and perhaps head injuries such as concussion, additional examination with a 19-channel qEEG may be required. The 19-channel qEEG records at 19 locations instead of five. Instead of the electrodes being attached with individual sensors, the client wears a cap with 19 different sensors inside it that will measure the entire brain. The technician will analyze the report after artifacting out muscle movements. The report will provide data on brain wave amplitudes, ratios between brain wave bandwidths, possible topological brain maps, and comparisons from normative databases. Both the five-channel and 19-channel qEEGs may offer information about possible neurofeedback treatments. The 19-channel qEEG offers a more comprehensive assessment of neurological dysregulation.

The results of the 19-channel qEEG were more comprehensive, showing overall trauma and dysregulation throughout the other brain sensor sites. Cognitive inefficiency and memory problems were discovered, with low amplitudes of alpha brain waves. A decrease in the brain's slow wave activity has been found with posttraumatic stress disorder. This suppresses traumatic experiences, resulting in emotional, behavioral, and physiological consequences (Chapin & Russell-Chapin, 2014). Low alpha waves at the midline and occipital lobes are often seen in EEGs and are related to emotional trauma (Swingle, 2008). The two EEG test results assisted in the development of a thorough treatment plan.

Assessment-Based Treatments

Several treatment approaches exist that use assessment data as part of the treatment process. In Carrie's case, potentially useful approaches included diaphragmatic breathing, biofeedback (skin temperature control, heart rate variability [HRV] training), and neurofeedback.

Diaphragmatic Breathing

Carrie was trained in diaphragmatic breathing. As an assessment tool, diaphragmatic breathing provides the counselor with beginning data on the client's possible problems. Taking a baseline of a client's breathing patterns offers clues about anxiety and other lifestyle issues in general, such as body tension and muscle rigidity.

The average normal adult breathing rate is 12–15 breaths per minute (Schwartz & Andrasik, 2003). Young children ages 2 to 5 years breathe faster, around 25–30 breaths per minute, and children ages 5 to 12 take 20–25 breaths per minute. Adolescents typically breathe 15–20 times per minute. Obtaining a baseline breathing measurement for a client helps establish a goal for treatment that is often targeted at a relaxed, peak performance state of 6 breaths per minute. Relaxed breathing not only alleviates anxiety but assists the brain in absorbing glucose and oxygen (Schwartz & Andrasik, 2003).

Biofeedback

Because Carrie had already been through several residential treatment programs, biofeedback was recommended as the first treatment strategy. Her biofeedback interventions were skin temperature control and HRV training.

A skin temperature control assessment is important to include as part of evaluation and treatment. Obtaining a baseline pretreatment value is simple to achieve. Ask the client to gently hold a thermometer for approximately 1 minute. Record the temperature.

HRV is a biofeedback technique that teaches heart rhythm feedback or beat-to-beat changes in heart rate to enhance self-regulation (Chapin & Russell-Chapin, 2014; McCraty et al., 2009). HRV is a measure of the variation in time between heartbeats that can be used to determine the amount of stress the body is under. A lower HRV signifies predominant activation of the sympathetic nervous system. There are many HRV software and hardware packages on the market. A monitoring system called *emWave* (www.heartmath.org) uses a plethysmograph that slides into a small cuff on the big finger of the nondominant hand or an ear sensor clip to measure HRV. In addition, RESPeRATE (www.resperate.com) is a blood pressure control device that allows clients to practice breathing by watching visual corrective feedback.

Neurofeedback

Neurofeedback is a noninvasive brain-based treatment for a variety of mental health concerns, including reducing symptoms of ADHD, depression, anxiety, and other clinical issues. Neurofeedback can also be used to enhance peak performance. It uses EEG technology to collect brain wave data on treatment response at the sites identified in Figure 7.2. The head map of functions in Figure 7.2 illustrates the diverse and specific functions of the differing brain site locations, assisting the counselor in correlating possible brain sites and functions with actual treatment activity and goals.

Clients receive feedback about their brain activity and can adjust their performance accordingly. If neurofeedback is warranted on the basis of client evaluation, then the counselor obtains brain wave data at pre- and posttest measurements. Neurofeedback software calculates numerous statistical compilations for brain wave amplitude, illustrating objective data from the differences in a single session or over multiple sessions.

Carrie's Treatment Response

Carrie was seen twice per week, with careful monitoring of symptoms and measurable biweekly goals. Her progress was gradual but optimistic. Carrie was taught HRV using the emWave software. She became proficient at the low levels, and she was able to practice this heart rate coherence during her everyday living activities. She was able to reduce her breathing rate to 15 breaths per minute. Because of her low body weight, Carrie was usually cold. Teaching Carrie skin temperature activities helped to bring her skin temperature to at least 86 degrees. The goal is 90 degrees for a relaxed state of peak performance. As her trust began to grow, Carrie agreed to slowly increase her calorie consumption to 1,200 calories and decrease her running to 5 miles per day. Her weight increased by 10 pounds. By our 10th neurofeedback session, Carrie began to look me directly in the eyes. Carrie stated that she felt safe in my office. I asked her to look in the mirror and describe what she saw. Carrie turned to me and exclaimed, "There is life in my eyes!" Truly, there was vibrancy in Carrie for the first time.

> **Reflective Question**
>
> Why might it be helpful to combine several assessment-based treatment tools to help Carrie rather than selecting only one?

My Brain-Based Approach to the Case of Carrie

Using the case of Carrie, I have demonstrated my brain-based approach to assessment and neurocounseling throughout the chapter. Each assessment offered additional information that assisted in developing a carefully crafted treatment plan based on all the data and patterns. The four steps for formulating a neurocounseling treatment planning were followed. According to Chapin (2021), the steps are "prioritizing the presenting problem, determining the brain locations that are implicated from the assessments of neurological dysregulation, selecting a neurocounseling intervention and identifying behavioral goals for coaching clients on ways to improve their neuroplasticity" (p. 70). In this case, the client's symptoms were very severe. Ethically responsible principles and judgments must be considered at every step of the treatment process.

> **Reflective Question**
>
> If only a few assessments could be used with Carrie, what essential evaluations would you choose?

Conclusion

The case of Carrie demonstrates how useful a thorough and comprehensive assessment battery can be. Each assessment was a unique yet interdependent aspect of the treatment. Many of the evaluations discussed in this chapter are consistently used in regular counseling. However, when the goal is neurocounseling, or bridging brain and behavior, assessments that address body perception, EEG, baseline skin temperature and HRV, and insomnia are routinely used as part of the overall assessment battery. These neurocounseling tests offer additional information about the client's physiology and provide baseline measures to begin customized goal development (Chapin & Russell-Chapin, 2022). Although it is not necessary for every counselor to be an expert in all of these neurocounseling measurements, all counselors should aim to be "neurowise" rather than "neuronaïve" about the importance of neurocounseling information to the overall assessment, treatment, and health of the client.

> **Reflective Question**
>
> How might the use of neurocounseling assessments for Carrie result in more effective treatment?

Quiz

1. Which statement about validity and reliability is most true?
    a. The test needs to be valid and reliable.
    b. The test only needs to be valid.
    c. The test only needs to be reliable.
    d. It is more important to be reliable than valid.
2. The main goal of a neurocounseling assessment is:
    a. A comprehensive evaluation.
    b. Physiological, emotional, and behavioral self-regulation.
    c. An evaluation for conceptualization and treatment.
    d. None of the above.
3. Which of the following are possible sources of brain dysregulation?
    a. Genetic predisposition.
    b. Substance abuse.
    c. High fever.
    d. All of the above.

References

Amen, D. (2001). *Healing ADD: The breakthrough program that allows you to see and heal the six types of ADD.* Putnam.

American Psychological Association. (n.d.). *Assessment.* APA dictionary of psychology. https://dictionary.apa.org/assessment

Beck, A. T., Steer, R. A., & Brown, G. K. (1996). *Manual for the Beck Depression Inventory–II.* Psychological Corporation.

Brannigan, G. G., & Decker, S. L. (2003). *Bender Visual-Motor Gestalt Test, Second Edition.* Pearson.

Burns, D. D. (1993). *Ten days to self-esteem.* Quill.

Chapin, T. (2021). Integrating neurocounseling with assessments, treatment planning, and outcome evaluation. In L. Russell-Chapin, N. Pacheco, & J. DeFord (Eds.), *Practical neurocounseling: Connecting brain functions to real therapy interventions* (pp. 57–84). Routledge.

Chapin, T., & Russell-Chapin, L. (2014). *Neurotherapy and neurofeedback: Brain-based treatment for psychological and behavioral problems.* Routledge.

Chapin, T., & Russell-Chapin, L. (2022, May). Brain-based assessment in neurocounseling: Part 2. *Counseling Today, 64*(11), 19–23.

Council for Accreditation of Counseling and Related Educational Programs. (2024). *2024 CACREP standards.* https://www.cacrep.org/wp-content/uploads/2023/06/2024-Standards-Combined-Version-6.27.23.pdf

Demos, J. N. (2005). *Getting started with neurofeedback.* Norton.

Greenberg, L. M., & Waldmant, I. D. (1993). Developmental normative data on the Test of Variables of Attention (T.O.V.A.). *Journal of Child Psychology and Psychiatry, 34*(6), 1019–1030. https://doi.org/10.1111/j.1469-7610.1993.tb01105.x

McCraty, R., Atkinson, M., Tomasino, D., & Bradley, R. T. (2009). The coherent heart: Heart–brain interactions, psychophysiological coherence, and the emergence of system-wide order. *Integral Review, 5*(2), 10–115.

Millon, T., Millon, C., & Grossman, S. (2015). *Millon Clinical Multiaxial Inventory–IV.* Pearson.

Morin, C. M., Belleville, G., Belanger, L., & Ives, H. (2011). The Insomnia Severity Index: Psychometric indicators to detect insomnia cases and evaluate treatment response. *Sleep, 34*(5), 601–608. https://doi.org/10.1093/sleep/34.5.601

Porges, S. (2015). *Body perception questionnaire.* Traumatic Stress Research Consortium. https://www.traumascience.org/body-perception-questionnaire

Russell-Chapin, L. (2021). Demonstrating neurocounseling and neuroanatomy with the case of Patrice: Struggling with stability and independence. In L. Russell-Chapin, N. Pacheco, & J. DeFord (Eds.), *Practical neurocounseling: Connecting brain function to real therapy interventions* (pp. 22–56). Routledge.

Schwartz, M., & Andrasik, F. (2003). *Biofeedback: A practitioner's guide* (3rd ed.). Guilford Press.

Swingle, P. (2008). *Basic neurotherapy: The clinician's guide.*

Weathers, F. W., Litz, B. T., Keane, T. M., Palmieri, P. A., Marx, B. P., & Schnurr, P. P. (2013). *The PTSD Checklist for DSM-5 (PCL-5).* https://www.ptsd.va.gov/professional/assessment/documents/PCL5_Standard_form.PDF

Chapter 8

Neurocounseling Approaches to Wellness and Optimal Performance

Theodore J. Chapin

The battle for the hearts, minds, and souls of those seeking health is ever present in modern society. On one side stand those who favor a disease and illness model that often reflects a cost-conscious, depersonalized, and medication-based approach to diagnosis and treatment. On the other side are those who favor a holistic, mind-body-spirit model that is individualized and wellness focused and that seeks optimal functioning. It is against this backdrop that this chapter explores the concept of wellness and extends this discussion to focus on the foundational role of several neurocounseling-based strategies for optimal performance. Also reviewed is the role of wearables for self-monitoring of physiological measures such as activity, sleep, and stress. These concepts are illustrated and applied through the case study of William, a man who struggles with excess weight, anxiety, compulsiveness, trauma, and depressed mood. To conclude, the chapter briefly discusses future trends in wellness and optimal performance.

2024 CACREP Standards

This chapter addresses 2024 Council for Accreditation of Counseling and Related Educational Programs (CACREP) Standards pertinent to the Foundational Counseling Curriculum (Section 3) areas of Social and Cultural Identities and Experiences (Standard B), Lifespan Development (Standard C), and Counseling Practice and Relationships (Standard E):

- The role of religion and spirituality in clients' and counselors' psychological functioning (Standard B.11.)
- Models of resilience, optimal development, and wellness in individuals and families across the life span (Standard C.7.)
- Principles and strategies of caseload management and the referral process to promote independence, optimal wellness, empowerment, and engagement with community resources (Standard E.17.)

Clinical Case Study: William

William is a 51-year-old man. He grew up as the oldest of three boys in his family. William was an above-average student and was active in scouting, sports, and a conservative Christian church. His mother was a devoted but compulsive homemaker, and his father was an emotionally constricted disciplinarian who used excessive corporal punishment. William experienced his first bout of depression during his freshman year of college. He was reportedly overwhelmed by the academic and social demands of an independent college life. He was later married and divorced, which seemed to increase his anxiety and compulsiveness. He then remarried and had two children. He pursued a successful career, although he was once asked to resign after repeated criticism from managers. This triggered a second episode of depression. Over the past decade, William's body has gone from muscular and athletic to moderately obese. Recently, William suffered severe physical and emotional trauma after elective surgery. William sought help to better manage his emotions and regain his physical fitness.

Wellness

For decades, counseling has been at the forefront in considering the importance of physical and spiritual health when providing mental health services. The wellness movement was spurred when Halbert Dunn (1961), a physician and consultant to the World Health Organization, challenged the then-prevailing belief that health was solely the absence of disease. Dunn instead suggested that health was a state of physical, mental, and social well-being. Dunn described wellness as integrated functioning of maximized individual potential.

Hettler (1984) was among the first to propose a wellness model, followed by a wellness model that emerged from the counseling profession in 1988 titled the *wheel of wellness* (Myers et al., 2000) and later the *indivisible self* (Myers & Sweeney, 2005). Subsequently, a brain-based wellness model emerged called the *healthy mind platter* (Rock et al., 2012). Rock et al. (2012) developed the model depicted in Figure 8.1 to explain how optimal brain functioning can be maintained through daily healthy mental habits.

Copyright © 2011 by David Rock and Daniel J. Siegel. All rights reserved.

FIGURE 8.1

The Healthy Mind Platter for Optimal Brain Matter™

Note. Healthy Mind Platter for Optimal Brain Matter © 2011 by David Rock and Daniel J. Siegel, MD. Used with permission. All rights reserved.

The three models described here all propose similar lifestyle areas deemed important to a person's well-being. These lifestyle areas were called *wellness dimensions* by Hettler (1984), *wellness factors* by Myers and Sweeney (2005), and *mental activities* by Rock et al. (2012). The following section uses Hettler's six wellness dimensions as an organizing framework for the healthy mind platter's mental activities to advance an integrative brain-based approach to wellness assessment and psychoeducation. Together, these wellness dimensions and mental activities provide a template and guide for the development of a healthy lifestyle. The following section also includes a wellness assessment for the case of William to elucidate the importance of screening for lifestyle habits during intake. It is important to mention that lifestyle habits are an often unrecognized yet debilitating influence on a person's overall health.

Occupational and Intellectual Wellness: Focus Time, Playtime, Downtime

Occupational wellness involves personal satisfaction and enrichment in one's life work. Intellectual wellness recognizes the value of learning and cognitive stimulation. These wellness domains are consistent with the healthy mind platter's mental activities of focus time, playtime, and downtime.

Focus time pertains to goal-directed brain states in the completion of tasks. Individuals may experience what Mihaly Csikszentmihalyi (1990) called the *flow state,* defined as full immersion in the task at hand with energized focus and enjoyable involvement. His initial observations of the flow state were with artists, particularly painters, who became so deeply involved in their work that they would often go significant periods of time without food, water, and sleep. In describing their experiences, they used the metaphor of being carried away in a current of water (Csikszentmihalyi, 1975). Today, many sports psychologists describe a similar phenomenon when they talk about being "in the zone" (Strack et al., 2011). Flow can occur across many types of human endeavors, including education, music, sports, gaming, spiritual and religious practice, and work. Along with physical exercise, focused attention may help to prevent the development of Alzheimer's disease (Baumgart et al., 2015).

Playtime refers to engagement in creative and spontaneous activity that results in new neural connections in the brain and reduces the potential for aggressive and emotionally dysregulated behavior (van den Berg et al., 1999).

Downtime refers to relaxation that lacks a specific goal or focus, rather than to active leisure activities. Downtime helps the brain to recharge by activating different regions—namely, the default mode network for task-negative states (Rock et al., 2012). Put simply, downtime is about being in the moment rather than doing tasks (Rock et al., 2012, p. 8). Downtime provides space for preconscious processing in the right hemisphere of the brain, leading to the development of new insights (Segal, 2004).

William reported that intellectual wellness was a strength area. He felt actively engaged in work-related activities (focus time). He was also an avid reader. William noted that he frequently reverted to passive activities when returning home from work, such as watching television for long periods. He rarely engaged in defocused downtime and wanted to engage in more meaningful leisure activities (playtime).

Physical Wellness: Physical Time, Sleep Time

Physical wellness includes engaging in healthy behavior through sleep, exercise, diet, and self-care. This wellness domain is consistent with two mental activities in the healthy mind platter: physical time and sleep time.

Physical time is defined as physical activity, preferably aerobic, that strengthens the brain, such as by stimulating neuronal growth (known as neurogenesis). Ratey (2008) found that 40 minutes of aerobic exercise three times a week stimulates neurological production of brain-derived neurotrophic factor (BDNF). He described this as Miracle-Gro for the brain and said that it is important for both energy metabolism and synaptic plasticity because it activates glutamate, increases antioxidant production, and grows new brain cells. Although exercise is widely known to be as effective as antidepressant medication in reducing symptoms of major depression (Blumenthal et al., 1999), it is less known that exercise can also increase the size of the hippocampus and improve memory in older adults (Erickson et al., 2011). In a study by Erickson et al. (2011), 40 minutes of brisk walking resulted in a 2% increase in hippocampal volume in the intervention group compared with a 1.4% decrease in the control group, which had engaged in nonaerobic yoga and resistance-band training.

Sleep time is the mental activity that facilitates the brain's consolidation of learning and memory; the brain's resting state, sleep helps people recover from activities. Sleep has a restorative function; during sleep, glial cells clean the brain of toxins and debris (Xie et al., 2013). Sleep appears to be related to neuroplasticity via synaptic potentiation and cellular gene and protein translation, as well as efficient memory processing from initial learning (encoding) to long-term memory (consolidation) across whole brain networks (Abel et al., 2013). Thus, people who lack quality sleep are unable to recover fully from the day, and their consolidation of learning and memories is impaired. Inadequate sleep has also been linked to the development of depression (Nakata, 2011).

William reported receiving inconsistent quality sleep. On some nights, he was able to fall asleep quickly with no nighttime or early morning awakening. On other nights, he stayed up late to watch television, or he drank alcohol during the evening, both of which had an impact on the quality of his sleep. Most of all, William wanted to regain his athletic figure and seemed highly motivated to restart an active workout regimen.

Social Wellness: Connecting Time

Social wellness reflects the quality of interpersonal relationships and community interaction. This wellness domain is comparable to the mental activity of connecting time in the healthy mind platter. *Connecting time* refers to engagement in face-to-face relationships and with the natural world, which activates circuitry in the brain. In an older landmark study, social relationships predicted mortality over a 9-year period even when physical health, health behaviors, and socioeconomic status were controlled for (Berkman & Syme, 1979). In their review of 81 studies, Uchino et al. (1999) found that social support has positive effects on blood pressure in addition to the cardiovascular, endocrine, and immune systems.

William reported having lifelong friends and a strong marriage, although he admitted to neglecting parts of these relationships. For example, he often lacked energy to go out and socialize on the weekends. He felt connected to his children.

Emotional and Spiritual Wellness: Time-In

Emotional wellness values awareness and acceptance of feelings as well as an optimistic approach to life. Spiritual wellness involves one's spiritual practice, search for meaning and purpose, and tolerance of others' beliefs and traditions. Engagement in spiritual practice and acceptance of experience are comparable to the healthy mind platter's mental activity of time-in.

Time-in involves quiet internal reflection and meditation. Many spiritual meditation techniques exist, all of which involve a quiet environment, a comfortable posture, a focus of attention, and an open and reflective attitude. Meditation has been linked to neurological benefits associated with attention, mind wandering, retrieval of episodic memory, and emotional processing. These include increased cortical thickness (gyrification) of gray matter, which promotes improved information processing (Luders et al., 2012); increased gray matter of the hippocampus, which is involved in the formation of new memories (Holze et al., 2011); slowing of age-related neurological decline (Luders, 2014); and increased activation of the default mode, salience (selection of stimuli deserving of attention), and executive control (high-level cognitive tasks) networks (Ganesan et al., 2020; Xu et al., 2014). An early meta-analysis of 47 mindfulness trials with 3,515 participants found moderate evidence for improved anxiety, depression, and pain; some evidence for reduced stress and improved quality of life; and low or insufficient evidence for improved mood, attention, eating habits, sleep, weight, and reduced substance use (Goyal et al., 2014). A more recent meta-analysis of 191 studies (Schlechta Portella et al., 2021) found that meditation improves mental health (anxiety, depression, stress, and mindfulness); physical and metabolic health (blood pressure, cancer symptoms, and chronic pain); and overall vitality, well-being, and quality of life (decreased stress, higher quality of life, and better work health). Meditation appears to have comparable outcomes to other active treatments such as medication, exercise, and behavioral therapies. Clearly, time-in and meditation should be part of a comprehensive wellness plan and not a stand-alone strategy.

William was dissatisfied with his current religious and spiritual involvement. He felt that his religious experience was dogmatic and did not leave space for exploration. William engaged in little time-in activity beyond attending church. He had little interest in engaging in a more traditional, quiet, and still meditative process. William was open to meditating while being physically active, such as when walking, gardening, or biking, or while being immersed in a creative project.

> **Reflective Question**
>
> How might William's level of engagement in different mental activities be related to his primary emotional and behavioral problems?

Neurocounseling Strategies to Enhance Wellness

Neurocounseling strategies involve the use of brain-based interventions to facilitate client change (Montes, 2013). They use both traditional talk therapy methods and a variety of specialized neurologically based techniques. Several neurocounseling strategies were applied to the case of William, including therapeutic lifestyle changes, biofeedback, and neurofeedback.

Therapeutic Lifestyle Changes

The daily habits people form and the lifestyle choices they make play an important role in their overall health and functioning. Ivey et al. (2022) identified 17 stress management strategies, or *therapeutic lifestyle changes*, that enhance wellness. Three of the five most important lifestyle changes were identified as nutrition and weight management, sleep, and exercise. Along with screen time, these factors are known to facilitate or hinder healthy self-regulation (Chapin & Russell-Chapin, 2014). When the brain and body are in a constant state of distress (sympathetic fight, flight, or freeze) as a result of bad habits and poor lifestyle choices, wellness and optimal performance are not possible. Problematic habits must be ameliorated before healing recovery (parasympathetic response) can occur.

Nutrition and Weight Management

Excessive consumption of simple carbohydrates such as sugar, pasta, and other white foods (e.g., potatoes, flour, white rice) has been found to have an adverse impact on the brain's neurological functioning, affecting memory and neuronal plasticity (Molteni et al., 2002). In a recent study, Mujica-Parodi et al. (2020) found that neurobiological marker changes associated with aging can be seen much earlier—in their 40s—among people who consume a high-carbohydrate diet but might be reversible by minimizing the consumption of simple carbohydrates. Such a diet has also been identified as a significant factor in the development of metabolic disease, cardiovascular disease, obesity, and Type 2 diabetes (Stanhope et al., 2013). The best diet for neurological regulation and optimal production of neurotransmitters consists of protein (e.g., lean meat, fish, organic dairy, nuts), complex carbohydrates (e.g., fresh vegetables, fruit), and both saturated fats (e.g., dairy, coconut oil) and monounsaturated fats (e.g., omega-3 fatty acids, olive oil; Amen, 2001). A healthy diet has been associated with larger total brain volume; greater gray matter, white matter, and hippocampal volume; improved attention and memory; reduced depression and anxiety; better physical health; and decreased inflammation, autoimmune disorders, and diabetes (Moh, 2020).

Although nothing substitutes for the nutritional value of a healthy diet, careful use of selective dietary supplements can improve neurological and immunological functioning (Balch, 2010). Recent research has confirmed the value of dietary supplements in neurological disease and aging. Naureen et al. (2022) noted the beneficial effects of polyphenols, omega-3 fatty acids, B vitamins, and several ayurvedic herbs. Supplements found to reduce and manage neurological inflammation that interferes with optimal brain functioning include omega-3 fatty acids found in fish oil, curcumin from the spice turmeric, vitamin D, and N-acetylcysteine. The fatty acids in omega-3s are converted into oxylipins that initiate and terminate the immune response and promote neuronal repair (Zivkovic et al., 2011). Curcumin also serves an anti-inflammatory function that reduces oxidative stress caused by various diseases, including diabetes, cancer, arthritis, cardiovascular disease, and Alzheimer's disease (Ghosh et al., 2015). Vitamin D, usually acquired with exposure to sunlight, has been found to promote immune functioning and general physical and mental health (Aranow, 2011). Finally, N-acetylcysteine acts to support the production of glutathione, which is essential for synaptic plasticity (Dean et al., 2011) and has been found to be helpful in the treatment of addictions, bipolar disorder, obsessive-compulsive disorder, trichotillomania, and schizophrenia. Counselors should either acquire specialized training before suggesting dietary supplements to clients or consider referring their client to a nutritionist or a functional medicine professional.

Sleep

The National Sleep Foundation recommends 9 to 11 hours of sleep for school-age children (6–12 years), 8 to 10 hours for teenagers (13–18 years), and 7 or more hours for adults (18-plus years; Suni & Singh, 2023). Sleep duration does not always equate to sleep quality. Education in sleep hygiene can help ensure ready sleep onset, deep quality rest and recovery, and refreshed awakening (Chen et al., 2010). General suggestions for healthy sleep hygiene include limiting the use of alcohol, caffeine, and nicotine; avoiding eating, drinking, or exercising before bedtime; creating a cool, quiet, and dark sleep environment; keeping a regular sleep routine; using the bedroom only for sleep or intimacy; using natural light when one wakes and during the day to replenish natural melatonin; and limiting naps. If a person experiences ongoing problems with sleep, further treatment options include melatonin supplementation, biofeedback, and neurofeedback.

Screen Time

Screen time on smartphones, computers, video games, and television continues to increase as the use of electronic devices is more deeply integrated into education, work, and leisure activities. Use of these devices has become so pervasive that average use of screen time has surpassed sleep and other vital life activities. Screen time has rocketed to 6 hours a day for tweens, 9 hours a day for teens, and 7 hours a day for adults (Swingle, 2015). Currently, 12% of all U.S. users would meet the criteria for addiction with significant neurological impact (Swingle, 2015).

Wellness and optimal performance are best achieved through a lifestyle that takes advantage of the benefits of technology without allowing it to supplant other healthy interests and activities. To maintain optimal health and performance, Swingle (2015) recommended limiting screen time to 1 to 2 hours a day beyond use involved in school and work; stopping screen time (including television) 1 to 2 hours before bed; avoiding toddler use of screen time before age 2 years; restricting child use of screen time between ages 2 and 6 years; and maintaining engagement in other activities, including face-to-face social interaction and outdoor recreation.

Researchers documented dramatic increases in screen time during the COVID-19 pandemic (Pandya & Lodha, 2021). Time on social media, messaging applications, and videoconference platforms increased an average of 5 hours a day, with heavy users logging 17.5 hours a day and light users 30 hours per week. Young adults showed the highest increase to 8.8 hours a day compared with those 65 or older, who logged 5.2 hours a day (Pandya & Lodha, 2021). Wagner et al. (2021) investigated the increase in recreational screen time during the coronavirus pandemic, attributing it to boredom and the desire to connect with others and noting its association with worsened mental health. Hedderson et al. (2023) found that overall adolescent screen time nearly doubled as a result of the pandemic, from an average of 3.8 hours a day to 7.7 hours a day. Prompted by the data, U.S. Surgeon General Dr. Vivek Murthy issued an advisory on social media use, warning of a profound risk of harm to adolescents in areas such as body image; engagement with hate-based, suicide, and self-harm content; loss of sleep; lack of exercise; and depression (Office of the Surgeon General, 2023). Murthy advised families to develop a family media plan, to establish tech-free zones at meals and bedtime, to teach kids responsible online behavior and model that behavior themselves, and to limit sharing of personal information online.

The consequences of increased screen time and decreased physical and otherwise usual activity levels on overall health and wellness have been well documented. In a systematic review of 13 studies, Stiglic and Viner (2019) found evidence for a variety of harmful effects, including problems with obesity, higher energy intake, less healthy diet, poorer quality of life, behavior problems, anxiety, depression, hyperactivity and inattention, lower self-esteem, poorer well-being, diminished psychosocial health, metabolic syndrome, decreased cardiovascular fitness, inhibited cognitive development, lower educational attainment, and decreased sleep.

Exercise

One relatively inexpensive and effective way to maintain wellness and optimal performance is daily aerobic exercise (Chapin & Russell-Chapin, 2014). Not only does exercise provide obvious physical and long-term health benefits, but it also promotes neurological, cognitive, and mental health. As previously noted, exercise increases both hippocampal volume and serum levels of BDNF (Erickson et al., 2011). Exercise also stimulates the release of irisin, an exercise hormone released during moderate aerobic activity when the cardiovascular system is engaged and muscles are being exerted (Fu et al., 2021). Irisin has also been found to brown the body's white adipose tissue (body fat), to regulate insulin use, and to have anti-inflammatory and antioxidative properties. Its dysfunction has been associated with cardiovascular disease and hypertension. Gibbons et al. (2023) found that when exercise was combined with intermittent fasting, such as by skipping breakfast, both exercise and fasting were neuroprotective and upregulated BDNF. Prolonged fasting for 20 hours had no effect on BDNF, and neither did prolonged low-intensity cycling. However, 90 minutes of interval cycling was found to increase BDNF by 4 to 5 times.

> **Reflective Question**
>
> Which therapeutic lifestyle changes could help to enhance William's performance, based on the results of his wellness assessment?

Biofeedback and Neurofeedback

In a compilation of work on the application of biofeedback and neurofeedback in sports psychology, optimal self-regulation is described as involving both bottom-up and top-down strategies (Strack et al., 2011).

Bottom-up strategies may include the biofeedback training techniques of (a) peripheral skin temperature and (b) heart rate variability. Elevated peripheral skin temperature (warm hands and feet) indicates relaxed muscles and a slowed, steady heart rate. Healthy heart rate variability, accomplished through slow, diaphragmatic breathing, indicates the achievement of a parasympathetic response or physiological state of calm recovery. The zone for optimal performance can be achieved by rhythmically paced, smooth breathing; attention focused on the heart; and thoughts directed toward a positive outcome. Combined, these promote calm focus and smooth recovery. For example, consider a golfer who hits an errant shot and becomes overly upset and anxious, resulting in loss of composure and several more shanked shots. By relaxing one's muscles, maintaining even blood flow throughout the body, breathing evenly, and steadying one's overall level of arousal, one can more readily restore recovery and focus and better realize optimal performance.

Top-down strategies for optimal performance involve neurofeedback training in optimal brain wave functioning. Neurofeedback is a form of biofeedback applied to the brain's electrical activity. It uses the electroencephalogram, computer technology, and principles of reinforcement to identify, target, increase, or decrease certain helpful or unhelpful brain wave patterns or to improve communication across brain wave networks that affect performance (Chapin & Russell-Chapin, 2014). Most often, the first step in neurofeedback treatment is to evaluate current brain wave functioning and any abnormal brain wave patterns related to attention-deficit/hyperactivity disorder, anxiety, trauma, or obsessive-compulsive disorder with the initial goal of modifying these brain waves toward more normal functioning. Next, optimal brain wave patterns for a given behavior are strengthened. In sports, this may involve strengthening alpha brain waves for improved calm focus. In the arts, it may involve strengthening theta and alpha waves for improved creativity. In business and public speaking, it may involve training alpha waves for calm focus and low beta waves for improved cognitive efficiency (Chapin & Russell-Chapin, 2014). Even the process of aging, which is characterized by a steady decline in alpha waves, has seen the application of neurofeedback to help forestall cognitive decline. This technique is called *brain brightening* and involves alpha wave training four times a year to maintain cognitive efficiency (Budzynski, 1996).

Wearables

The use of wearable electronic devices, or *wearables*, to monitor personal physiological indicators of health and performance has exploded in recent years. These devices, which may be styled as watches or rings, are typically worn near the skin and can monitor many physiological functions, including physical activity, heart rate, breathing patterns or heart rate variability, blood oxygen levels, skin temperature, blood pressure, and sleep patterns. They can track the wearer's physical performance over time, providing guidance on cardio fitness, energy levels, and sleep. They can alert one to stressful states, early signs of illness, and menstrual cycle schedules. They can track other health information, including diet, exercise, recovery, and sleep. Some of the more popular devices include the Fitbit, the Apple Watch, and the Oura Ring. Prices range from $60 to several hundred dollars, depending on the features and purchase plan options.

These devices permit both individual and group monitoring of healthy functioning. They have been applied in various group settings, including health care and cancer treatment, insurance wellness programs, corporate wellness programs, and large university student health studies. They have also been applied in individual pursuits, including sports performance, running, personal fitness, yogic breathing, and sleep. In a study of 997 Americans across 46 states, Solino-Fernandez et al. (2019) found that two out of three participants were willing to wear a health insurance wearable device if they believed it would improve their health, prevent disease, or reduce their insurance premiums. Results of a larger national survey of 4,551 U.S. adults indicated that the potential of wearable devices was underutilized, with only one third of the sample participants currently and actively using them (Chandrasekaran et al., 2020). The survey results also showed that current users of wearables were younger (18–34 years old), wealthier (household incomes over $75,000 a year), more educated (college graduates), and more technology literate.

Many clients interested in personal fitness, wellness, and optimal performance can receive immediate and ongoing biofeedback on their physiological functioning. This feedback not only permits awareness of physiological states but empowers the individual with information to readily modify their behavior (Miller, 1969). By its nature, biofeedback helps individuals make subtle but important changes in their body to achieve new ways to effectively control their physiological responses, thus improving health, performance, and overall wellness.

William did not use any wearable devices in his neurocounseling experience. If he had, he might have benefited from many of their features. Skin temperature, heart rate, and breathing feedback would have helped him modify his anxiety. Tracking his activity levels, diet, and sleep patterns might have also helped him increase his exercise, lose weight, and improve his sleep. Even further, these changes could have helped him relax, feel calm in his interpersonal interactions, and gain confidence in his newfound health and wellness.

Reflective Question

You have now provided William with psychoeducation on therapeutic lifestyle changes and collaboratively formed a wellness plan.

As a next step in the treatment process, how might you use biofeedback and neurofeedback with William?

My Brain-Based Approach to the Case of William

Overall assessment of William's current level of wellness found his primary strength area to be his intellectual wellness and the most prominent areas of concern to be physical, emotional, and spiritual wellness. The problem areas detected in William's wellness assessment led to intervention through therapeutic lifestyle changes. William began to consume a diet of protein, fruit, and vegetables. He limited his intake of sugar and simple carbohydrates. He also began taking several dietary supplements, including fish oil and curcumin, vitamin D, and N-acetylcysteine. William began working out 1 hour a day. His exercise routine consisted of performing a variety of exercises, stretching, lifting light weights, and biking. He maintained a steady sleep routine of 7 to 8 hours in a dark, cool room. Over time, William lost 80 pounds and gained more energy to attend to neglected aspects of his work relationships, friendships, and marriage.

Counseling, neurocounseling, biofeedback, and neurofeedback were integrated into William's treatment. Through these interventions, William searched for meaning in life events and began to change the way he perceived them. William converted his outrage over receiving corporal punishment into motivation to learn about parenting. William realized that his previous marriage and divorce had spurred him to become fully responsible and to not be stuck in the life that he and others thought he should live. Although his forced resignation felt demoralizing at the time, it motivated him to start his own business. In his trauma after elective surgery, William had learned about the fragility of life and the fallibility of others.

William developed improved self-regulation through biofeedback and neurofeedback training. His biofeedback consisted of heart rate variability and peripheral skin temperature training to increase his ability to generate a calming, parasympathetic response. His neurofeedback consisted of four protocols, one intended to reduce his tendency toward overactivation, another to process his traumatic experiences, a third to reduce his obsessive-compulsive tendencies, and a fourth to reduce negative self-talk and associated depressed mood.

William's improved performance led to several significant life achievements. Over time, William built a strong professional reputation; owned his own business; became a professional author; and provided trainings locally, nationally, and occasionally in other countries.

Reflective Questions

After reading through this chapter, which threats to William's optimal performance might the counselor need to continue attending to?
What else needs to be done to help William?

Conclusion

A healthy, meaningful, and high-performing life cannot be achieved solely through amelioration of illness, disease, or pathology. By harnessing the mind-body connection implicit in neurocounseling interventions such as therapeutic lifestyle changes, biofeedback, wearable devices, neurofeedback, and meditation, clients can more fully hope to overcome dysfunctional physical, emotional, or behavioral states and realize the promise of wellness and optimal performance.

The prevalence of science-backed wellness trends in popular media speaks to an ever-increasing public consciousness around improved wellness and optimal performance. An article in NBC News BETTER (DiGiulio, 2020) predicted nine wellness trends likely to continue in this decade:

1. Prioritizing self-care from tireless career and personal accomplishments to activities and experiences that create happiness and enjoyment in life
2. Rebranding sleep from number of hours to quality sleep
3. Lifetime strength training for physical stability and healthier aging
4. Using muscle recovery tools to promote blood flow and decrease soreness to boost recovery after a workout
5. Reducing stigma and increasing attention to mental and emotional health
6. Spending quiet time at home with close others instead of pursuing leisure activities focused on entertainment and material consumption
7. Increasing plant-based, ecofriendly eating to decrease inflammation and prevent autoimmune and other metabolic diseases
8. Focusing greater attention on the gut-brain connection and following a diet rich in fiber, fruits, vegetables, and nuts to promote both physical and mental health
9. Improving conflict resolution skills that promote dialogue, apology, forgiveness, and compromise toward constructive conflict negotiation

Perhaps everyone could benefit from having at least one new resolution over the next decade for wellness and optimal performance.

Reflective Questions

What might have been missed had the wellness assessment not been conducted? What wellness trend or trends might you wish to take up?

Quiz

1. Which of the following is not one of the six dimensions of Hettler's (1984) model of wellness?
   a. Physical.
   b. Emotional.
   c. Spiritual.
   d. Health.
2. Which of the following is not a healthy lifestyle strategy?
   a. Moderate use of alcohol.
   b. Dietary supplements.
   c. Exercise.
   d. Screen time.
3. Which of the following mental activities does not help clients to regulate their emotions?
   a. Time-in.
   b. Focus time.
   c. Playtime.
   d. Sleep time.
4. Which of the following functions is not typically measured by wearable devices?
   a. Physical activity.
   b. Heart rate.
   c. Sleep quality.
   d. Cortisol secretion.

References

Abel, T., Havekes, R., Saletin, J., & Walker, M. (2013). Sleep, plasticity, and memory from molecules to whole-brain networks. *Current Biology, 23*(17), 774–788.

Amen, D. (2001). *Healing ADD*. Putnam.

Aranow, C. (2011). Vitamin D and the immune system. *Journal of Investigative Medicine, 59*(6), 881–886.

Balch, P. A. (2010). *Prescription for nutritional healing: A practical A-Z reference to drug-free remedies using vitamins, minerals, herbs and food supplements* (5th ed.). Avery.

Baumgart, M., Snyder, H. M., Carrillo, M. C., Fazio, S., Kim, H., & Johns, H. (2015). Summary of the evidence on modifiable risk factors for cognitive decline and dementia: A population-based perspective. *Alzheimer's & Dementia, 11*(6), 718–726. https://doi.org/10.1016/j.jalz.2015.05.016

Berkman, L. F., & Syme, L. S. (1979). Social networks, host resistance, and mortality: A nine-year follow-up study of Alameda County residents. *American Journal of Epidemiology, 185*(11), 1070–1088. https://doi.org/10.1093/aje/kwx103

Blumenthal, J., Babyak, M., Moore, K., Craighead, W., Herman, S., Khatri, P., & Ranga Krishnan, K. (1999). Effects of exercise training on older adults with major depression. *Archives of Internal Medicine, 159*(19), 2349–2356. https://doi.org/10.1001/arch-inte.159.19.2349

Budzynski, T. (1996). Brain brightening: Can neurofeedback improve cognitive process? *Biofeedback, 24*(2), 14–17.

Chandrasekaran, R., Katthula, V., & Moustakas, E. (2020). Patterns of use and key predictors for the use of wearable health care devices by US adults: Insights from a national survey. *Medical Internet Research, 22*(10), Article e22443. https://doi.org/10.2196/22443

Chapin, T., & Russell-Chapin, L. (2014). *Neurotherapy and neurofeedback: Brain-based treatment for psychological and behavioral problems.* Routledge.

Chen, P.-H., Kuo, H.-Y., & Chueh, K.-H. (2010). Sleep hygiene education: Efficacy on sleep quality in working women. *Journal of Nursing Research, 18*(4), 282–289. https://doi.org/10.1097/JNR.0b013e3181fbe3fd

Council for Accreditation of Counseling and Related Educational Programs. (2024). *2024 CACREP standards.* https://www.cacrep.org/wp-content/uploads/2023/06/2024-Standards-Combined-Version-6.27.23.pdf

Csikszentmihalyi, M. (1975). *Beyond boredom and anxiety.* Jossey-Bass.

Csikszentmihalyi, M. (1990). *Flow: The psychology of optimal experience.* Harper & Row.

Dean, O., Giorlando, F., & Berk, M. (2011). N-acetylcysteine in psychiatry: Current therapeutic evidence and potential mechanisms of action. *Journal of Psychiatry & Neuroscience, 36*(2), 78–86. https://doi.org/10.1503/jpn.100057

DiGiulio, S. (2020, January 5). *9 science-backed wellness trends from the last decade that we're taking into the 2020s.* NBC News BETTER. https://www.nbc.com/better/lifestyle/9-science-backed-wellness-trends-last-decade-we-re-taking-ncna1110096

Dunn, H. (1961). *High level wellness.* Beatty.

Erickson, K., Voss, M., Prakash, R., Basak, C., Szabo, A., Chaddock, L., Kim, J., Heo, S., Alves, H., White, S., Wojcicki, T., Mailey, E., Viera, V., Martin, S., Pence, B., Woods, J., McAuley, E., & Kramer, A. (2011). Exercise training increases size of hippocampus and improves memory. *Proceedings of the National Academy of Sciences, USA, 108*(7), 3017–3022.

Fu, J., Li, F., Tang, Y., Cai, L., Zeng, C., Yang, Y., & Yang, J. (2021). The emerging role of irisin in cardiovascular diseases. *Journal of the American Heart Association, 10*(20), Article e022453. https://doi.org/10.1161/jaha.121.022453

Ganesan, S., Beyer, E., Moffat, B., Van Dam, N. T., Loranzetti, V., & Zalesky, A. (2020). Focused attention meditation in healthy adults: A systemic review and meta-analysis of cross-sectional functional MRI studies. *Neuroscience and Biobehavioral Reviews, 141*(10), Article 104846. https://doi.org/10.1016/j.neubiorev.2022.104846

Ghosh, S., Banerjee, S., & Sil, P. C. (2015). The beneficial role of curcumin on inflammation, diabetes and neurodegenerative disease: A recent update. *Food Chemical Toxicology, 83*(9), 111–124. https://doi.org/10.1016/j.fct.2015.05.022

Gibbons, T. D., Cotter, J. D., Ainslie, P. N., Abraham, W. C., Mockett, B. G., Campbell, H. A., Jones, E. M. W., Jenkins, E. J., & Thomas, K. N. (2023). Fasting for 20 h does not affect exercise-induced increases in circulating BDNF in humans. *Journal of Physiology, 601*(11), 2121–2137. https://doi.org/10.1113/jp283582

Goyal, M., Singh, S., Sibinga, E., Gould, N. F., Rowland-Seymour, A., Sharma, R., Berger, Z., Sleicher, D., Maron, D. D., Shihab, H. M., Ranasinghe, P. D., Linn, S., Saha, S., Bass, E. B., & Haythornthwaite, J. A. (2014). Meditation programs for psychological stress and well-being: A systematic review and meta-analysis. *JAMA Internal Medicine, 174*(3), 357–368. https://doi.org/10.1001/jamainternmed.2013.13018

Hedderson, M., Bekelman, T., Li, M., Knapp, E., Palmore, M., Dong, Y., Elliot, A., Friedman, C., Galarce, M., Gilbert-Diamond, D., Glueck, D., Hockett, C., Lucchini, M., McDonald, J., Sauder, K., Zhu, Y., Karagas, M., Dabelea, D., & Ferrara, A. (2023). Trends in screen time use among children during the COVID-19 pandemic, July 2019 through August 2021. *JAMA Network Open, 6*(2), Article e2256157. https://doi.org/10.1001/jamanetworkopen.2022.56157

Hettler, B. (1984). Wellness: Encouraging a lifetime of pursuit of excellence. *Health Values, 8*(4), 13–17.

Holze, B., Carmody, J., Vangel, M., Congleton, C., Yerramsetti, S., Gard, T., & Lazar, S. (2011). Mindfulness practice leads to increases in regional brain gray matter density. *Psychiatry Research: Neuroimaging, 191*(1), 36–43. https://doi.org/10.1016/j.pscychresns.2010.08.006

Ivey, A. E., Ivey, M. B., & Zalaquett, C. (2022). *Intentional interviewing and counseling* (10th ed.). Cengage.

Luders, E. (2014). Exploring age-related brain degeneration in meditation practitioners. *Annuls of the New York Academy of Sciences, 1307*(1), 62–72. https://doi.org/10.1111/nyas.12217

Luders, E., Kurth, F., Mayer, E., Toga, A., Narr, K., & Gaser, C. (2012). The unique brain anatomy of meditation practitioners: Alteration in cortical gyrification. *Frontiers in Human Neuroscience, 6*(2), Article 34. https://doi.org/10.3389/fnhum.2012.00034

Miller, N. E. (1969). Learning of visceral and glandular responses: Recent experiments on animals show the fallacy of an ancient view of the autonomic nervous system. *Science, 163*(3866), 434–445. https://doi.org/10.1126/science.163.3866.434

Moh, Y. S. (2020, September). The impact of diet and nutrition on mental health, Part 1. *Counseling Today, 63*(3), 10–12.

Molteni, R., Barnard, R., Ying, Z., Roberts, C. K., & Gomez-Pinilla, F. (2002). A high-fat, refined sugar diet reduces hippocampal brain-derived neurotrophic factor, neuronal plasticity, and learning. *Neuroscience, 112*(2), 803–814. https://doi.org/10.1016/S0306-4522(02)00123-9

Montes, S. (2013, December). The birth of the neurocounselor. *Counseling Today, 56*(6), 32–40.

Mujica-Parodi, L. R., Amgalan, A., Sultan, S. F., Antal, B., Sun, X., Skiena, S., Lithen, A., Adra, N., Ratai, E.-M., Weistuch, C., Govindarajan, S. T., Strey, H. H., Dill, K. A., Stufflebeam, S. M., Veech, R. L., & Clark, K. (2020). Diet modulates brain networks stability, a biomarker for brain aging, in young adults. *Proceedings of the National Academy of Sciences, USA, 117*(11), 6170–6177. https://doi.org/10.1073/pnas.1913042117

Myers, J. E., & Sweeney, T. J. (2005). The indivisible self: An evidence-based model of wellness. *Journal of Individual Psychology, 60*(3), 234–245.

Myers, J. E., Sweeney, T. J., & Witmer, J. M. (2000). The wheel of wellness counseling for wellness: A holistic model for treatment planning. *Journal of Counseling & Development, 78*(3), 251–266. https://doi.org/10.1002/j.1556-6676.200.tb01906.x

Nakata, A. (2011). Work hours, sleep sufficiency, and prevalence of depression among full-time employees: A community-based cross-sectional study. *Journal of Clinical Psychiatry, 72*(5), 605–614. https://doi.org/10.4088/JCP.10m06397gry

Naureen, Z., Dhuli, K., Medori, M. C., Caruso, P., Manganotti, P., Chiurazzi, P., & Bertelli, M. (2022). Dietary supplements in neurological diseases and brain aging. *Journal of Preventive Medicine and Hygiene, 63*(2S3), E174–E188. https://doi.org/10.15167/2421-4248/jpmh2022.63.2S3.2759

Office of the Surgeon General. (2023). *Social media and youth mental health: The U.S. Surgeon General's advisory*. https://www.hhs.gov/sites/default/files/sg-youth-mental-health-social-media-advisory.pdf

Pandya, A., & Lodha, P. (2021). Social connectedness, excessive screen time during COVID-19 and mental health: A review of current evidence. *Frontiers in Human Dynamics, 3*, Article 684137. https://doi.org/10.3389/fhumd.2021.684137

Ratey, J. (2008). *Spark: The revolutionary new science of exercise and the brain*. Little, Brown.

Rock, D., Siegel, D. J., Poelmans, S. A. Y., & Payne, J. (2012). The healthy mind platter. *NeuroLeadership Journal, 4*, 1–23.

Schlechta Portella, C. F., Ghelman, R., Abdala, V., Schveitzer, M. C., & Afonso, R. F. (2021). Meditation: Evidence map of systemic reviews. *Frontiers in Public Health, 9*(12), Article 742715. https://doi.org/10.3389/fpubh.2021.742715

Segal, E. (2004). Incubation in insight problem-solving. *Creativity Research Journal, 16*(1), 141–148. https://doi.org/10.1207/s15326934crj1601_13

Solino-Fernandez, D., Ding, A., Bayro-Kaiser, E., & Ding, E. L. (2019). Willingness to adopt wearable devices with behavioral and economic incentives by health insurance wellness programs: Results of a US cross-sectional survey with multiple consumer health vignettes. *BMC Public Health, 19*(1), Article 1649. https://doi.org/10.1186/s12889-019-7920-9

Stanhope, K., Schwarz, J.-M., & Havel, P. J. (2013). Adverse metabolic effects of dietary fructose: Results from the recent epidemiological, clinical, and mechanistic studies. *Current Opinions in Lipidology, 24*(3), 198–206. https://doi.org/10.1097/MOL.0b013e3283613bca

Stiglic, N., & Viner, R. M. (2019). Effects of screentime on the health and well-being of children and adolescents: A systematic review of reviews. *BMJ Open, 9*(1), Article e023191. https://doi.org/10.1136/bmjopen-2018-023191

Strack, B., Linden, M., & Wilson, V. (2011). *Biofeedback and neurofeedback applications in sport psychology.* Association for Applied Psychophysiology and Biofeedback.

Suni, E., & Singh, A. (2023, November 3). *How much sleep do you need?* Sleep Foundation. https://www.sleepfoundation.org/how-sleep-works/how-much-sleep-do-we-really-need

Swingle, M. (2015). *I-mind: How cell phones, computers, gaming and social media are changing our brains, our behavior and the evolution of the species.* Inkwater.

Uchino, B. N., Holt-Lunstad, J., Uno, D., Betancourt, R., & Garvey, T. S. (1999). Social support and age-related differences in cardiovascular function: An examination of potential mediators. *Annual Behavioral Medicine, 21*(2), 135–142. https://doi.org/10.1007/BF02908294

van den Berg, C. L., Hol, T., Van Ree, J. M., Spruijt, B. M., Everts, H., & Koolhaas, J. M. (1999). Play is indispensable for an adequate development of coping with social challenges in the rat. *Developmental Psychobiology, 34*(2), 129–138. https://doi.org/10.1002/(SICI)1098-2302(199903)34:2%3C129::AID-DEV6%3E3.0.CO;2-L

Wagner, B. E., Folk, A. L., Hahn, S. L., Barr-Anderson, D., Larson, N., & Neumark-Sztainer, D. (2021). Recreational screen time behaviors during the COVID-19 pandemic in the U.S.: A mixed-methods study among a diverse population-based sample of emerging adults. *International Journal of Environmental Research and Public Health, 18*(9), Article 4613. https://doi.org/10.3390/ijerph18094613

Xie, L., Kang, H., Xu, Q., Chen, M. J., Liao, Y., Thiyagarajan, M., O'Donnell, J., Christensen, D., Nicholson, C., Iliff, J., Takano, T., Deane, R., & Nedergaard, M. (2013). Sleep drives metabolic clearance from the adult brain. *Science, 342*(6156), 373–377. https://doi.org/10.1126/science.1241224

Xu, J., Vik, A., Groote, I., Lagopoulos, J., Holen, A., Ellingsen, O., Haberg, A. K., & Davanger, S. (2014). Nondirective meditation activates default mode network and areas associated with memory retrieval and emotional processing. *Frontiers in Human Neuroscience, 8,* Article 86. https://doi.org/10.3389/fnhum.2014.00086

Zivkovic, A., Telis, N., German, J. B., & Hammock, B. D. (2011). Dietary omega-3 fatty acids aid in the modulation of inflammation and metabolic health. *California Agriculture, 65*(3), 106–111. https://doi.org/10.3733/ca.v065n03p106

Chapter 9

Neuroscience of Attention: Empathy and Counseling Skills

Carlos P. Zalaquett, Ravza N. Aksoy, Allen E. Ivey, Mary Bradford Ivey, and Thomas Daniels

Counseling changes the brains of both client and counselor. Interpersonal interactions have the potential to strengthen existing neuronal connections and build new neural networks. Learning, practicing, and remembering activate the brain. When a client learns a new skill, repeatedly practices a new behavior, or accesses memories, their neural networks fire in concert, creating electrochemical routes that shape into long-term memories and possible new patterns of behavior. This process is known as *neuroplasticity*—the selective organizing of connections among neurons in the brain. With practice, connections become stronger and more efficient, and they establish connections with other parts of the brain. The old view that brain cells are irreplaceable and that if a brain cell dies you are out of luck has changed. The idea that people only lose neurons as they age is a myth. Older adults, through learning and new experiences, have the potential to increase the gray matter in their brains. Recent studies have shown that the efficiency of cognitive abilities, such as orienting and executive inhibitory functions, increases with age (Veríssimo et al., 2022).

However, learning and memory are not possible without attention and without suppressing irrelevant distractors. These are main functions of attention (Weksler et al., 2021). Attentional control and executive functioning allow us to navigate efficiently the complexities of everyday life, and deficits in these processes may produce negative consequences (Hopfinger & Slotnick, 2020). Ivey et al. (2023, 2024) added that attentional processes and the microskill of attending to behavior make possible and enhance learning, memory, and change. Attention is essential for counseling and therapy.

As a counselor, you cannot help clients cope with stress and change behavior without *attending* empathically to their stories and goals (Ivey et al., 2024; Zalaquett et al., 2019). Current research in neuroscience and neurobiology has reinforced and supported what counselors and therapists have been doing for years. Research into artificial intelligence and virtual learning software that mimics the input and output function of the neurons that are involved in the brain's attentional process (Lindsay, 2020) also contributes support. It is time to draw on this new information and apply it further to counseling and clinical practice. Advances in neuroscience and artificial intelligence will enhance the mind-brain-body conceptualization and enable counselors to practice more effectively.

This chapter explains the neuroscientific foundation of attention, and how it serves as a basis for counseling practice, with a focus on four main areas: neuroscience of attention; attention and empathic understanding; counseling skills; and rewiring memories into positive, resilient action. A fictional case study illustrates how neuroscience helps the counselor to both conceptualize and act in the counseling and therapy session.

2024 CACREP Standards

This chapter addresses 2024 Council for Accreditation of Counseling and Related Educational Programs (CACREP) Standards pertinent to the Foundational Counseling Curriculum (Section 3) area of Counseling Practice and Relationships (Standard E):

- Culturally sustaining and responsive strategies for establishing and maintaining counseling relationships across service delivery modalities (Standard E.7.)
- Counselor characteristics, behaviors, and strategies that facilitate effective counseling relationships (Standard E.8.)
- Interviewing, attending, and listening skills in the counseling process (Standard E.9.)
- Strategies for adapting and accommodating the counseling process to client culture, context, abilities, and preferences (Standard E.11.)

Clinical Case Study: Gloria

Microaggressions can affect mind, brain, and body (Luke et al., 2020). Let us look at a counselor and client from a brain-based skills approach. Gloria is a 15-year-old high school student in her junior year who migrated from Honduras 2 years ago. Gloria, who is fluent in both English and Spanish, achieved high grades in her first year at the high school. Gloria is referred to counseling by the high school counselor because of anxiety and panic symptoms that are interfering with her academic performance. Gloria has experienced declining grades since this school year began. Gloria reports to you that other students have made "rude" comments about her accent and mockingly told her, "You speak good English." Gloria has trouble concentrating in the school environment because, she says, "the stress is too much."

Reflective Questions

> Before proceeding, consider that this is a microaggression in action. It may seem mild to many of us and as though it should be ignored, but you see the immediate and powerful reaction it has brought out in Gloria. Continued racist, sexist, or homophobic microaggressions gradually make the world feel unsafe and dangerous, alienating the recipient.
>
> Can you think of microaggressions you and your clients may have received in the past? Bullying is one common example, as is sexist language and even being ignored.
>
> What is the counselor's responsibility? How can the counselor handle these events to facilitate client empowerment and resilience?

Neuroscience of Attention

Gloria is attending to a strong stimulus: a microaggression. Such negative experiences rapidly lock themselves into long-term memory and can change behavior. There can be no counseling and therapy without attention. The counselor needs to draw out Gloria's story, her thoughts and feelings. This chapter is designed to reveal how attentional processes, microskills, and neurobiological structures and mechanisms underlie counseling and the therapy process.

The nervous system and its main branches, the central nervous system and the peripheral nervous system, play a pivotal role in active attending. Impairments to the nervous system can reduce one's capacity to attend (e.g., toxins or diseases can reduce attention). What one attends to can be traumatic, leading to fight-or-flight reactions by the sympathetic nervous system, or it can be soothing ("rest and digest") through action in parasympathetic nervous system and its vagus nerve. In Gloria's case, the negative interaction has awakened the sympathetic nervous system, and there is a need for calming via the parasympathetic nervous system and the vagus nerve.

Both the sympathetic and the parasympathetic nervous systems are part of the autonomic nervous system, a branch of the peripheral nervous system. A balance between the sympathetic and parasympathetic systems is critical for both mental and physical health. Key to achieving balance are three basic networks identified by Petersen and Posner (2012): alerting, orienting, and executive attention.

Alerting

The wakeup call of the nervous system is a whole brain-body activity that starts with perceptions—visual, auditory, tactile, olfactory, and taste. One or more of these perceptions wakes up the brain stem to vigilance and arousal, producing norepinephrine. In turn, the spinal cord, thalamus, amygdala, hypothalamus, and prefrontal and parietal areas of the brain become involved. Maintaining vigilance after perception relies on the prefrontal cortex.

Alerting is most easily described as anything that draws one's immediate attention, such as a car horn; a shout; a thunderclap; or a microaggression, such as that experienced by Gloria. Alerting can occur any time one's emotions are evoked or one has motivation—for example, when one sees a friend out of the corner of one's eye, is suddenly drawn to a brilliant sunset, senses danger, or feels one is not approved of. Alerting represents the focal point of the beginning of attention and the counseling relationship.

The process of alerting can take considerable brain energy, devouring glucose necessary for cell functioning. Maintaining vigilance and awareness appears to gradually wear down alerting and brain efficiency. For example, consider Transportation Security Administration staff, truck drivers, and prison guards—anyone with a stressful but important job. Reaction times vary throughout the day (Ghassemzadeh et al., 2019). Moreover, people of color can become weary and lose attention and alertness as a result of necessary vigilance in situations that are unwelcoming or that demonstrate White privilege (Golden, 2021). Needless to say, alerting becomes an issue not only for clients but for counselors. Social racism creates significant individual and institutional racial biases that provide preferential treatment to people on the basis of their race or ethnicity, negatively affecting the attention and functioning of people of color (Stevens & Abernethy, 2018).

A counselor's attention can decline during the day and even during a session, and thus they may miss important things that clients say in the here and now. Furthermore, a counselor can emphasize or minimize their client's cultural characteristics or worldview. The obvious point is that if counselors are to be with clients in the moment, they need to increase their efforts to be alert to them and discover what patterns alert clients both to danger (sympathetic nervous system) and to comfort (parasympathetic nervous system, vagus nerve).

Orienting

Once the counselor has been alerted to the client through the senses, the counselor must navigate all incoming information to the brain. The orienting network is involved in directing attention to a specific stimulus and location in space. The function of the orienting network can be stimulus driven (exogenous, automatic, or bottom-up) and goal directed (endogenous, voluntary, or top-down). So far, the counselor and Gloria have been alerted, and both have oriented their attention to the classroom story. The next attentional issue occurs once the counselor has all the information. This action of the mind—that is, executive functioning—is, somewhat surprisingly, what counseling is all about.

Executive Attention

Executive attention is more popularly known as self-regulation, executive functioning, emotion regulation, and self-efficacy. Counselors attend to clients; they listen for cognitions and underlying emotions. They try to understand what is going on in the client's mind. They make decisions. This is executive functioning in action.

A critical part of executive functioning is cognitive and emotion regulation. At the beginning of counseling, the counselor discovers that Gloria's emotions have allowed her to take on the microaggression internally and that she believes that she is the one who is inadequate and should not speak up. Emotion often rules cognition. Counselors need to have competence in controlling and regulating themselves, because it is their goal to facilitate client executive functioning and emotion regulation and build resilience through appropriate attentional patterns. Executive functioning, like alerting and orienting, requires widespread brain connections. The prefrontal cortex, insula, and anterior cingulate cortex are particularly important in balancing emotion and cognition.

Posner and Rothbart (2018) discussed three attentional networks: The first obtains and maintains the alert state, the second orients to sensory stimuli, and the third provides executive control of voluntary behavior. This last network monitors behavioral acts (and was the original definition of executive control). Such executive functioning is critical to maintaining focus on a topic or task. The second attention network operates differently, through monitoring, error detection, and correction. Every counselor makes mistakes in counseling, and the ability to self-reflect and correct them is essential. With Gloria, there is a need to strengthen her cognitive executive control but also to help her understand her emotional error in accepting false and disruptive behavior from others.

Executive functioning is also related to mentalizing, located primarily in the prefrontal cortex, anterior cingulate cortex, and temporal parietal junction (Mahy et al., 2014). Mentalizing, also termed *theory of mind* (ToM), is the process of making sense of others (clients) and oneself. It includes subjective processes, such as the thoughts occurring in others and one's own self-reflections (Arioli et al., 2021; Wade et al., 2018). Mentalizing requires the counselor to enter the client's world, but with an awareness that it is the client's and not the counselor's world. Given these many important functions, understanding executive functioning, emotion regulation, and self-regulation is vital to becoming an effective counselor and therapist. ToM is basic to empathy, and an important part of the counselor's task in the case study is to generate a theory of how Gloria is processing information in her mind.

Reflective Questions

> How do you work to balance emotion and cognition?
> Does the concept of ToM make sense to you?
> What is going on in Gloria's mind?
> Can you identify the three aspects of alerting, orienting, and executive attention as they occurred in the interview?
> What is the place of the nervous system in counseling work?

Attention and Empathic Understanding

The psychologist Carl Rogers made empathy central to the helping professions (Ivey et al., 2024). Neuroscience, however, has taken our understanding of the meaning and value of empathy to a new level (Levy et al., 2019). Extensive functional MRI (fMRI) studies of the brain have both theoretical and applied implications for the counseling field. The following discussion illustrates the complexity of understanding and experiencing the world of others.

Spunt (2013) spoke to the importance of the brain's mirror system as basic to understanding empathy's place in the communication process. The basics of the mirror system are most easily understood in terms of observational learning. Think of a youth who wants to learn how to throw a boomerang. Watching someone who is already adept at throwing the boomerang activates portions of the youth's brain.

The counseling relationship facilitates neuroplasticity, and practice (homework and action plans) outside of the therapeutic relationship strengthens the development of new neural networks. The conversation in counseling helps "rewire" the brain for more effective living. Neurophysiology has shown that empathy is indeed a necessary (and sometimes sufficient) condition for the relationship, which can itself enable client change.

Empathy consists of various different components, such as cognitive empathy, affective empathy, and mentalizing (Arioli et al., 2021; Stevens & Taber, 2021; Zebarjadi et al., 2021). Affective empathy helps counselors understand the emotions experienced by others, and cognitive empathy enables them to understand client verbalizations, thought patterns, and behaviors. Advances in neuroscience have confirmed that empathy is a multicomponent construct essential in navigating the social environment and developing deeper intellectual and emotional understanding. A meta-analysis conducted by Fan et al. (2011) across 40 fMRI studies showed that affective empathy is associated with increased activity in the insula, whereas the right supramarginal gyrus recognizes a lack of empathy and autocorrects (Engen & Singer, 2013). Cognitive empathy is associated with higher activity in the midcingulate cortex and the dorsomedial prefrontal cortex. Eres et al. (2015) found greater gray matter density in both places.

Speaking to the importance of self-correction in listening, psychologist Tania Singer said that "we use ourselves as a yardstick" and project our emotional state onto others (Bergland, 2013, "The Neuroscience of Empathy," para. 4). This "emotional egocentricity" can distort how we understand the emotions of other people, particularly if they are different from our own. There is relatively little research in this area.

Neuroimaging studies by Lamm et al. (2011) have also shown that emotional components are shared vicariously. When people experience direct pain themselves (firsthand sensation), the somatosensory motor cortex, insula, and anterior cingulate cortex are activated. When they watch others experience pain (secondhand pain), the insula and anterior cingulate cortex are activated, but not the somatosensory cortex. The insula integrates visceral and autonomic information with salient stimuli, acting as an infrastructure for the representation of subjective bodily feelings of positive and negative emotions.

Listening is the building block of the relationship. The microskills approach has made the underlying behaviors of listening clear. Research using fMRI has revealed that listening literally lights up the brain. Kawamichi et al. (2015) found that the Rogerian microskills of attending to behavior, paraphrasing, reflecting feelings, and summarizing create the foundation for a strong therapeutic relationship and the benefits that stem from this alliance.

Counselors activate key brain structures when they listen. The ventral striatum becomes active when encountering abstract positive communication. Counselors think of Rogerian positive regard, authenticity, and being with the client as key aspects of listening. The microcounseling approach (Ivey et al., 1968) identified the concrete behaviors of listening and developed the term *attending behavior*. Culturally appropriate eye contact, body language, vocal tone, and verbal following have since become standard basic counseling practice.

The right anterior insula has been identified as key in emotional appraisal. Ghahremani et al. (2015) pointed out that the insula identifies what is salient and has a close connection with the anterior cingulate cortex, which is concerned with empathy, emotion, and reward anticipation. Together, these structures are important in inhibitory control, a critical factor in dealing constructively with emotion. The superior temporal sulcus is involved in the perception of where others are gazing (joint attention) and thus is important in determining where others' emotions are being directed (Beauchamp, 2015).

Polyvagal Theory and Safety

Physiological needs and safety are the foundation of Maslow's (1943) hierarchy of needs (see Figure 9.1). Counselors too often focus on self-actualization in counseling and therapy, paying insufficient attention to Maslow's foundation of physiological and safety needs. In developing his polyvagal theory, Porges (2011) clarified the importance of Maslow's original ideas. He identified the vagus nerve and its connections as the physiological basis for safety (Porges, 2022). To conduct effective counseling and therapy, counselors need to provide a safe environment in an effective relationship.

Reflective Questions

> Notice that neuroscience brings us back to counseling's emphasis on relationship. Safety, both cognitive and physiological, is foundational to any relationship. Think through the different challenges in providing safety and compassion when you are culturally similar to your client. Now, change the client's gender to be different from yours. How might safety and relationship change?
>
> Next, assume a cultural difference between you and your client in race, ethnicity, sexual orientation, spirituality, or economic background. How prepared are you to provide a safe relationship?

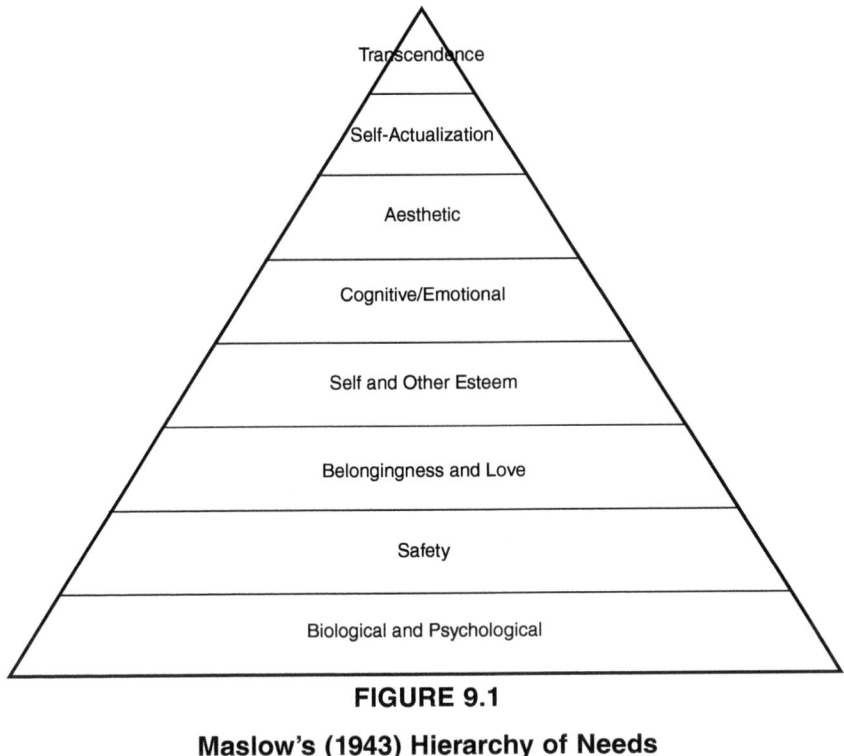

FIGURE 9.1

Maslow's (1943) Hierarchy of Needs

Note. Copyright 2016 by Carlos P. Zalaquett, Allen E. Ivey, and Mary B. Ivey. All rights reserved. Adapted with permission.

Safety is a major issue for Gloria in the counseling relationship. Creating a trusting relationship is necessary. Then, the interview needs to address the lack of feelings of safety that Gloria experiences in the classroom. These are cognitive and emotional issues, but they are based on physiological foundations. Unless the client feels safe, cognition and even emotional recognition are not enough.

Porges (2022) focused on four factors of compassionate relationship that will sound familiar to those who have worked with microskills. Porges gave considerable attention to prosodic vocal tone, facial expression and body language, and eye contact as key to building feelings of safety. He also spoke of socialization, in which verbalizations provide content to these key nonverbal dimensions. Furthermore, he provided additional specifics that can help guide the client to a calming state (e.g., taking a deep breath). Attending, listening skills, and empathy help to substantiate the relationship and thus help to engender feelings of safety and trust between counselor and client. In this way, the relationship itself produces pleasurable dopamine in the client's brain, thus enabling the client to be more open to discussing their issues fully.

Counseling Skills, Calming, and Activating

From 1966 to 1968, the microskills framework focused on listening skills and attending behavior (Ivey et al., 1968). At that time, the basics of attending behavior—visuals, vocal tone, verbal following, body and facial language—were identified through the study of videotaped sessions parallel to Porges's (2022) observations of physiological safety. Through an authentic and safe relationship and listening, counselors can encourage feelings in the body that support cognitive and emotional change. Counselors are not only counseling with words; rather, they are working with the whole of the client's body, brain, and mind. An axiom of today is that microcounseling is the mind-brain-body axis operating in an atmosphere of people and the environment.

Listening skills provide a calming and safe foundation for relationship. If heard, clients can learn and develop in a compassionate counseling relationship and prepare to take risks. One way of fostering growth is through empathy. Listening microskills communicate to clients the three components of empathy (i.e., affective, cognitive, mentalizing). Deficits in empathic listening may lead to ineffective counselors, deficient counseling relationships, and damaging interventions. The microskill of reflection of feeling is central to affective empathy, the emotional dimension of empathy, which helps one experience, in a conscious way, the emotional state of the other person. At the same time, affective empathy will often enable one to feel at least some of the client's emotions in one's own body. Affective empathy implies a self-other distinction as well as an understanding of the origin of the other person's emotional experience. Why is this important? Because it increases one's general sensitivity to the emotions of others, enhances one's capacity to fully understand their emotional experience, and facilitates helping behavior.

Paraphrasing and clarifying clients' language and thought processes is critical to the counselor's ability to understand the minds of others and predict their behavior without necessarily sharing their emotions. Cognitive empathy with minimal affective empathy could also facilitate competitive, antagonistic, and deceptive behavior.

The microskill of summarization is closely associated with ToM and mentalizing. Human behavior is based on fluid mental states, which makes it difficult to understand others. Everyone's actions are driven by needs, feelings, desires, beliefs, or reasons. When people interact with another person, they automatically (often preconsciously) read that person's underlying mental states and base their responses on what underlies the other person's behavior. At times, people understand others quite inaccurately, often because they do not listen fully. Counselors make serious errors in mentalizing when they unconsciously mix up their own experiences with the thoughts and feelings of the client.

The basic listening sequence encompasses the empathic skills of active listening, using open and closed questions, encouraging, paraphrasing, reflecting feelings, and summarizing to help clients explore and find ways to address their issues (Ivey et al., 2023). The purpose of the basic listening sequence is to not only draw out the client's cognitive and affective worlds but also understand the client's internal mental state. Most summaries are primarily cognitive but include client emotional and feeling tone. Because emotions are often first reactions and may occur before cognition regulates them, counselors need to consistently think about (mentalize) the possible underlying unsaid emotions. Clients feel reassured when they sense that they are in tune with the counselor.

Through the basic listening sequence, the counselor draws out Gloria's story, develops increased empathy, and begins to understand what is going on in Gloria's mind (ToM). Equally important, perhaps even more so, is the basis of trust in the relationship that enables Gloria to explore issues more fully and move later to behavioral change and action.

Because this chapter focuses primarily on the empathic listening portion of the microskills framework, the activating influencing skills receive only secondary attention. Influencing skills such as empathic confrontation, feedback, self-disclosure, interpretation, psychoeducation, and therapeutic confrontation were added to the framework over the years. All these specifically affect the sympathetic nervous system, leading first to new thoughts and behaviors and later to activities that will enable Gloria to be more proud, culturally aware, and self-confident. Throughout the activating, influencing change process, which involves risk for the client, there is a constant need for safety and relationship. Thus, the newly rewired brain learns to balance activation and calming.

We have seen how the counselor used attending and listening skills early in the session to develop feelings of safety and trust, thus enabling him to draw out Gloria's story of the microaggression and its impact on her.

Figure 9.2 presents the microskills hierarchy, outlining the systematic step-by-step communication skill units. Note that multicultural understanding, neuroscience, ethics, and positive psychology form the foundation, in recognition that communication style varies with individual cultural background and history. Professionally, it is essential that interpersonal communication be bounded by ethical practice. Finally, microskills practice is based on what we have called the *positive asset search,* and the positive psychology–resilience foundation speaks to the issue that counseling and psychotherapy have too often focused on problems rather than on positive ways to build health and resilience.

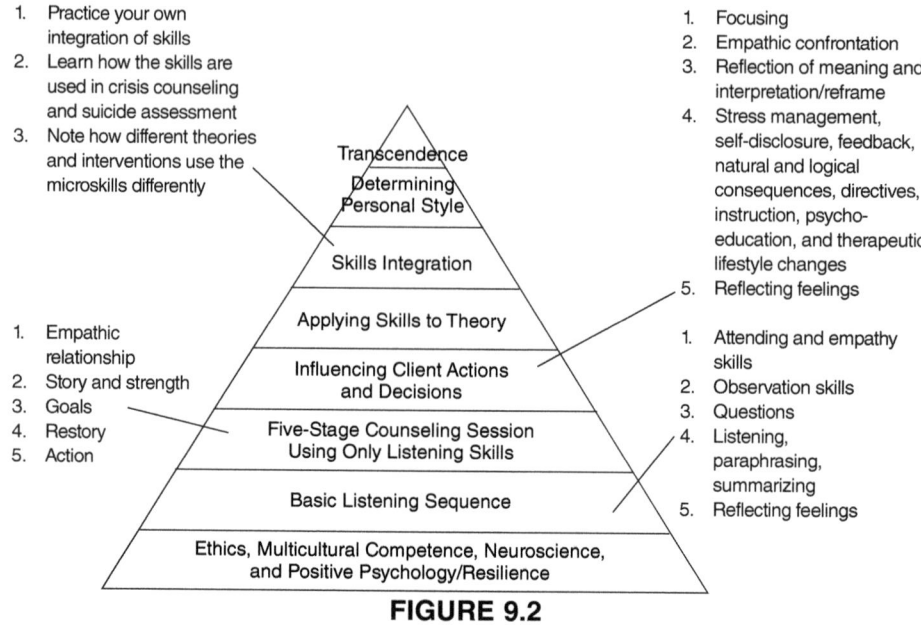

FIGURE 9.2

**The Microskills Hierarchy:
A Pyramid for Building Cultural Intentionality**

Note. Copyright 2012, 2014, 2017 by Allen E. Ivey. All rights reserved. Adapted with permission.

Reflective Questions

Consider that listening is obviously central in providing a safe relationship. How would you personally define listening?

How might different theories—cognitive behavior, humanistic, narrative, multicultural, feminist, and so forth—define listening differently?

Research has shown that the listener inevitably influences the way in which clients construct and think about their issues. In other words, there is no immaculate listening. What type of listener are you?

Rewiring Memories Into Positive, Resilient Action

Counseling and therapy are typically thought of as changing the meaning of issues and thus moving toward action. This action is typically thought of as oriented externally in terms of new behavior, thoughts, and feelings. However, these changes also affect the body through blood flow, heart rate, neurotransmitters, hormones, and onward to the gut, cell, and DNA. These are powerful agents of change, and a strictly traditional cognitive behavior or humanistic orientation is limiting. Counseling and therapy are deeply intertwined with biological processes; being aware of this will enhance counselors' effectiveness.

Let us look at a counselor and client from a brain-based skills approach. What are some specifics that lead to the changes described in the preceding paragraph? The client, Gloria, shares her story; the counselor listens, reflects, and seeks to help restory the memory and its meaning. Two brains are active in the session, and each person's brain, including short- and long-term memory, may change during the interaction. Two sets of memories in the hippocampus meet in the here and now of conscious conversation. Working memory brings life and the possibility of change to the session. Ultimately, counseling change is an interactive process of influencing clients' working memory in positive ways, helping the client rewire old memories and establish new ones. Counselors use working memory to access long-term memory, and significant change in long-term memory leads to changes in thoughts, feelings and emotions, and behaviors. Counseling is not a one-way process, and counselors also learn and change as they work with clients.

Working memory is the integrated centerpiece of action in counseling conversation. Working memory can be defined as the area in which people store high-speed data from here-and-now consciousness as well as information from short- and long-term memory. Gloria and the counselor can each likely store, at most, 18 items in working memory. However, the amount of information in working memory can change at any moment. For example, a highly emotional experience may leave the client with as few as one or two items in working memory. Similarly, counselors can become fixated on one slightly distorted view of what the client is saying, thus disrupting the therapeutic process.

Concerning the change process that occurs through the interaction of client and counselor, counselors are also dealing with the relationship of the "executive CEO" prefrontal cortex with the amygdala and the limbic hypothalamic-pituitary-adrenal hormonal axis. Attending behavior (attentional processes are heavily controlled by the thalamus and the prefrontal areas) remains foundational in determining whether declarative long-term memories are solidified in the hippocampus. However, emotional involvement through the energizing amygdala is necessary for working memory to function.

The listening process is ultimately aimed at understanding and calming, which highlights the importance of the vagus nerve and of developing new memories and stories that help promote a habit of calming under stress. With an understanding of their client's issues and concerns, counselors can move toward influencing skills and activating change processes.

With practice, long-term memory becomes automated in procedural memory, thus allowing the person to be fully in the here and now. This is also a goal of counseling and therapy—that is, to help clients learn new ways of being that eventually become so much a part of them that they seldom have to think about their actions.

Our Brain-Based Approach to the Case of Gloria

Microaggressions can rest deeply in the soul, and their emotional effect can be long lasting. Changing long-term memories is a challenging task. Counselors begin this process by listening to client stories in a safe space. In the case of Gloria, the counselor started by exploring her long-term, embedded memories.

As Gloria reexamined her memories of the microaggression, the counselor encouraged her to externalize her perceptions and view the challenging situation as it might be seen by others, and particularly by her Latinx friends. Through counseling, Gloria came to realize that she was not at fault for having an accent and should not have blamed herself.

The counselor also brought out positive stories and memories of Gloria's parents, who left Honduras to escape gang-related violence in their community. Her father worked in Tegucigalpa as a high-level supervisor at a well-known multinational food company. He sought and was granted a transfer to work in the United States with the same company. Gloria viewed him as a hero and began to take pride in her cultural heritage. With support from the counselor's use of both influencing and listening skills, and the strength of her father, she reframed her classroom experiences as oppression and racism while simultaneously expressing anger toward the classmates who had spoken mockingly to her—a dramatic change in language and worldview in a relatively short time.

Gloria moved from passive acceptance of microaggressions to active awareness of them and of the need to change her behavior, cognitions, and emotions. A new form of executive functioning and emotion regulation occurred. The development of new neural connections and a rewiring led to a change in memory and resulted in immediate changes in behavior.

Approaching the end of high school, Gloria began to contemplate further study in college. Her thoughts and feelings of cultural inferiority were replaced by cultural pride. Gloria became equipped for a more active stance toward the world. Her executive functioning changed, and her emotions were regulated in new ways. In an atmosphere of safety and trust, Gloria was able to reframe the difficult high school classroom experience in a new way.

Reflective Questions

> Consider how in this case study microaggression was reframed as oppression and Gloria externalized her cognitive and emotional issues. Before counseling, Gloria's microaggressions were internalized, lowering her self-confidence and leading to her feeling unsafe in the classroom. At the conclusion, her brain was rewired for her to take active resilient action when she encounters microaggression in the future.
> What were the specific steps and microskills used to restore safety with Gloria? What skills relate to reframing and action?
> What can you take away from this session?

Conclusion

One key task of counseling is to help the client restory their past experience and develop new memories and connections (e.g., behaviors, thoughts, feelings, meanings). Successful counseling is a collaborative endeavor that can change the client and their long-term memory in significant ways and even build new neural networks in the brain (neuroplasticity). The attending microskills discussed in this chapter provide the cognitive and emotional "charge" to promote understanding and change. The influencing skills both start and solidify the change process and improve action.

Quiz

1. Which of the following is *not* one of the things that Porges (2022) suggested is helpful to consider in establishing safety?
    a. Prosodic vocal tone (quality of intonation).
    b. Gestures and body language.
    c. Style of eye contact.
    d. Open-ended questions.
2. Which of the following is *not* a form of empathy?
    a. Cognitive.
    b. Situational.
    c. Affective.
    d. Mentalization.

3. In helping a client deal with microaggressions, the central goal is to:
   a. Listen carefully and fully.
   b. Provide a safe environment so that the client can talk easily.
   c. Use neuroscientific concepts to understand what is going on inside the client.
   d. Build resilience by strengthening the client so that the client can work with future insults and challenges more effectively and not internalize what others say about them.

References

Arioli, M., Cattaneo, Z., Ricciardi, E., & Canessa, N. (2021). Overlapping and specific neural correlates for empathizing, affective mentalizing, and cognitive mentalizing: A coordinate-based meta-analytic study. *Human Brain Mapping, 42*(14), 4777–4804. https://doi.org/10.1002/hbm.25570

Beauchamp, M. (2015). The social mysteries of the superior temporal sulcus. *Trends in Cognitive Science, 19*(9), 489–490.

Bergland, C. (2013, October 10). The neuroscience of empathy: Neuroscientists identify specific brain areas linked to compassion. *Psychology Today.* https://www.psychologytoday.com/blog/the-athletes-way/201310/the-neuroscience-empathy

Council for Accreditation of Counseling and Related Educational Programs. (2024). *2024 CACREP standards.* https://www.cacrep.org/wp-content/uploads/2023/06/2024-Standards-Combined-Version-6.27.23.pdf

Engen, H. G., & Singer, T. (2013). Empathy circuits. *Current Opinion in Neurobiology, 23*(2), 275–282. https://doi.org/10.1016/j.conb.2012.11.003

Eres, R., Decety, J., & Molenberghs, P. (2015). Individual differences in local gray matter density are associated with differences in affective and cognitive empathy. *NeuroImage, 117,* 305–310.

Fan, Y., Duncan, N., de Greck, M., & Northoff, G. (2011). Is there a core neural network in empathy? An fMRI based quantitative meta-analysis. *Neuroscience and Biobehavioral Reviews, 35*(3), 903–911.

Ghahremani, A., Rastogi, A., & Lam, S. (2015). The role of right anterior insula and salience processing in inhibitory control. *Journal of Neuroscience, 35*(8), 3291–3292.

Ghassemzadeh, H., Rothbart, M. K., & Posner, M. I. (2019). Anxiety and brain networks of attentional control. *Cognitive and Behavioral Neurology, 32*(1), 54–62. https://doi.org/10.1097/WNN.0000000000000181

Golden, K. B. (2021). White privilege: Unconscious racism, Freud, and neuroscience of implicit bias. *Critical Philosophy of Race, 9*(2), 295–322.

Hopfinger, J. B., & Slotnick, S. D. (2020). Attentional control and executive function. *Cognitive Neuroscience, 11*(1–2), 1–4. https://doi.org/10.1080/17588928.2019.1682985

Ivey, A., Ivey, M. B., & Zalaquett, C. P. (2023). *Intentional interviewing and counseling: Facilitating client development in a multicultural society* (10th ed.). Cengage.

Ivey, A., Ivey, M. B., & Zalaquett, C. P. (2024). *Essentials of intentional counseling and psychotherapy in a multicultural world* (4th ed.). Cengage.

Ivey, A., Normington, C., Miller, D., Morrill, W., & Haase, R. (1968). Microcounseling and attending behavior: An approach to pre-practicum training. *Journal of Counseling Psychology, 15*(5, Pt. 2), 1–12.

Kawamichi, H., Yoshihara, K., Sasaki, A. T., Sugawara, S. K., Tanabe, H. C., Shinohara, R., Sugisawa, Y., Tokutake, K., Mochizuki, Y., Anme, T., & Sadato, N. (2015). Perceiving active listening activates the reward system and improves the impression of relevant experiences. *Social Neuroscience, 10*(1), 16–26. https://doi.org/10.1080/17470919.2014.954732

Lamm, C., Decety, J., & Singer, T. (2011). Meta-analytic evidence for common and distinct neural networks associated with directly experienced pain and empathy for pain. *NeuroImage, 54*(3), 2492–2502.

Levy, J., Goldstein, A., & Feldman, R. (2019). The neural development of empathy is sensitive to caregiving and early trauma. *Nature Communications, 10*(1), Article 1905. https://doi.org/10.1038/s41467-019-09927-y

Lindsay, G. W. (2020). Attention in psychology, neuroscience, and machine learning. *Frontiers in Computational Neuroscience, 14*, Article 29. https://doi.org/10.3389/fncom.2020.00029

Luke, C., Redekop, F., & Moralejo, J. (2020). From microaggressions to neural aggressions: A neuro-informed counseling perspective. *Journal of Multicultural Counseling and Development, 48*(2), 120–129. https://doi.org/10.1002/jmcd.12170

Mahy, C., Moses, L., & Pfeifer, J. (2014). How and where: Theory-of-mind in the brain. *Developmental Cognitive Neuroscience, 9*, 68–81.

Maslow, A. H. (1943). A theory of human motivation. *Psychological Review, 50*(4), 370–396. https://doi.org/10.1037/h0054346

Petersen, S. E., & Posner, M. I. (2012). The attention system of the human brain: 20 years after. *Annual Review of Neuroscience, 35*, 73–89. https://doi.org/10.1146/annurev-neuro-062111-150525

Porges, S. W. (2011). *The polyvagal theory: Neurophysiological foundations of emotions, attachment, communication, and self-regulation.* Norton.

Porges, S. W. (2022). Polyvagal theory: A science of safety. *Frontiers in Integrative Neuroscience, 16*, Article 871227. https://doi.org/10.3389/fnint.2022.871227

Posner, M. I., & Rothbart, M. K. (2018). Temperament and brain networks of attention. *Philosophical Transactions of the Royal Society B: Biological Sciences, 373*(1744), Article 20170254. https://doi.org/10.1098/rstb.2017.0254

Spunt, R. (2013). Mirroring, mentalizing, and the social neuroscience of listening. *International Journal of Listening, 27*(2), 61–72. https://doi.org/10.1080/10904018.2012.756331

Stevens, F. L., & Abernethy, A. D. (2018). Neuroscience and racism: The power of groups for overcoming implicit bias. *International Journal of Group Psychotherapy, 68*(4), 561–584. https://doi.org/10.1080/00207284.2017.1315583

Stevens, F., & Taber, K. (2021). The neuroscience of empathy and compassion in pro-social behavior. *Neuropsychologia, 159,* Article 107925. https://doi.org/10.1016/j.neuropsychologia.2021.107925

Veríssimo, J., Verhaeghen, P., Goldman, N., & Weinstein, M. T. (2022). Evidence that ageing yields improvements as well as declines across attention and executive functions. *Nature Human Behavior, 6*(1), 97–110. https://doi.org/10.1038/s41562-021-01169-7

Wade, M., Prime, H., Jenkins, J. M., Yeates, K. O., Williams, T., & Lee, K. (2018). On the relation between theory of mind and executive functioning: A developmental cognitive neuroscience perspective. *Psychonomic Bulletin & Review, 25*(6), 2119–2140. https://doi.org/10.3758/s13423-018-1459-0

Weksler, A., Jacobson, H., & Bronfman, Z. Z. (2021). The transparency of experience and the neuroscience of attention. *Synthese, 198,* 4709–4730. https://doi.org/10.1007/s11229-019-02366-8

Zalaquett, C. P., Ivey, A., & Ivey, M. B. (2019). *Essential theories of counseling and psychotherapy: Everyday practice in our diverse world.* Cognella Academic.

Zebarjadi, N., Adler, E., Kluge, A., Jääskeläinen, I. P., Sams, M., & Levy, J. (2021). Rhythmic neural patterns during empathy to vicarious pain: Beyond the affective-cognitive empathy dichotomy. *Frontiers in Human Neuroscience, 15,* Article 380. https://doi.org/10.3389/fnhum.2021.708107

Chapter 10

Leveraging the Neuroeducation Process to Enhance Outcomes

Eric T. Beeson, Thomas A. Field, Chad Luke, Raissa Miller, Laura K. Jones, and Isaac Burt

In this chapter, we will provide readers with an overview of neuroeducation as a counseling intervention, outline ethical principles for the most effective delivery to clients, discuss the benefits and contraindications of neuroeducation, and provide examples showing how neuroeducation can be integrated into practice.

2024 CACREP Standards

This chapter addresses 2024 Council for Accreditation of Counseling and Related Educational Programs (CACREP) Standards pertinent to the Foundational Counseling Curriculum (Section 3) area of Counseling Practice and Relationships (Standard E):

- Critical thinking and reasoning strategies for clinical judgment in the counseling process (Standard E.2.)
- Ethical and legal issues relevant to establishing and maintaining counseling relationships across service delivery modalities (Standard E.6.)
- Counseling strategies and techniques used to facilitate the client change process (Standard E.10.)

This chapter also addresses the following Doctoral Standards for Counselor Education and Supervision (Section 6) in the area of Doctoral Curriculum (Standard B):

- Legal and ethical issues and responsibilities in counseling across multiple settings and across service delivery modalities (Standard B.1.f.)

Clinical Case Study: Enzo

You have been working with Enzo weekly for 3 months and have a complete biopsychosocial assessment and treatment plan focusing on (a) reducing angry outbursts at work and (b) increasing the quality of his nonromantic interpersonal relationships. You are looking at his intersecting identities using the ADDRESSING (age/generation, disability developmental or acquired, religion/spirituality, ethnicity/race, socioeconomic status, sexual orientation, Indigenous heritage, national origin, and gender) model (Hays, 1996, 2008). Enzo is a 45-year-old first-generation Italian American cisgender male who uses the pronouns he/him. He is fluent in both English and Italian and has no developmental or acquired disabilities. He identifies as an atheist and is of high socioeconomic status. He is sexually attracted to and engaged in sexual activity with women only.

During your next session, Enzo begins by saying, "I just saw on TikTok that trauma is stored in my body and that I need to take all these supplements to rid my body of the memories." You observe Enzo exhibiting more psychomotor agitation, rapid speech, and tangential thought processes than usual when discussing this advice. He asks, "Is this true? What should I do?"

As the counselor in this situation, how would you respond to Enzo's question? Would you be excited because you just read a book on the topic and have new information to share? Would you get defensive and blame your graduate program for not including more neuroscience training? Would you feel out of your depth and freeze? If you are like we were early in our careers, you would probably share, with good intentions, all the wonderful neuroscience information you know about memory systems, trauma, and the stress response system. However, your client's attention might wane, or they could tell you that you remind them of their elementary school science teacher. (Yes, both situations happened to us.)

Despite our good intentions and enthusiasm for the study of neuroscience, we often found ourselves relying on the neuroscience information rather than the contextual and relational components that surround that information. Our research and collective experiences have elevated the art and science of exploring neuroscience in therapeutic work. One method of integrating neuroscience is by using neuroeducation—that is, psychoeducation about neuroscience. Neuroeducation can be a tremendously powerful method of integrating neuroscience into our work, when provided with intentionality. Throughout this chapter, we will review the history and models of neuroeducation, highlight a 10-step process to guide the effective use of neuroeducation, and apply these steps to support Enzo and other clients whom you will serve in the future.

Introduction to Neuroeducation

In their model of neuroscience-informed counseling, Beeson and Luke (2023) described *neuroeducation* (Miller & Beeson, 2021) as a Level 2 skill, meaning that neuroscience is used to extend and refine our existing skills related to psychoeducation. Miller and Beeson (2021) expanded on previous descriptions of neuroeducation (Fishbane, 2013; Miller, 2016) to describe neuroeducation as a transdiagnostic and transtheoretical process that is grounded in the real therapeutic relationship and uses neuroscience information not to explain but to explore past, current, and future conceptualizations of the human experience, expectations for change, and actualized changes in counseling and psychotherapy. The process of neuroeducation can be informed by decades of research into the more general practice of psychoeducation (e.g., Magill et al., 2021) as well as by emerging models of neuroeducation in physical therapy (Louw et al., 2016), psychology (De Raedt, 2020; Kryza-Lacombe et al., 2021), addiction (Ekhtiari et al., 2017), and counseling (Miller, 2016; Miller & Beeson, 2021). Despite differences in focus, the models have similar threads and emphasize client centeredness, individualization, and creativity. Although neuroeducation includes teaching elements, these elements are delivered in an exploratory and interactive manner that engages the client rather than lectures at them.

In their neuroeducation model, Miller and Beeson (2021) presented a common factors (Wampold, 2015) approach to the neuroeducation process. This process is grounded in client/counselor coregulation, which is the foundation of an authentic relationship. The counselor uses microskills and shares neuroscience information with the intent to collaboratively explore the client's past, present, and future in a manner that is culturally and developmentally responsive. Miller and Beeson described how the neuroeducation process can be used to support positive narratives, increase compassion and empathy for self and others, reduce blame and shame, normalize the client experience and change process, support regulation, and enhance client engagement and motivation. They also asserted that neuroeducation is best delivered when it is embodied, contextualized, individualized, holistic, intentional, hypothesis driven, collaborative, emergent, appreciative, and compassionate. Counselors should use their basic microskills to guide the meaning-making process throughout, with an emphasis on the relationship over the content.

As shown in the case of Enzo, neuroscience information may be indicated or even requested by some clients during the counseling process. Counselors need to have specific skills in order to use neuroscience information to help clients explore their current experiences and to strengthen (rather than rupture) the therapeutic relationship.

10-Step Process for Providing Neuroeducation

We recommend a 10-step process to elevate the neuroeducation process during counseling, as follows:

Step 1. Attend to the client and the relationship.
Step 2. Explore the client's theory and motivations.
Step 3. Identify the neuroscience concepts, or neuroconcepts, relevant to the client's story and developmental and cultural contexts.
Step 4. Reflect on your theory and motivations and on ethical considerations.
Step 5. Consider the influence of social positions and power differentials.
Step 6. Assess client knowledge of and interest in the neuroconcept.
Step 7. Determine your method for introduction and exploration.
Step 8. Deliver information ethically and with curiosity and humility.
Step 9. Use information as a vehicle for exploration.
Step 10. Coconstruct a plan for what to do next.

Step 1: Attend to the Client and the Relationship

A professional counselor's best first response to a client's implicit or explicit request for neuroscience information is to do what we do best—that is, attend to the client with active listening and therapeutic presence. This engagement sends a message of safety and elevates the client's expertise (rather than the counselor's) concerning their own experience, which serves as a foundation for clarifying the intent and meaning of the client's request. By slowing down to focus on engagement, the counselor is also able to gauge the client's current state of regulation (e.g., high or low arousal) and support the client in obtaining an optimal state for learning. Sharing neuroscience information too early in the dialogue can shift the focus away from the client and the therapeutic relationship, thus sending a message that the client's experiences are unimportant or too uncomfortable for the counselor, which does not model how to attend to and tolerate one's experiences.

Step 2: Explore the Client's Theory and Motivations

Clients will hold their own theories to explain their life experiences. The counselor can seek to learn the client's perspective while considering how neuroscience information might support, expand, or challenge that perspective. Most, if not all, client theories have an adaptive origin that has helped the client to cope and regulate. If the client's existing theory continues to be adaptive in their current context (e.g., "My past gave me the opportunity to develop resilience and healthy boundaries in relationships."), the counselor should use neuroscience information that lends credence to the client's theories. Although a theory may be initially adaptive (as with trauma), it can sometimes be overgeneralized to future life experiences and lead to challenges (e.g., "My brain must not work right, because I keep sabotaging all relationships."). After attending and connecting to the client, the counselor can use neuroscience information to propose an alternative theory for exploration.

Clients may seek neuroscience information that connects to their current theories either explicitly (e.g., "What is going on in my brain?" "Why is my brain so broken?"), as in the case of Enzo, or implicitly (e.g., "There must be something wrong with me that can't change."). Some of these questions remain core to counselors' scope of practice, whereas others may require medical referral. For example, a client who is having sensory experiences such as extreme derealization is wondering whether they are experiencing seizures or have a brain tumor. Before exploring neuroscience information, you should try to decipher what the client is hoping to gain from the experience.

With curiosity, you might explore a variety of conceptualization questions: What does the question imply about the client's existing theories? What does the client hope to gain? What does the client hope the counselor will provide? How will this impact the counseling relationship? Exploring the meaning behind the client's inquiry will clarify the client's motivations and desired response. In many cases, directly exploring neuroscience information is not crucial to the client's presenting issue or concern and may even have unintended consequences. In the case of Enzo, immediately providing neuroscience information about implicit memory, the hypothalamic-pituitary-adrenal (HPA) axis, or the effects of certain supplements on physiological functioning might cause the counselor to miss his preexisting theory (e.g., "My career is so important to me. Why are they making me so angry?"), the dissonance of potential new theories (e.g., "I used to think it was all their fault, and now I am learning that it might be mine."), and potential motivations (e.g., confirming the external causes, exculpating from biogenically derived behaviors). Taking time to first attend to the client's experiences and explore their theories and motivations increases the likelihood that any information that follows will be beneficial to therapeutic outcomes. Without this step, the counselor may provide information that holds little worth for the client, distracts from the counseling process, and/or creates distance in the therapeutic relationship. Even if the counselor has an initial idea of relevant neuroscience information, it is more helpful to first consider this information in the context of the client's theories and motivations. The counselor, when providing neuroeducation, can also be mindful to not cross a boundary of competence related to providing information about concepts that are outside their scope of professional knowledge. In the case of Enzo, the counselor should not provide information to Enzo about the use of supplements unless the counselor has the professional training and credentialing to do so (e.g., naturopathic training and credentialing).

Step 3: Identify Neuroconcepts Relevant to the Client's Story and Developmental and Cultural Contexts

Integrating neuroscience into your work requires you to be actively engaged with an ever-evolving body of research. Examples of search terms you can try in your library database or Google Scholar are listed in the glossary. The next step in providing neuroeducation to your client is to carefully select the information you will collaboratively explore together. You may wonder how much neuroscience you need to know to integrate neuroeducation into your practice. Ultimately, you are responsible for ensuring your competence in any intervention you use. We recommend that counselors complete training in basic neuroanatomy and physiology, which may be attained through undergraduate biopsychology coursework, and stay current in basic and translational neuroscience research.

At this point, you might also be wondering what type of neuroscience information you could include when working with Enzo and what neuroscience information would be relevant to the clients you serve. These are great questions, and the answers are as vast as neuroscience itself and should be as developmentally and culturally responsive as possible to meet the needs of those we serve. Neuroeducation is less about having a predetermined script of information to deliver and more about using that information as a collaborative vehicle for exploration. We encourage you to tap into your research skills as you identify those neuroconcepts that might be most relevant and responsive to the needs of those clients you serve.

To begin, think about your passions. Given the expanse of neuroscience publications, there is likely a "neuroscience of" publication on virtually any topic. You can start with what you are interested in and do a quick search on "neuroscience of [insert your topic]." For instance, if you are interested in racial trauma, you could search for "neuroscience of racial trauma." A quick Google Scholar search would provide peer-reviewed articles and other resources that could become the basis for a neuroeducation exercise with a client.

You could also take any construct that you already know about (e.g., emotions, cognitions, empathy) and find out whether there is neuroscience information that expands your understanding. This approach could extend to a variety of therapeutic processes and interventions. For instance, you could read about the neuroscience of the change process as well as cognitive behavioral, psychodynamic, and/or Adlerian counseling. This material could not only serve as the foundation for neuroeducation interventions but also enhance your general skill delivery, even if you never mention neuroscience in a counseling session.

Before integrating any neuroscience information, it is essential to evaluate the quality of that information, just as with any other body of research (Kryza-Lacombe et al., 2021; Luke et al., 2020; Miller & Beeson, 2021). Give preference to sources with the most primary citations from basic neuroscience research rather than secondary citations or translations of translations. Because most counselors are not formally trained as neuroscientists, websites and other resources may be helpful for building a scope of competence. The use of infographics, imagery, and other creative means of exploring neuroscience is recommended to strengthen your practice.

Step 4: Reflect on Your Theory and Motivations and on Ethical Considerations

When considering neuroeducation, reflect on your motivations. Are you personally excited about the topic? Does the topic have some personal meaning to your own recovery journey? Did you just attend a related training? Does the topic align with your complete biopsychosocial understanding of the client's world? We have found ourselves a little overzealous at times when learning new concepts, and we encourage restraint to ensure that motives for using neuroeducation align with the goals and desires of the client being served. Using restraint can limit some of the ethical concerns that are inherent in the haphazard delivery of neuroeducation.

As we think about the use of neuroeducation in the counseling process, it is important to consider potential concerns and cautions that exist. Luke et al. (2020) identified five categories of ethical concerns about the integration of neuroscience among counselors as follows:

- It may not fit with counselor identity and the counseling profession's humanistic ethos.
- It may push counselors' scope of practice too far, given the lack of training standards and neuroscience background among counselors.
- There are limits in understanding of neuroscience research and the ability to apply it accurately and effectively to the counseling relationship.
- It might be misused and unintentionally cause clients to be harmed.
- It is unethical to not integrate neuroscience into counseling.

Despite the noted benefits of neuroeducation, such as reduced blame, treatment optimism, reduced stigma, and treatment engagement (see Lebowitz et al., 2021, for a review), some problematic forms of neuroeducation, such as biogenic or neurogenic explanations of mental health concerns, have been linked to suboptimal client outcomes. Luke et al. (2020) summarized the literature on biogenic explanations, and we extend their discussion here with a focus on the perceptions of laypersons and the public, clients or people who might seek help, and providers of health and mental health care.

Layperson/Public Perception

Neuroscience information is said to have a *seductive allure*, meaning that people tend to believe and assign superiority to information that is tied to neuroscience, even if that information is superfluous or false (e.g., Fernandez-Duque et al., 2015; Weisberg et al., 2008). Although biogenic explanations of mental disorders do in fact seem to reduce blaming these individuals for their health concerns, this reduction in blame is paired with a perception that these individuals are more dangerous (Larkings & Brown, 2018) or inherently flawed and beyond repair (Berent & Platt, 2021).

Client Self-Perception

Biogenic explanations seem also to influence clients' self-perceptions. Whereas early on there was the belief—a hope, really—that explaining a diagnosis to a client in neurobiological terms would free them to dissociate their disorder from themselves (e.g., Badenoch, 2008), the evidence to support this belief has remained anecdotal. Again, research has demonstrated a reduction in self-blame but also increased pessimism in prognosis (Lebowitz et al., 2021), inflation of symptoms (Lebowitz & Ahn, 2018), as well as decreased confidence (Lebowitz et al., 2021) and investment in psychosocial interventions (Lebowitz & Appelbaum, 2017). However, exploring the nondeterministic nature of genes, and how individuals can change their biology through treatment before neuroeducation, could mitigate the prognostic pessimism. Hence, we can see the need to contextualize what neuroscience does and does not tell us before introducing neuroscience information in our work.

Provider Perception

Health care providers are not immune to the stigma attached to biogenic explanations of mental health concerns. Despite a counterintuitive assumption that practitioners of the medical model might hold non-essentialist views of patients and clients, research results are mixed. While biogenic explanations were found to reduce stigma among physicians in one study (Knapp et al., 2014), other studies have found that biogenic explanations of psychological concerns led to reduced empathy (Lebowitz & Ahn, 2014) and increased stigma (Larkings & Brown, 2018). The glaring limitation in these studies is that professional counselors were not included in the samples, and we hypothesize that our unique training and values paired with a sound neuroeducation process can limit the risks and elevate the value.

Counselors have advocated that it is the delivery and discussion of biogenic factors that influences how such information is received (Beeson & Luke, 2023; Luke, 2020). Our presence and manner of delivery matter. Counselors must provide neuroeducation with an awareness of how each of the aspects of the biopsychosocial model have relevance to conceptualizing client problems. Counselors have the potential to cause harm if they provide neuroeducation in a manner that reduces human experience strictly to neurobiological explanations; such an approach gives no space for client subjectivity, intersubjectivity, psychological processes, and social/cultural heritage and experiences. Reducing client experience to biogenic explanations only is known as *neuroessentialism* (Schultz, 2018). Counselors should thus always check in about the client's subjective experiences and provide neuroeducation in an inquisitive and nondefinitive manner that gives space for clients to explore whether the information fits their subjective experience. Furthermore, when counselors are intentional in using neuroscience information to strengthen a client's nonjudgmental self-compassion and belief in their capacity to change (prognostic optimism), rather than provide biogenic explanations that foster self-blame and hopelessness of change (prognostic pessimism), clients are more likely to benefit.

Step 5: Consider the Influence of Social Positions and Power Differentials

Given our role as counselors, we hold inherent power positions within the therapeutic relationship that can be further exaggerated when we also hold positions of social privilege related to race, gender, sexuality, and any other identity status. If done haphazardly, providing neuroeducation could amplify power differentials whereby the counselor is viewed as the knowledge holder and expert.

You can use the Multicultural and Social Justice Counseling Competencies (Ratts et al., 2015) to reflect on your own identity statuses and their intersections, as well as those of the client, as you consider your susceptibility to this power imbalance during the neuroeducation process. To limit this vulnerability, use basic microskills to facilitate exploration, rather than explanation, of neuroscience information. Be honest about what you do or do not know while elevating what the client knows—that is, their personal lived experience and expertise. Elevate common factors like congruence, cultural responsiveness, collaboration, empathy, and positive regard. Ground the neuroscience information in the client's current understanding, theory, and worldview. When possible, use neuroscience information to create a shared language for exploring the client's past, present, and future experiences rather than to provide a definitive answer to their current experiences. Any exploration of information should be done with an emphasis on client agency, autonomy, and empowerment. In all situations, the counselor should consider how neuroscience information could be culturally sustaining to a client, meaning how it can preserve, support, and affirm the client's heritage, values, language, and cultural identity.

Neuroscience has the potential to affirm and support people with migrant and Black, Indigenous, or people of color (BIPOC) identities in their experiencing of persisting environmental stressors and support the importance of addressing systemic barriers and adversities. Counselors providing neuroeducation can attend to stories and narratives the client shares about adversity they face in their social environment as well as current and historical challenges faced by the communities to which they belong. This information can inform how to fit neuroscience education to the client's current situation. Taking an *emic* approach (i.e., individualizing presented information) rather than an *etic* approach (i.e., sharing the same information with all clients, regardless of context; Myers et al., 2022) to educating clients about concepts such as stress and trauma is much more likely to make neuroscience palatable to historically and currently disenfranchised and marginalized populations (Burt & Pankey, 2023). Taking such an approach can be restorative and healing to the client. Some clients may have encountered counselors who neither broached nor affirmed their social and cultural identities and heritages; who reduced their experiences to diagnostic criteria; or who, by dismissing adverse experiences such as racism, disaffirmed their identity, experiences, and perceptions. Counselors therefore need to effectively attend to the client's past experiences in counseling in addition to their current, historical, and ongoing experiences of adversity in their environment.

Step 6: Assess Client Knowledge of and Interest in the Neuroconcept

Another strategy to reduce the negative impact of power differentials is to connect with the client's existing knowledge and interest in exploring neuroscience information. This knowledge and interest can be influenced by their developmental stage and stage of change (Prochaska & DiClemente, 1982). Younger children might need simplified explanations supported by visuals or play, whereas adults with some background in neuroscience may be able to enter a higher order discussion, although adults can certainly benefit from visuals and play as well.

Using a stage-of-change approach to exploring neuroscience information can help connect to the client's current state and pace exploration in a way that is formative. For instance, Enzo already expressed contemplation about the value of neuroscience, so it will be important to start there, using neuroscience information to prompt further contemplation, preparation, and action/maintenance (e.g., "Sounds like this video had quite the impact, and concepts like memory, trauma, and supplements might be important to you. Let's start by exploring your current experience with each topic."). However, if the client has never contemplated the impact of neuroscience on their lived experience, then the introduction of that material will have to be more tentative and exploratory (e.g., "I know you have mentioned these angry outbursts happen before you are even aware of them and you feel hopeless. Some people say this might be related to complex memory systems throughout our brain and body. Would you like to explore more?" "What do you already know about . . . ?"). In either case, obtain consent before connecting the neuroscience information to their current stage of change and preexisting knowledge, using each as a foundation for exploration.

Step 7: Determine Your Method for Introduction and Exploration

The final step before delivering the neuroscience information is to use the previous points of reflection to determine how you will introduce and explore the information with those you serve. Whenever possible, include the client in deciding what method of exploration to take. For example, will you describe the information verbally? Or will you draw it out or use a video, infographic, or some other means? Knowing the method, you can make sure you have all the necessary materials to be responsive to your client's desire. When introducing and exploring the neuroscience information, consider the following principles that we have consolidated from several authors (e.g., Field et al., 2019; Field & Ghoston, 2020; Magill et al., 2021; Miller & Beeson, 2021):

- *Constrain selection.* Select one or two concepts to explore. In our experience, sharing more than two concepts can be overwhelming to clients and leave them disengaged. This means counselors must carefully select which information to include and which to exclude, which can be prioritized with the client.
- *Less is more.* Information should be delivered in concise fashion, avoiding repetition or being overly verbose. Spend no more than 1–2 minutes on your introduction. Regardless of the length of time, infuse check-ins throughout the process to promote collaborative exploration (e.g., "What is this bringing up for you?" "How does this compare to what we have already talked about?").
- *Distill without diluting.* We recommend that counselors provide information from primary sources, whenever possible, rather than relying on highly translated secondary sources. Counselors should be thoughtful in how to deliver this information in a digestible manner. A challenging but crucial skill to learn for delivering neuroeducation is how to convey the essential aspects of the informational content (i.e., distill) without making the response vague, unclear, or overgeneralized (i.e., dilute). One strategy is to use experiential methods, such as exercises that explain neurophysiological functioning to help the client experience the concept before exploring the information underneath it.
- *Ensure relevancy.* During the collaborative exploration of neuroscience information, some clients may find it helpful for the counselor to discuss connections to their current experiences. Connect the neuroscience information to what you already know about the client. Lead with a reflection of the client's experience to elevate relevancy (e.g., "I have heard you describe this cycle of diminished motivation and rewards over the duration of your substance use.") and then connect their lived experience to the neuroscience information (e.g., "I wonder if this connects to the sensitization of the dopaminergic systems we have talked about. Would you like to explore this more?").

In addition, we suggest that counselors do the following:

- Deliver small units of information in plain language at a moderate pace and using the client's language, symbols, metaphors, and so on, whenever possible.
- Begin with concepts that are straightforward and approachable before scaffolding to more difficult concepts.
- Attend to the client's nonverbal indicators of inattention or disinterest and check their degree of engagement.
- Explore the client's reactions to the information and the process.

Just as with basic counseling microskills, neuroeducation should be provided in a nonjudgmental tone and manner, with a calming vocal tone and pitch that indicates care and safety for the client. The relational context and delivery of the information matter just as much as, if not more than, the information itself.

Ideally, you will have intentionally practiced your delivery several times and in several different ways so that you can be responsive to each client in language, preferred mode of processing, and cultural values. This exploration could take many forms, including having traditional therapeutic conversations facilitated with sound microskills, watching a video, using a handout, creating a new artifact, or using any other method that facilitates exploration. Whenever possible, connect the method and form of exploration to the existing worldview and capacity of the client.

Step 8: Deliver Information Ethically and With Curiosity and Humility

While exploring neuroconcepts, continue to be curious and humble in what you know and what you do not know. Pay attention to limits on your scope of practice when providing neuroeducation. Using tentative language such as "Some research suggests there is a complex brain network related to dopamine that begins even before the substance is taken" is preferable to making definitive statements such as "You are doing this because your brain is hijacked by dopamine." Consider exploring the client's reaction and meaning making (e.g., "What does this say to you?" "How does this compare with how you have experienced the world?") rather than telling them what it means from your perspective. This is especially important when introducing conversations related to medication. Although general information can be shared and the client's experience with the medication can be processed, specific questions and guidance should be referred to the attending prescriber.

Especially when working with migrant and BIPOC populations, it is imperative for the counselor to not only appear knowledgeable but possess humility (Johnson et al., 2022). This entails being open to correction by the client, acknowledging when the counselor does not know the answer to a client's question, and speaking to the limitations and scientific biases of neuroscience research (Day-Vines et al., 2022). Counselors can consider that neuroscience research can sometimes reinforce a deficit approach by lacking any discussion of cultural strengths and perpetuating erroneous perceived weaknesses of an already historically disenfranchised and marginalized population (Moore et al., 2021). When providing neuroeducation, counselors can help clients understand how both environmental and neurobiological factors impact cognitions, emotions, behavior, and physiological responses. If done properly, neuroeducation can aid the client in understanding how their social environment is impacting them and provide strategies for addressing structural and systemic forms of oppression in addition to providing intrapersonal strategies to reduce the stress and trauma caused by adversities in their environment.

Step 9: Use Information as a Vehicle for Exploration

As you have likely gleaned from reading this chapter, neuroscience information is secondary to relationship factors. Instead of offering neuroscience information as a rigid, deterministic explanation or as "the answer," use the information as a vehicle for exploring a client's past, present, and future, and compare this information to their previous theories. Asking clients to situate, or compare, the neuroconcept within their past and present life as well as their future goals can help you to connect, contextualize, and collaborate on what comes next. For instance, asking open questions (e.g., "How have you seen that concept in your life?" "How does this concept compare to other ways you have viewed your experiences?" "How might that concept be playing out in your current situation?" "How might this concept influence your next steps?") could help drive further exploration, rather than explanation, and meaning making.

Step 10: Coconstruct a Plan for What to Do Next

One goal of neuroeducation is to help the client gain insight into who they have been in the past, who they are in the present, and who they are becoming in the future. However, insight alone does not predict outcomes. As you close the neuroeducation process, be mindful of what happens next with your client and coconstruct a path forward. You could explore what the client's initial takeaways were from the information and how the information may be helpful to understanding their experiences. You could elevate change talk and help them craft a coherent narrative about their experiences that drives change (even if that is acceptance) in their life. You could check in about any concepts that were unclear and welcome any follow-up questions the client has in response. In doing so, you are making the provision of neuroeducation a relational experience that reestablishes the balance of power within the relationship.

Our Brain-Based Approach to the Case of Enzo

Given that Enzo's mental status appears to be different than usual, you should first connect to his current state, assist in coregulation, and use reflections and immediacy to ground him via the working therapeutic relationship (Step 1). Because you have already been working with Enzo for 3 months, you might be able to ground him in previous exercises or notice changes (e.g., "I notice you seem a little more animated today than usual; it seems like this TikTok video really had an impact."). Using your observation skills and naming the experience within the context of your existing working relationship should promote some degree of coregulation.

Using your basic counseling microskills, you can attend to Enzo's expressed interest in how trauma is stored in the body and the use of supplements to address traumatic memories (Step 2). Consider Enzo's theory for the genesis of his trauma experiences and his motivations for sharing this information with you. You can also explore his use of TikTok and technology in general. Although his statement was focused on TikTok, there may be important context around his use of technology in general that could also fit into a neuroscience-informed case conceptualization. You can use encouragers (e.g., "I'd like to hear more about . . .") and open questions (e.g., "How did it feel when you learned this?") to explore his account of the current experiences brought to light by this TikTok video as well as his theory for who he has been, who he is, and who he is becoming.

At this point, you can identify potential neuroconcepts that might serve as vehicles of exploration (Step 3). In this case, the options include neuroscience of trauma, memory formation and reconsolidation, impact of technology use, embodied cognition, and impact of supplements. Invite Enzo into the decision-making process and prioritize the potential topics. In the process of choosing, reflect on your own reason for exploring each concept and consider ethical principles of neuroeducation application (Step 4). In facilitating neuroeducation, how might you foster rather than diminish (i.e., prevent deterministic interpretation) Enzo's hope for change? Be mindful of your position as the counselor, your intersecting identities, and the intersecting identities of the client and work to minimize power differentials (Step 5). Using tentative language (e.g., "Some researchers have found . . .") can be helpful here.

Let's assume that your engagement and coregulation have calmed Enzo. He is less agitated and more present. He notes that he was really intrigued by the idea of trauma being stored in the body and really wants to know more about that concept. He says that although he does not really know much about it, his instincts are telling him that taking supplements is probably not the answer (Step 6). You reflect on your own theory, knowledge, and position of power and determine that you have useful and accurate neuroscience information that you can explore with Enzo. You decide that a brief discussion about the impact of trauma on biological systems is a good place to start (Step 7). You use an original "Embodied Impact of Trauma" infographic, which you created based on your close review of primary science literature, to guide the discussion (Step 8). After briefly introducing the information, you check in with Enzo to ask what sense he makes of the infographic (Step 9). He seems intrigued and shares an example of often feeling "on edge" despite not being able to identify anything dangerous in his environment. Together, you decide that this hypervigilance is something you want to explore in more depth to get a better sense of when it is happening and how Enzo might be able to prevent or effectively respond to such situations (Step 10).

Conclusion

Neuroeducation is a potentially useful therapeutic process that is most effective when delivered with careful consideration of various client and counselor factors. The 10-step process outlined in this chapter offers detailed guidance on navigating these interrelated factors in an ethical and humanistic manner. We hope that this chapter has inspired your curiosity about engaging in neuroeducation and offered helpful ideas for how to do so with intentionality and connectedness.

Quiz

1. Which of the following statements best describes neuroeducation?
   a. Telling a client what is wrong in their brain.
   b. A process of exploring neuroscience information with a client.
   c. Teaching a client about their entire neuroanatomy and nervous system functioning.
   d. Giving a client a brief lecture on the neuroscience of [insert presenting concern].
   e. Referring a client to a neuropsychologist.
2. When exploring neuroscience information with clients, it is important to:
   a. Select only one or two concepts and deliver them as concisely as possible.
   b. Share as many concepts as possible because the more you can share, the better.
   c. Rely mostly on translational theorists rather than primary science research.
   d. Focus more on the content than the process.
   e. Stress the superiority of neuroscience over other biopsychosocial factors.

3. Current research on providing biogenic or neurobiological explanations suggests:
    a. There are no risks, only benefits, to providing biogenic and neurological explanations to clients.
    b. Clients and health care providers both consider biogenic and neurological explanations less stigmatizing.
    c. Biogenic and neurobiological explanations can be delivered successfully without attention to the therapeutic relationship.
    d. Biogenic and neurobiological explanations are quick ways to help clients feel more hopeful about their prognosis.
    e. There are potential benefits to sharing biogenic and neurological explanations with clients but also certain risks related to stigma, prognostic pessimism, and neuroessentialism.

References

Badenoch, B. (2008). *Being a brain-wise therapist: A practical guide to interpersonal neurobiology*. Norton.

Beeson, E. T., & Luke, C. (2023). Neuroscience-informed counseling. Another lens to look at the human experience. In R. Fulmer (Ed.), *Counseling and psychotherapy: Theory and beyond* (pp. 254–310). Cognella.

Berent, I., & Platt, M. (2021). Essentialist biases toward psychiatric disorders: Brain disorders are presumed innate. *Cognitive Science, 45*(4), Article e12970. https://doi.org/10.1111/cogs.12970

Burt, I., & Pankey, B. (2023). Critically analyzing the field of neuroscience and its therapeutic application with Black populations. *Journal of Multicultural Counseling and Development, 51*(3), 174–182. https://doi.org/10.1002/jmcd.12274

Council for Accreditation of Counseling and Related Educational Programs. (2024). *2024 CACREP standards*. https://www.cacrep.org/wp-content/uploads/2023/06/2024-Standards-Combined-Version-6.27.23.pdf

Day-Vines, N. L., Bryan, J., Brodar, J. R., & Griffin, D. (2022). Grappling with race: A national study of the broaching attitudes and behavior of school counselors, clinical mental health counselors, and counselor trainees. *Journal of Multicultural Counseling and Development, 50*(1), 25–34. https://doi.org/10.1002/jmcd.12231

De Raedt, R. (2020). Contributions from neuroscience to the practice of cognitive behaviour therapy: Translational psychological science in service of good practice. *Behaviour Research and Therapy, 125*, Article 103545. https://doi.org/10.1016/j.brat.2019.103545

Ekhtiari, H., Rezapour, T., Aupperle, R. L., & Paulus, M. P. (2017). Neuroscience-informed psychoeducation for addiction medicine: A neurocognitive perspective. In T. Calvey & W. M. U. Daniels (Eds.), *Progress in brain research: Vol. 235. Brain research in addiction* (pp. 239–264). Elsevier. https://doi.org/10.1016/bs.pbr.2017.08.013

Fernandez-Duque, D., Evans, J., Christian, C., & Hodges, S. D. (2015). Superfluous neuroscience information makes explanations of psychological phenomena more appealing. *Journal of Cognitive Neuroscience, 27*(5), 926–944. https://doi.org/10.1162/jocn_a_00750

Field, T. A., Beeson, E. T., Luke, C., Ghoston, M., & Golubovic, N. (2019). Counselors' neuroscience conceptualizations of depression. *Journal of Mental Health Counseling, 41*(3), 260–279. https://doi.org/10.17744/mehc.41.3.05

Field, T. A., & Ghoston, M. R. (2020). *Neuroscience-informed counseling with children and adolescents*. American Counseling Association.

Fishbane, M. D. (2013). *Loving with the brain in mind: Neurobiology and couple therapy*. Norton.

Hays, P. A. (1996). Addressing the complexities of culture and gender in counseling. *Journal of Counseling & Development, 74*(4), 332–338. https://doi.org/10.1002/j.1556-6676.1996.tb01876.x

Hays, P. A. (2008). *Addressing cultural complexities in practice: Assessment, diagnosis, and therapy* (2nd ed.). American Psychological Association.

Johnson, K. F., Ieva, K., & Byrd, J. (2022). Introduction to the special issue on school counselors addressing education, health, wellness, and trauma disparities. *Professional School Counseling, 26*(1b). https://doi.org/10.1177/2156759X221105451

Knapp, S., Marziliano, A., & Moyer, A. (2014). Identity threat and stigma in cancer patients. *Health Psychology Open, 1*(1). https://doi.org/10.1177/2055102914552281

Kryza-Lacombe, M., Richards, E., Hansen, N., & Goldin, P. (2021). Integrating neuroeducation into psychotherapy practice: Why and how to talk to patients about the brain. *The Behavior Therapist. 44*(7), 361–370.

Larkings, J. S., & Brown, P. M. (2018). Do biogenetic causal beliefs reduce mental illness stigma in people with mental illness and in mental health professionals? A systematic review. *International Journal of Mental Health Nursing, 27*(3), 928–941. https://doi.org/10.1111/inm.12390

Lebowitz, M. S., & Ahn, W. K. (2014). Effects of biological explanations for mental disorders on clinicians' empathy. *Proceedings of the National Academy of Sciences, USA, 111*(50), 17786–17790. https://doi.org/10.1073/pnas.1414058111

Lebowitz, M. S., & Ahn, W. K. (2018). Blue genes? Understanding and mitigating negative consequences of personalized information about genetic risk for depression. *Journal of Genetic Counseling, 27*(1), 204–216. https://doi.org/10.1007/s10897-017-0140-5

Lebowitz, M. S., & Appelbaum, P. S. (2017). Beneficial and detrimental effects of genetic explanations for addiction. *International Journal of Social Psychiatry, 63*(8), 717–723. https://doi.org/10.1177%2F0020764017737573

Lebowitz, M. S., Dolev-Amit, T., & Zilcha-Mano, S. (2021). Relationships of biomedical beliefs about depression to treatment-related expectancies in a treatment-seeking sample. *Psychotherapy, 58*(3), 366–371. https://doi.org/10.1037/pst0000320

Louw, A., Zimney, K., Puentedura, E. J., & Diener, I. (2016). The efficacy of pain neuroscience education on musculoskeletal pain: A systematic review of the literature. *Physiotherapy Theory and Practice, 32*(5), 332–355. https://doi.org/10.1080/09593985.2016.1194646

Luke, C. (2020). *Neuroscience for counselors and therapists: Integrating the sciences of mind and brain* (2nd ed.). Cognella.

Luke, C., Beeson, E. T., Miller, R., Field, T. A., & Jones, L. K. (2020). Counselors' perceptions of ethical considerations for integrating neuroscience with counseling. *The Professional Counselor, 10*(2), 204–219. https://doi.org/10.15241/cl.10.2.204

Magill, M., Martino, S., & Wampold, B. (2021). The principles and practices of psychoeducation with alcohol or other drug use disorders: A review and brief guide. *Journal of Substance Abuse Treatment, 126*, Article 108442. https://doi.org/10.1016/j.jsat.2021.108442

Miller, R. M. (2016). Neuroeducation: Integrating brain-based psychoeducation into clinical practice. *Journal of Mental Health Counseling, 38*(2), 103–115. https://doi.org/10.17744/mehc.38.2.02

Miller, R., & Beeson, E. T. (2021). *The neuroeducation toolbox: Practical translations of neuroscience in counseling and psychotherapy*. Cognella.

Moore, J. L., III, Hines, E. M., & Harris, P. C. (2021). Introduction to the special issue: Males of color and school counseling. *Professional School Counseling, 25*(1_part_4). https://doi.org/10.1177/2156759X211040045

Myers, L. J., Lodge, T., Speight, S. L., & Haggins, K. (2022). The necessity of an emic paradigm in psychology. *Journal of Humanistic Psychology, 62*(4), 488–515. https://doi.org/10.1177/00221678211048568

Prochaska, J. O., & DiClemente, C. C. (1982). Transtheoretical therapy: Toward a more integrative model of change. *Psychotherapy, 19*(3), 276–278. https://doi.org/10.1037/h0088437

Ratts, M. J., Singh, A. A., Nassar-McMillan, S., Butler, S. K., & McCullough, J. R. (2015). *Multicultural and social justice counseling competencies*. Association for Multicultural Counseling and Development. https://www.counseling.org/docs/default-source/competencies/multicultural-and-social-justice-counseling-competencies.pdf?sfvrsn=20

Schultz, W. (2018). Neuroessentialism: Theoretical and clinical considerations. *Journal of Humanistic Psychology, 58*(6), 607–639. https://doi.org/10.1177/0022167815617296

Wampold, B. E. (2015). How important are the common factors in psychotherapy? An update. *World Psychiatry, 14*(3), 270–277. https://doi.org/10.1002/wps.20238

Weisberg, D. S., Keil, F. C., Goodstein, J., Rawson, E., & Gray, J. R. (2008). The seductive allure of neuroscience explanations. *Journal of Cognitive Neuroscience, 20*(3), 470–477. https://doi.org/10.1162/jocn.2008.20040

Chapter 11

Neuroscience-Informed Counseling Theory

Carlos P. Zalaquett, SeriaShia Chatters, Ravza N. Aksoy, and Allen E. Ivey

This chapter provides an overview of the intersection between counseling theories and neuroscientific research. The interaction between neuroscience and counseling offers strong evidence of the biological substrates of behavior and the effects of different therapies on the brain (Ivey et al., 2023, 2024; Ivey & Zalaquett, 2011; Zalaquett et al., 2019). Neuroscience provides unique approaches to case conceptualization, assessment strategies, diagnosis, and therapy planning (Beijan et al., 2022; Luke et al., 2019).

In this chapter, we review a few select theories to demonstrate how they benefit from the integration of neuroscience. The case of Natasha is presented to explore an integrative approach to neuroscience-informed counseling theory and neurocounseling intervention.

2024 CACREP Standards

This chapter addresses 2024 Council for Accreditation of Counseling and Related Educational Programs (CACREP) Standards pertinent to the Foundational Counseling Curriculum (Section 3) area of Counseling Practice and Relationships (Standard E):

- Theories and models of counseling, including relevance to clients from diverse cultural backgrounds (Standard E.1.)
- Evidence-based counseling strategies and techniques for prevention and intervention (Standard E.15.)

This chapter also addresses the following Doctoral Standards for Counselor Education and Supervision (Section 6) in the area of Doctoral Curriculum (Standard B):

- Integration of theories relevant to counseling (Standard B.1.b.)

Clinical Case Study: Natasha

Natasha is a 24-year-old Puerto Rican American woman who attended a counseling center at a midsize university in the southwestern United States because of stress and academic concerns. She previously served for 5 years in the U.S. Army. During that time, she was deployed on three occasions and saw action in both Gulf Wars, including Operation Iraqi Freedom and Operation Enduring Freedom. Natasha was recently divorced after a 3-year marriage that was emotionally and physically abusive. She was afraid of being physically attacked by her ex-husband, who had made a number of threats during the divorce. She also reported bouts of excruciating abdominal pain, which she attributed to the stressful situation with her ex-husband. Natasha had originally been attending a college on the East Coast. One day, she abruptly stopped attending all of her classes at this college and moved to the opposite side of the continental United States to avoid a potential encounter with her ex-husband and the reminders of the stress and fear triggered by the abusive relationship. Because she had moved and neglected to drop her classes at the East Coast college, Natasha failed all of her courses for the spring semester. During the initial interview, Natasha reports, "I am a failure," and says she feels hopeless.

Reflective Question

What would you focus on first with Natasha—her cognitive distortions related to her academic failure, her current physical symptoms, her lack of social supports, or her narrative of the divorce?

Neuroscience-Informed Theoretical Orientation

Neuroscience research is revealing the biological underpinnings of various therapeutic orientations, offering ways to enrich the interventions with neuroscience-informed techniques and provide more personalized and effective treatments. For example, the finding of specific and encompassing neural paths underlying different disorders (Taylor et al., 2021) leads to a question: Which theoretical orientation or intervention is most effective at managing which problematic neural pathway? Given the brain's tendency toward negative affective bias in clients with mood disorders (Zhang et al., 2022), which theoretical orientation or intervention would best reduce their bias, increase their positive experiences, and encourage thickening of their preferred neural pathways? Although these questions have yet to be fully answered, research findings have had a significant impact on the understanding of the neurobiological underpinnings of human functioning and how mental health practitioners might be able to influence the brain's functions. In fact, practitioners are beginning to understand the neurological processes behind new, positive therapeutic learning experiences that can lessen the impact of a client's prior negative learning (De Raedt, 2020). New counselors are demonstrating a clear interest in neuroscience findings (Beeson et al., 2019; S. Kim & Zalaquett, 2019; Russo et al., 2021).

In this section, we briefly review how neuroscience informs three theoretical orientations: a cognitive approach, cognitive behavior therapy (CBT); a psychodynamic approach, interpersonal psychotherapy (IPT); and a postmodern approach, narrative therapy.

CBT

CBT is a broad category encompassing several therapeutic approaches, including rational emotive behavior therapy, cognitive therapy, cognitive behavior modification, and dialectical behavior therapy, among others. In general, CBT is a therapeutic approach that emphasizes the role of thinking in inducing and perpetuating client distress. CBT treatment strategies involve counselors and therapists assisting clients to intentionally direct their thought patterns from dysfunctional to functional processes (David & Szentagotai, 2006; Paquette et al., 2003; Ressler & Mayberg, 2007). The use of CBT has considerable empirical support as an intervention for various disorders.

Techniques used to examine the neurological correlates of CBT include MRI and positron emission tomography (PET), which can be used to map brain structure and functions and demonstrate the connectivity and interactions between various brain regions (Månsson et al., 2021). Neurocognitive functioning is necessary for successful performance of several CBT techniques, including Socratic dialogue, guided discovery, self-monitoring, self-reflection, emotion regulation, and behavioral homework. Poor executive functioning and memory processes might limit the effect of these cognitive techniques (De Raedt, 2020).

Next, we will briefly review the neuroscience associated with CBT in the treatment of phobias, trauma, and depression.

CBT and Phobias

Over the past decade, researchers have studied the impact of psychotherapy on brain activation primarily in the context of phobias. *Phobias,* or extreme or irrational fears of or aversions to something or a specific situation, cause the client to elicit a specific or a conditioned fear response to a stimulus. Empirical studies have indicated that exposure therapy, a behavioral intervention rooted in systematic desensitization (Wolpe, 1958), is successful in relieving phobia symptoms.

Paquette et al. (2003) studied brain activation and the use of CBT in the treatment of spider phobia. During the pretest, study participants experienced a fear response to photographic stimuli that was correlated with activation of the right dorsolateral prefrontal cortex (PFC) and parahippocampal gyrus. During postintervention, a significant reduction in activation was noted in those same areas. The authors maintained that initial parahippocampal activation had led to the development of avoidance toward spiders, reinforcing the phobia. Paquette et al. hypothesized that reduced activation of the right dorsolateral PFC might have been related to study participants' use of metacognitive strategies aimed at self-regulating the fear triggered by the spider film excerpts.

In addition to such parahippocampal activity, researchers have found activation in the right and left amygdala, periamygdaloid, and medial PFC in the process of extinction (e.g., the reduction in the fear of spiders after repeated exposure to a stimulus). Álvarez-Pérez et al. (2021) conducted a study evaluating exposure therapies for specific phobias to small animals. Participants who received treatment reported decreases in anxiety responses with associated decreases in PFC and increases in precuneus brain activity associated with less anxiety. However, participants also had no change to amygdala brain activity, suggesting that their fear response persisted even if their anxiety was reduced. The authors proposed that the PFC has a bidirectional relationship with limbic regions such as the amygdala. This frontolimbic emotional processing pathway allows for humans to voluntarily appraise emotional experiences such as fear and plan responsive actions. A meta-analysis examining differential activations between human subjects and people with a specific phobia in various regions of the limbic circuit also showed a greater convergence of activations in the right amygdala, insula, and cingulate cortex of phobic patients compared with the controls (Rosenbaum et al., 2020).

All of these outcomes support the clinical efficacy of in-person and virtual CBT and their changes in determined areas of the brain.

CBT and Trauma

As described in Chapter 4, a traumatic event can have a significant impact on the brain that may permanently stimulate atypical structural plastic changes within the amygdala, PFC, and other structures. Once symptoms of posttraumatic stress disorder (PTSD) are present, neurons in the amygdala may have formed neural networks associated with abnormal flashbacks, easily triggered by stimuli (Sapolsky, 2003). Although CBT may not be able to deconstruct these permanent changes, counseling may stimulate the formation of alternative or competing neural pathways that may offset maladaptive circuitry associated with PTSD symptoms (Maguschak & Ressler, 2008; Quirk & Mueller, 2007). CBT may induce neuroplastic changes in structural interconnectedness in brain regions and neurons by promoting reduced expression of learned fear and moderating the underlying neuronal pathways (De Raedt, 2006; Martin & Kandel, 1996; Navalta et al., 2018; Ressler & Mayberg, 2007). A study of male and female Vietnam veterans found decreased activation in the medial PFC and increased activation in the left and right amygdala and the periamygdaloid after veterans therapeutically processed traumatic experiences (Shin et al., 2004). This landmark study was followed by several studies that subsequently confirmed that CBT produced changes in brain activation during the treatment of anxiety disorders (see Beauregard, 2007; Frewen et al., 2010; Makinson & Young, 2012; Straube et al., 2006). Furthermore, CBT has been associated with amelioration of trauma-related symptoms through shifts in activation of the left dorsal striatal and frontal networks that alter the inhibitory control network (Brooks & Stein, 2015; Navalta et al., 2018).

CBT and Depression

The psychological and somatic symptoms of depression have been correlated with several types of biological dysfunction. Depression increases the volume of the amygdala (Roozendaal et al., 2009), decreases the volume of the hippocampus (MacQueen et al., 2008) and the PFC (Joëls & Baram, 2009), and changes the connectivity between the amygdala and the PFC (de Almeida et al., 2009) and the amygdala and the hippocampus (Fu et al., 2008). Symptoms of mood dysregulation may be associated with decreased metabolism of the PFC (Galynker et al., 1998). Symptoms of apathy may be associated with elevated levels of cortisol in the bloodstream that affect reward circuits in the limbic system (Arnsten, 2009; Lieberman, 2006). Symptoms of excessive guilt and hopelessness are correlated with dysfunction of the hypothalamic-pituitary-adrenal axis, which is related to arousal of the sympathetic branch of the autonomic nervous system (Du et al., 2009; Fiocco et al., 2006; Gillespie & Nemeroff, 2005; Segerstrom & Miller, 2004).

In addition to the aforementioned impact on the brain, Fu et al. (2008) found that amygdala-hippocampal activity was associated with depression and that dorsal anterior cingulate activity was a predictor of a positive treatment response to CBT. This is an important finding because of its potential translational implications. Dorsal anterior cingulate activity may be used to identify those individuals who are most and least likely to benefit from CBT, leading to more effective selection of treatment and utilization of resources. Jones et al. (2022) studied a sample of 83 adult participants with major depressive disorder and a history of childhood trauma. Using PET and MRI scans, they confirmed the presence of a reduced hippocampus and amygdala. This study provides additional identification of biological targets for counseling interventions aimed at preventing major depressive disorders following early trauma. For example, given that CBT produces increases in hippocampus volume in clients with PTSD (Hanson et al., 2015; E. J. Kim et al., 2015), it may be useful for normalizing the structural and functional changes triggered by childhood trauma and treating ensuing mental disorders such as trauma and depression.

Biofeedback can be added as a component of cognitive behavior treatment and has been used in counseling for many years. *Biofeedback* provides clients with feedback on certain aspects of their biology and body, such as information about breathing patterns, skin temperature, galvanic skin responses, and muscle tension. Each specific biofeedback technique is selected on the basis of the client's needs and concerns.

Studies investigating the efficacy of using biofeedback, or adding it to traditional therapeutic interventions, to treat anxiety (Karavidas et al., 2007; Siepmann et al., 2008) and depression (Karavidas et al., 2007) have produced promising results. Forms of biofeedback that appear to be effective in the treatment of panic disorder and PTSD include respiratory sinus arrhythmia, electromyographic feedback, heart rate variability training (Gevirtz & Dalenberg, 2008; Hahusseau et al., 2022; Schuman et al., 2023), LORETA Z-score neurofeedback (Bell et al., 2019), and infra-low frequency neurofeedback training (Kirk & Dahl, 2022). Neurofeedback is a form of biofeedback that has indications for use with children, adolescents, and adults experiencing a variety of issues, including anxiety, brain injury, depression, trauma, substance use, and attention-deficit/hyperactivity disorder (ADHD; Simkin et al., 2014).

Learning Theory in Neurofeedback

Neurofeedback is a treatment method designed to alter brain functioning through the use of signals provided to the client via various illustrations (a thermometer, a flying plane, dimming or brightening the screen) that reflect changes in the client's real-time electroencephalogram (EEG). Neurofeedback is grounded in the behavioral learning theory of operant conditioning

(Collura, 2017; Russell-Chapin et al., 2020). During neurofeedback, clients receive positive reinforcement for reaching a state of optimal EEG activity in targeted brain regions. When a person displays desired brain activity in targeted regions, a video game, music, or a movie may play to reinforce optimal activity in that brain state. When the client is outside the limits of optimal arousal, the reinforcer is withdrawn, and the video game, music, or movie stops playing. This real-time positive feedback assists clients to train their brains to recognize optimal arousal and alter their approach in response. Neurofeedback has particular indications for ADHD and may be equally as effective as stimulant medication (Fuchs et al., 2003). In a functional MRI (fMRI) study, neurofeedback normalized functioning in the anterior cingulate cortex, which is associated with selective attention (Lévesque et al., 2006).

By using neurofeedback, clients get real-time feedback on their brain activity to achieve self-regulation via operant conditioning to support CBT (De Raedt, 2020). The most promising technological tools to support CBT techniques are fMRI neurofeedback, EEG neurofeedback, and wearable devices such as EEG headbands and heart rate smartwatches that are combined with applications (De Raedt, 2020).

Neurofeedback protocols have been created and used for reducing hyperarousal by training areas of resting rhythm in the sensorimotor region and alpha activity in the parietal and temporal regions, in conjunction with trauma-informed therapy (Askovic et al., 2017). In Australia, neurofeedback has been added to psychotherapy in the treatment of clients with refugee experiences, including torture, trauma, grief, resettlement, and separation problems (Holt & McLean, 2021). A recent study demonstrated that even brief exposure to feedback learning is more efficient than repetitive practice without evaluation, as evidenced by an improvement in task-dependent functional connectivity and neural pattern consistency during a mental motor execution task (Lee et al., 2019). Consistent with previous research results, this study also showed that neurofeedback training is effective for inducing changes in functional connectivity from the primary motor cortex (Lee et al., 2019; Yamashita et al., 2017). Furthermore, by utilizing top-down information on intention, neurofeedback may contribute to the development of knowledge for controlling functional circuitry at the neural level (Lee et al., 2019).

Reflective Question

How might you use CBT to address Natasha's cognitive distortions?

Psychodynamic Therapy

Psychodynamic therapy has been found to be an effective treatment for numerous mental disorders, most notably depression (Leichsenring et al., 2004). Counseling in general seems to be helpful for depression by altering the density of serotonin (5-HT) receptors (Bhagwagar et al., 2004; Drevets et al., 1999; Sargent et al., 2000). Karlsson et al. (2010) conducted a study comparing the effects of psychotherapy and fluoxetine (Prozac) on the density of serotonin 5-HT1A receptors. Using binding potential values, a crucial measure in PET studies to establish the density of available receptors (Laruelle et al., 2002), Karlsson et al. estimated the ratio of specific and nondisplaceable binding in the white matter of participants' brains. Pre- and posttreatment symptom improvements were similar in both groups, but only those participants receiving psychotherapy showed an increase in serotonin 5-HT1A binding in the dorsolateral PFC, ventrolateral PFC, ventral anterior cingulate cortex, inferior temporal gyrus, insular cortex, and angular gyrus.

Psychodynamic approaches encourage the rediscovery of repressed emotions, often known as *corrective emotional experience*. This might encourage memory reconsolidation, a different neurobiological process with the capacity to undo the initial unhelpful emotional learning. Studies on the neurobiology of acceptance, a tactic used in experiential dynamic therapies, support subcortical modulation (bottom-up process; Grecucci et al., 2020).

Interpersonal Therapy and Depression

A specific form of psychodynamic therapy known as *interpersonal psychotherapy* (IPT) was developed specifically to treat depression (Weissman et al., 2000). IPT is a short-term, time-limited method based on psychiatrist and psychoanalyst Harry Stack Sullivan's interpersonal theory (Weissman & Markowitz, 1994). In IPT, clients appraise their current relationships and seek to improve interpersonal connectedness and social supports while coping with stress. Evidence for IPT's alteration of neurological functioning in the treatment of depression is emerging. In a study comparing the effects of IPT and paroxetine (Paxil; an antidepressant), a PET scanner was used to analyze clients' metabolic glucose levels (Brody et al., 2001). Posttreatment, glucose metabolism was reduced in the bilateral PFC for clients treated with the antidepressant and in the right PFC for clients treated with IPT. PET scans of clients in both groups showed increased metabolism in the left temporal cortex. These results demonstrate that the depressed brain may exhibit increased glucose metabolism in the PFC and decreased metabolic activity in the left temporal cortex. Such findings may indicate that treatment for depression may require some normalization of metabolic activity in these areas. Cera et al. (2022) summarized existing findings that psychodynamic therapy effects changes in the right superior and inferior frontal gyri, with a small cluster in the putamen, whereas mindfulness CBT affects the dorsal anterior insulae and the medial superior frontal gyrus.

Narrative Therapy

In the 1990s *narrative therapy*, a postmodern approach developed by Michael White and David Epston (1990), gained popularity (Beaudoin & Zimmerman, 2011; White, 2007). The primary impetus of narrative therapy is to assist clients in making meaning out of their experiences. In narrative therapy, clients explore opportunities to access preferred experiences or reauthor their narrative, and they learn how to separate their identities from their experiences or issues, better known as *deconstruction* or externalization (Beaudoin & Zimmerman, 2011; White, 2007).

> **Reflective Question**
>
> How might you enhance Natasha's social supports using an IPT framework?

A client's narrative is a story the client tells the therapist about their issues. The therapist, working from a narrative approach, will apply various narrative techniques in the session, such as collaboratively reauthoring the client's narrative through a process called *coconstruction*. From a neurobiological perspective, the client's current perspective of their narrative could have neural networks associated with these problem-related experiences. These neural networks are significantly more developed than those associated with preferred experiences because of the brain's tendency toward conservatism (Beaudoin & Zimmerman, 2011). Problem-saturated stories are the result of the PFC attempting to create meaning from repetitive generation of negative affect. As mentioned earlier, the infrequency of positive experiences results in neural networks that are mainly associated with emotionally laden problematic experiences. As these networks become more sophisticated, they have a higher likelihood of being activated.

Reauthoring is a process by which clients reexamine problematic situations they handled in undesirable ways to rewrite them (White & Epston, 1990). The narrative achieved during the process of reauthoring more closely represents clients' preferred ways of being and is more congruent with their values (White, 2007). In narrative therapy, these new narratives are called *problem-free narratives, problem-minimized events,* or *unique outcomes.* Through the process of reauthoring the narrative, more adaptive neural pathways are strengthened.

During reauthoring, counselors and therapists encourage conversations that may shift the client's affective experiences and preconceived narratives by revisiting the experience. Each revisited conversation can encourage alterations to memories that may be infused with new meaning and emotions or moods (Sousa, 2011). These memories can be altered in a negative or positive way, so therapists must be careful during the reauthoring process. LeDoux (2003) indicated that once an experience is retrieved in therapy, it may return to storage in an altered form. Major environmental events affect epigenetic processes that influence plasticity in gene regulation. Effective psychotherapy produces epigenetic changes such as DNA methylation, a potential biomarker of therapy success. Wilker et al. (2023) found that narrative exposure therapy produced a significant increase in cg25535999 methylation, supporting the central role of glucocorticoid signaling in trauma-focused therapy.

Reflective Question

What parts of Natasha's narrative could you help her to reauthor?

Future Directions in Neuroscience-Informed Theory

Counselors are using knowledge of neuroscience to inform conventional theory. For example, Ivey and colleagues (2023, 2024; Ivey & Zalaquett, 2011) advanced the application of neuroscience in counseling and social justice intervention using concepts such as neuroplasticity, attention and focus, wellness, and stress management. These researchers have used their neuroscience-based model to guide the application of counseling microskills to increase empathy, clarify emotions, and advance individual wellness and strengths in counseling and social justice interventions (Ivey et al., 2023, 2024; Ivey & Zalaquett, 2011; Zalaquett et al., 2019). Another example from the counseling field is the development of neuroscience-informed CBT (nCBT; Field et al., 2015, 2017, 2019; Miller et al., 2020). This approach modifies Ellis's (1962) ABC model, in which a person's beliefs about past events primarily determine the emotional and behavioral consequences. The goal of nCBT is to use conventional CBT techniques, such as cognitive restructuring or behavioral activation (identified as Wave2 in nCBT), while training the client to become aware of their physiological responses through mindfulness, neurofeedback, biofeedback, and healthy coping techniques (identified as Wave1 in nCBT; Field et al., 2015). Given the progressive intersection between neuroscience and counseling, more neuroscience-informed theories are expected to emerge from within the counseling profession.

Our Brain-Based Approach to the Case of Natasha

Natasha came to the southwestern college's counseling center because of concern about her academic performance and growing stress. During Natasha's initial visit, she completed an intake and several assessments to determine an initial diagnosis. Clients who met the criteria for anxiety- or stress-related disorders were offered psychotherapy, biofeedback, or both. During this first session, she was provided with assistance regarding her academic situation. This brought great relief to her because she was able to continue her studies at her new university. Being far away from her abusive ex-husband greatly reduced the anxiety she had felt for the past 3 years.

By the end of the first session, she agreed to use a combination of CBT, IPT, and narrative therapy along with biofeedback training. Natasha received weekly counseling appointments and weekly biofeedback appointments to reduce stress and physical tension. Some of Natasha's session goals were to increase her relaxation response using biofeedback-guided training. Natasha's skin response (through electromyography) and finger temperature were measured. Natasha worked to reduce all of the indicated feedback values. Feedback regarding her progress was communicated to her through the use of a light (with red indicating a high level of response and green indicating a low level), a numerical display (with high numbers indicating a high level of response), and sound (with a loud sound indicating a high level of response). Using all three forms of feedback allowed for individualized preferences because changes in each measure of feedback were correlated. Natasha responded well to the biofeedback training. During the next 6 weeks, she made great progress in reducing her tension and stress.

During counseling, CBT and IPT were first used to stabilize Natasha. CBT addressed core cognitive distortions such as her current identity as a failure and feelings of hopelessness. IPT addressed the need to increase Natasha's social support system. Natasha was connected to local resources as a means of helping her become more socially connected. Natasha also decided to reach out to a family member to whom she felt close. Narrative therapy was introduced later to help Natasha reauthor her relationship narrative with her ex-husband. This work was more challenging for Natasha because she came to realize that relationship had been informed by problematic attachment patterns in early childhood. Through reauthoring, Natasha overcame her reticence about deserving a healthy, nonabusive relationship that is marked by nonpossessive caring. At the conclusion of Natasha's sessions, she reported increased engagement in her academic activities and more connectedness in social activities.

Conclusion

A counselor's theoretical orientation is the lens through which they view their client and conceptualize the case. Case conceptualization ultimately informs treatment planning and the selection of interventions throughout the therapeutic process. A counselor's theoretical orientation can often be a part of their identity as a counselor and may determine how they believe client issues develop and how they believe issues are resolved. In the case of Natasha, her therapist decided to use narrative therapy and CBT with biofeedback training in the form of electromyography and finger temperature training. The use of these therapeutic interventions has empirical support as an adjunct to conventional therapeutic interventions. Neuroscience-based techniques such as biofeedback and neurofeedback can be used to complement or augment the counseling therapy. As you reflect on our approach to Natasha's case, consider the following question.

> **Reflective Question**
>
> If you were working with Natasha as a client, which neuroscience-informed therapeutic intervention would you have chosen and why?

Quiz

1. Researchers began studying the neurobiological outcomes of the use of therapeutic interventions by investigating:
    a. Depression.
    b. Anxiety.
    c. Phobias.
    d. Obsessions and compulsions.
2. The primary brain structure affected during the application of therapeutic interventions such as cognitive behavior therapy (CBT) and interpersonal psychotherapy (IPT) is the:
    a. Brain stem.
    b. PFC.
    c. Limbic system.
    d. b and c

References

Álvarez-Pérez, Y., Rivero, F., Herrero, M., Viña, C., Fumero, A., Betancort, M., & Peñate, W. (2021). Changes in brain activation through cognitive-behavioral therapy with exposure to virtual reality: A neuroimaging study of specific phobia. *Journal of Clinical Medicine, 10*(16), Article 3505. https://doi.org/10.3390/jcm10163505

Arnsten, A. F. T. (2009). Stress signaling pathways that impair prefrontal cortex structure and function. *Nature Reviews Neuroscience, 10*(6), 410–422.

Askovic, M., Watters, A. J., Aroche, J., & Harris, A. W. F. (2017). Neurofeedback as an adjunct therapy for treatment of chronic posttraumatic stress disorder related to refugee trauma and torture experiences: Two case studies. *Australasian Psychiatry, 25*(4), 358–363.

Beaudoin, M.-N., & Zimmerman, J. (2011). Narrative therapy and interpersonal neurobiology: Revisiting classic practices, developing new emphases. *Journal of Systemic Therapies, 30*(1), 1–13. https://doi.org/10.1521/jsyt.2011.30.1.1

Beauregard, M. (2007). Mind does really matter: Evidence from neuroimaging studies of emotional self-regulation, psychotherapy, and placebo effect. *Progress in Neurobiology, 81*(4), 218–236.

Beeson, E. T., Kim, S., Zalaquett, C. P., & Fonseca, F. D. (2019). Neuroscience attitudes, exposure, and knowledge among counselors. *Teaching and Supervision in Counseling, 1*(2), Article 1. https://doi.org/10.7290/tsc010201

Beijan, L. L., Prosek, E. A., Jones, L. D., Jackson, D., & Legacy, B. (2022). A consensual qualitative analysis of counselor educators' experiences incorporating neuroscience. *Counselor Education and Supervision, 61*(3), 247–261. https://doi.org/10.1002/ceas.12234

Bell, A. N., Moss, D., & Kallmeyer, R. J. (2019). Healing the neurophysiological roots of trauma: A controlled study examining LORETA Z-score neurofeedback and HRV biofeedback for chronic PTSD. *NeuroRegulation, 6*(2), 54–71. https://doi.org/10.15540/nr.6.2.54

Bhagwagar, Z., Rabiner, E. A., Sargent, P. A., Grasby, P. M., & Cowen, P. J. (2004). Persistent reduction in brain serotonin: A receptor binding in recovered depressed men measured by positron emission tomography. *Molecular Psychiatry, 9*(4), 386–392.

Brody, A. L., Saxena, S., Stoessel, P., Gillies, L. A., Fairbanks, L. A., Alborzian, S., Phelps, M. E., Huang, S.-C., Wu, H.-M., Hom, M. L., Ko, M. K., Au, S. C., Maidment, K., & Baxter, L. R. (2001). Regional brain metabolic changes in patients with major depression treated with either paroxetine or interpersonal therapy: Preliminary findings. *Archives of General Psychiatry, 58*(7), 631–640.

Brooks, S. J., & Stein, D. J. (2015). A systematic review of the neural bases of psychotherapy for anxiety and related disorders. *Dialogues in Clinical Neuroscience, 17*(3), 261–279. https://doi.org/10.31887/DCNS.2015.17.3/sbrooks

Cera, N., Monteiro, J., Esposito, R., Di Francesco, G., Cordes, D., Caldwell, J. Z. K., & Cieri, F. (2022). Neural correlates of psychodynamic and non-psychodynamic therapies in different clinical populations through fMRI: A meta-analysis and systematic review. *Frontiers in Human Neuroscience, 16*, Article 1029256. https://doi.org/10.3389/fnhum.2022.1029256

Collura, T. F. (2017). *Technical foundations of neurofeedback.* Routledge.

Council for Accreditation of Counseling and Related Educational Programs. (2024). *2024 CACREP standards.* https://www.cacrep.org/wp-content/uploads/2023/06/2024-Standards-Combined-Version-6.27.23.pdf

David, D., & Szentagotai, A. (2006). Cognitions in cognitive behavioral psychotherapy: Toward an integrative model. *Clinical Psychology Review, 26*(3), 284–298.

de Almeida, J. R. C., Versace, A., Mechelli, A., Hassel, S., Quevedo, K., Kupfer, D. J., & Phillips, M. L. (2009). Abnormal amygdala-prefrontal effective connectivity to happy faces differentiates bipolar from major depression. *Biological Psychiatry, 66*(5), 451–459. https://doi.org/10.1016/j.biopsych.2009.03.024

De Raedt, R. (2006). Does neuroscience hold promise for the further development of behavior therapy? The case of emotional change after exposure in anxiety and depression. *Scandinavian Journal of Psychology, 47*(3), 225–236.

De Raedt, R. (2020). Contributions from neuroscience to the practice of cognitive behaviour therapy: Translational psychological science in service of good practice. *Behaviour Research and Therapy, 125,* Article 103545. https://doi.org/10.1016/j.brat.2019.103545

Drevets, W. C., Frank, E., Price, J. C., Kupfer, D. J., Holt, D., Greer, P. J., Huang, Y., Gautier, C., & Mathis, C. (1999). PET imaging of serotonin 1A receptor binding in depression. *Biological Psychiatry, 46*(10), 1375–1387. https://doi.org/10.1016/s0006-3223(99)00189-4

Du, J., Wang, Y., Hunter, R., Wei, Y., Blumenthal, R., Falke, C., Khairova, R., Zhou, R., Yuan, P., Machado-Vieira, R., McEwen, B. S., & Manji, H. K. (2009). Dynamic regulation of mitochondrial function by glucocorticoids. *Proceedings of the National Academy of Sciences, USA, 106*(9), 3543–3548. https://doi.org/10.1073/pnas.0812671106

Ellis, A. (1962). *Reason and emotion in psychotherapy.* Stuart.

Field, T. A., Beeson, E. T., & Jones, L. K. (2015). The new ABCs: A counselor's guide to neuroscience-informed cognitive-behavior therapy. *Journal of Mental Health Counseling, 37*(3), 206–220.

Field, T. A., Beeson, E. T., Jones, L. K., & Miller, R. (2017). Counselor allegiance and client expectancy in neuroscience-informed cognitive-behavior therapy: A 12-month qualitative follow-up. *Journal of Mental Health Counseling, 39*(4), 351–365. https://doi.org/10.17744/mehc.39.4.06

Field, T. A., Miller, R., Beeson, E. T., & Jones, L. K. (2019). Treatment fidelity in neuroscience- informed cognitive-behavior therapy: A feasibility study. *Journal of Mental Health Counseling, 41*(4), 359–376. https://doi.org/10.17744/mehc.4l.4.06

Fiocco, A. J., Wan, N., Weekes, N., Pim, H., & Lupien, S. J. (2006). Diurnal cycle of salivary cortisol in older adult men and women with subjective complaints of memory deficits and/or depressive symptoms: Relation to cognitive functioning. *Stress, 9*(3), 143–152.

Frewen, P. A., Dozois, D. J., & Lanius, R. A. (2010). Neuroimaging studies of psychological interventions for mood and anxiety disorders: Empirical and methodological review. *Focus, 8*(1), 228–246.

Fu, C. H.-Y., Williams, S. C. R., Cleare, A. J., Scott, J., Mitterschiffthaler, M. T., Walsh, N. D., Donaldson, C., Suckling, J., Andrew, C., Steiner, H., & Murray, R. M. (2008). Neural responses to sad facial expressions in major depression following cognitive behavioral therapy. *Biological Psychiatry, 64*(6), 505–512. https://doi.org/10.1016/j.biopsych.2008.04.033

Fuchs, T., Birbaumer, N., Lutzenberger, W., Gruzelier, J. H., & Kaiser, J. (2003). Neurofeedback treatment for attention-deficit/hyperactivity disorder in children: A comparison with methylphenidate. *Applied Psychophysiology and Biofeedback, 28,* 1–12. https://doi.org/10.1023/A:1022353731579

Galynker, I. I., Cai, J., Ongseng, F., Finestone, H., Dutta, E., & Serseni, D. (1998). Hypofrontality and negative symptoms in major depressive disorder. *Journal of Nuclear Medicine, 39*(4), 608–612.

Gevirtz, R., & Dalenberg, C. (2008). Heart rate variability biofeedback in the treatment of trauma symptoms. *Biofeedback, 36*(1), 22–23.

Gillespie, C. F., & Nemeroff, C. B. (2005). Hypercortisolemia and depression. *Psychosomatic Medicine, 67*(Suppl. 1), S26–S28.

Grecucci, A., Sığırcı, H., Lapomarda, G., Amodeo, L., Messina, I., & Frederickson, J. (2020). Anxiety regulation: From affective neuroscience to clinical practice. *Brain Sciences, 10*(11), Article 846. https://doi.org/10.3390/brainsci10110846

Hahusseau, S., Baracat, B., Lebey, T., Laudebat, L., Valdez, Z., & Delorme, A. (2022). Heart rate variability biofeedback intero-nociceptive emotion exposure therapy for adverse childhood experiences [version 2; peer review; 2 approved]. *F1000Research, 9*, 326. https://doi.org/10.12688/f1000research.20776.2

Hanson, J. L., Nacewicz, B. M., Sutterer, M. J., Cayo, A. A., Schaefer, S. M., Rudolph, K. D., Shirtcliff, E. A., Pollak, S. D., & Davidson, R. J. (2015). Behavioral problems after early life stress: Contributions of the hippocampus and amygdala. *Biological Psychiatry, 77*(4), 314–323. https://doi.org/10.1016/j.biopsych.2014.04.020

Holt, R., & McLean, L. (2021). Australian psychotherapy for trauma incorporating neuroscience: Evidence- and ethics-informed practice. *Neuroethics, 14*(Suppl. 3), 295–309. https://doi.org/10.1007/s12152-019-09398-4

Ivey, A., Ivey, M. B., & Zalaquett, C. P. (2023). *Intentional interviewing and counseling: Facilitating client development in a multicultural society* (10th ed.). Cengage.

Ivey, A., Ivey, M. B., & Zalaquett, C. P. (2024). *Essentials of intentional counseling and psychotherapy: Counseling in a multicultural world* (4th ed.). Cengage.

Ivey, A. E., & Zalaquett, C. P. (2011). Neuroscience and counseling: Central issue for social justice leaders. *Journal for Social Action in Counseling and Psychology, 3*(1), 103–116. https://doi.org/10.33043/JSACP.3.1.103-116

Joëls, M., & Baram, T. Z. (2009). The neuro-symphony of stress. *Nature Reviews Neuroscience, 10*(6), 459–466. https://doi.org/10.1038/nrn2632

Jones, J. S., Goldstein, S. J., Wang, J., Gardus, J., Yang, J., Parsey, R. V., & DeLorenzo, C. (2022). Evaluation of brain structure and metabolism in currently depressed adults with a history of childhood trauma. *Translational Psychiatry, 12*(1), Article 392. https://doi.org/10.1038/s41398-022-02153-z

Karavidas, M. K., Lehrer, P. M., Vaschillo, E., Vaschillo, B., Marin, H., Buyske, S., Malinovsky, I., Radvanski, D., & Hassett, A. (2007). Preliminary results of an open label study of heart rate variability biofeedback for the treatment of major depression. *Applied Psychophysiology and Biofeedback, 32*, 19–30.

Karlsson, H., Hirvonen, J., Kajander, J., Markkula, J., Rasi-Hakala, H., Salminen, J. K., Nagren, K., Aalto, S., & Hietala, J. (2010). Research letter: Psychotherapy increases brain serotonin 5-HT 1A receptors in patients with major depressive disorder. *Psychological Medicine, 40*(3), 523–528. https://doi.org/10.1017/s0033291709991607

Kim, E. J., Pellman, B., & Kim, J. J. (2015). Stress effects on the hippocampus: A critical review. *Learning & Memory, 22*(9), 411–416. https://doi.org/10.1101/lm.037291.114

Kim, S., & Zalaquett, C. P. (2019). An exploratory study of prevalence and predictors of neuromyths among potential mental health counselors. *Journal of Mental Health Counseling, 41*(2), 173–187.

Kirk, H. W., & Dahl, M. G. (2022). Infra low frequency neurofeedback training for trauma recovery: A case report. *Frontiers in Human Neuroscience, 16*, Article 905823. https://doi.org/10.3389/fnhum.2022.905823

Laruelle, M., Slifstein, M., & Huang, Y. (2002). Positron emission tomography: Imaging and quantification of neurotransmitter availability. *Methods, 27*(3), 287–299.

LeDoux, J. (2003). The emotional brain, fear, and the amygdala. *Cellular and Molecular Neurobiology, 23*(4/5), 727–738.

Lee, D., Jang, C., & Park, H.-J. (2019). Neurofeedback learning for mental practice rather than repetitive practice improves neural pattern consistency and functional network efficiency in the subsequent mental motor execution. *NeuroImage, 188*, 680–693. https://doi.org/10.1016/j.neuroimage.2018.12.055

Leichsenring, F., Rabung, S., & Leibing, E. (2004). The efficacy of short-term psychodynamic psychotherapy in specific psychiatric disorders: A meta-analysis. *Archives of General Psychiatry, 61*(12), 1208–1216.

Lévesque, J., Beauregard, M., & Mensour, B. (2006). Effect of neurofeedback training on the neural substrates of selective attention in children with attention-deficit/hyperactivity disorder: A functional magnetic resonance imaging study. *Neuroscience Letters, 394*(3), 216–221.

Lieberman, A. (2006). Depression in Parkinson's disease—A review. *Acta Neurologica Scandinavica, 113*(1), 1–8.

Luke, C., Miller, R., & McAuliffe, G. (2019). Neuro-informed mental health counseling: A person-first perspective. *Journal of Mental Health Counseling, 41*(1), 65–79. https://doi.org/10.17744/mehc.41.1.06

MacQueen, G. M., Yucel, K., Taylor, V. H., Macdonald, K., & Joffe, R. (2008). Posterior hippocampal volumes are associated with remission rates in patients with major depressive disorder. *Biological Psychiatry, 64*(10), 880–883. https://doi.org/10.1016/j.biopsych.2008.06.027

Maguschak, K. A., & Ressler, K. J. (2008). β-catenin is required for memory consolidation. *Nature Neuroscience, 11*, 1319–1326. https://doi.org/10.1038/nn.2198

Makinson, R. A., & Young, J. S. (2012). Cognitive behavioral therapy and the treatment of posttraumatic stress disorder: Where counseling and neuroscience meet. *Journal of Counseling & Development, 90*(2), 131–140. https://doi.org/10.1111/j.1556-6676.2012.00017.x

Månsson, K. N. T., Lueken, U., & Frick, A. (2021). Enriching CBT by neuroscience: Novel avenues to achieve personalized treatments. *International Journal of Cognitive Therapy, 14*(1), 182–195. https://doi.org/10.1007/s41811-020-00089-0

Martin, K. C., & Kandel, E. R. (1996). Cell adhesion molecules, CREB, and the formation of new synaptic connections. *Neuron, 17*(4), 567–570.

Miller, R. M., Field, T. A., Beeson, E. T., Doumas, D. M., & Jones, L. K. (2020). The impact of neuroscience-informed cognitive-behavior therapy training on knowledge and interoceptive awareness. *Journal of Counselor Preparation and Supervision, 13*(2), Article 1. https://doi.org/10.7729/42.1348

Navalta, C. P., McGee, L., & Underwood, J. (2018). Adverse childhood experiences, brain development, and mental health: A call for neurocounseling. *Journal of Mental Health Counseling, 40*(3), 266–278. https://doi.org/10.17744/mehc.40.3.07

Paquette, V., Levesque, J., Mensour, B., Leroux, J.-M., Beaudoin, G., & Bourgouin, P. (2003). "Change the mind and you change the brain": Effects of cognitive-behavioral therapy on the neural correlates of spider phobia. *NeuroImage, 18*(2), 401–409.

Quirk, G. J., & Mueller, D. (2007). Neural mechanisms of extinction learning and retrieval. *Neuropsychopharmacology, 33*(1), 56–72. https://doi.org/10.1038/sj.npp.1301555

Ressler, K. J., & Mayberg, H. S. (2007). Targeting abnormal neural circuits in mood and anxiety disorders: From the laboratory to the clinic. *Nature Neuroscience, 10*(9), 1116–1124.

Roozendaal, B., McEwen, B. S., & Chattarji, S. (2009). Stress, memory and the amygdala. *Nature Reviews Neuroscience, 10*(6), 423–433. https://doi.org/10.1038/nrn2651

Rosenbaum, D., Leehr, E. J., Kroczek, A., Rubel, J. A., Int-Veen, I., Deutsch, K., Maier, M. J., Hudak, J., Fallgatter, A. J., & Ehlis, A. C. (2020). Neuronal correlates of spider phobia in a combined fNIRS-EEG study. *Scientific Reports, 10*(1), Article 12597. https://doi.org/10.1038/s41598-020-69127-3

Russell-Chapin, L. A., Pacheco, N. C., & DeFord, J. A. (Eds.). (2020). *Practical neurocounseling*. Routledge.

Russo, G. M., Schauss, E., Naik, S., Banerjee, R., Ghoston, M., Jones, L. K., Zalaquett, C. P., Beeson, E. T., & Field, T. A. (2021). Extent of counselor training in neuroscience-informed counseling competencies. *Journal of Mental Health Counseling, 43*(1), 75–93. https://doi.org/10.17744/mehc.43.1.05

Sapolsky, R. M. (2003). Stress and plasticity in the limbic system. *Neurochemical Research, 28*, 1735–1742.

Sargent, P. A., Kjaer, K. H., Bench, C. J., Rabiner, E. A., Messa, C., Meyer, J., Gunn, R. N., Grasby, P. M., & Cowen, P. J. (2000). Brain serotonin 1A receptor binding measured by positron emission tomography with [11C]WAY-100635: Effects of depression and antidepressant treatment. *Archives of General Psychiatry, 57*(2), 174–180.

Schuman, D. L., Lawrence, K. A., Boggero, I., Naegele, P., Ginsberg, J. P., Casto, A., & Moser, D. K. (2023). A pilot study of a three-session heart rate variability biofeedback intervention for veterans with posttraumatic stress disorder. *Applied Psychophysiology Biofeedback, 48*(1), 51–65. https://doi.org/10.1007/s10484-022-09565-z

Segerstrom, S. C., & Miller, G. E. (2004). Psychological stress and the human immune system: A meta-analytic study of 30 years enquiry. *Psychological Bulletin, 130*(4), 601–630.

Shin, L. M., Orr, S. P., Carson, M. A., Rauch, S. L., Macklin, M. L., Lasko, N. B., Peters, P. M., Metzger, L. J., Dougherty, D. D., Cannistraro, P. A., & Alpert, N. M. (2004). Regional cerebral blood flow in the amygdala and medial prefrontal cortex during traumatic imagery in male and female Vietnam veterans with PTSD. *Archives of General Psychiatry, 61*(2), 168–176. https://doi.org/10.1001/archpsyc.61.2.168

Siepmann, M., Aykac, V., Unterdörfer, J., Petrowski, K., & Mueck-Weymann, M. (2008). A pilot study on the effects of heart rate variability biofeedback in patients with depression and in healthy subjects. *Applied Psychophysiology and Biofeedback, 33*, 195–201. https://doi.org/10.1007/s10484-008-9064-z

Simkin, D. R., Thatcher, R. W., & Lubar, J. (2014). Quantitative EEG and neurofeedback in children and adolescents: Anxiety disorders, depressive disorders, comorbid addiction and attention-deficit/hyperactivity disorder, and brain injury. *Child and Adolescent Psychiatric Clinics, 23*(3), 427–464.

Sousa, D. A. (2011). *How the brain learns* (4th ed.). Corwin.

Straube, T., Glauer, M., Dilger, S., Mentzel, H. J., & Miltner, W. H. (2006). Effects of cognitive-behavioral therapy on brain activation in specific phobia. *NeuroImage, 29*(1), 125–135.

Taylor, J., Siddiqi, S., Lin, C., Talmasov, D., Goodkind, M., Etkin, A., & Fox, M. (2021). A transdiagnostic psychiatric brain circuit derived from atrophy patterns and brain lesions. *Biological Psychiatry, 89*(9 Suppl.), S257. https://doi.org/10.1016/j.biopsych.2021.02.643

Weissman, M. M., & Markowitz, J. C. (1994). *Interpersonal psychotherapy of depression.* Jason Aronson.

Weissman, M. M., Markowitz, J. W., & Klerman, G. L. (2000). *Comprehensive guide to interpersonal psychotherapy.* Basic Books.

White, M. (2007). *Maps of narrative practice.* Norton.

White, M., & Epston, D. (1990). *Narrative means to therapeutic ends.* Norton.

Wilker, S., Vukojevic, V., Schneider, A., Pfeiffer, A., Inerle, S., Pauly, M., Elbert, T., Papassotiropoulos, A., de Quervain, D., & Kolassa, I.-T. (2023). Epigenetics of traumatic stress: The association of NR3C1 methylation and posttraumatic stress disorder symptom changes in response to narrative exposure therapy. *Translational Psychiatry, 13*(1), Article 14. https://doi.org/10.1038/s41398-023-02316-6

Wolpe, J. (1958). *Psychotherapy by reciprocal inhibition.* Stanford University Press.

Yamashita, A., Hayasaka, S., Kawato, M., & Imamizu, H. (2017). Connectivity neurofeedback training can differentially change functional connectivity and cognitive performance. *Cerebral Cortex, 27*(10), 4960–4970.

Zalaquett, C. P., Ivey, A., & Ivey, M. B. (2019). *Essential theories of counseling and psychotherapy: Everyday practice in our diverse world.* Cognella Academic.

Zhang, Y.-H., Wang, N., Lin, X.-X., Wang, J.-Y., & Luo, F. (2022). Application of cognitive bias testing in neuropsychiatric disorders: A mini-review based on animal studies. *Frontiers in Behavioral Neuroscience, 16*, Article 924319. https://doi.org/10.3389/fnbeh.2022.924319

Chapter 12

Neuro-Informed Career-Focused Counseling

Chad Luke and Thomas A. Field

The purpose of this chapter is to explore how neuroscientific literature and principles offer an innovative perspective on self-knowledge, occupational knowledge, decision-making, and stress in the context of neuroscience-informed (or *neuro-informed*) career-focused counseling. We first applied neuro-informed counseling to career-related issues in the first edition of this text. We expanded the discussion later in Luke (2018) and Luke and Gibbons (2022). To effectively integrate neuroscience into career counseling, counselors may need to adjust their perspective to see their client's career issue as a clinical mental health counseling presenting problem, like and unlike other presenting problems, rather than as a stand-alone problem. Neuro-informed career-focused counseling (Luke, 2018) addresses career issues as a specific type of presenting problem, making them appropriate for integration with recent neuroscientific findings (Luke & Redekop, 2016). Although the field of neuroscience has yet to identify a "career region" of the brain, integration takes a broader perspective by looking at indirect and translational evidence. Translational approaches, generally speaking, "examine how basic biological (i.e., brain-based mechanisms) and behavioral factors interact in initiating and sustaining positive behavior change as a result of psychotherapy" (Feldstein-Ewing & Chung, 2013, p. 329). In other words, translational approaches assist with the process of applying basic scientific information to clinical practice (Woolf, 2008). In the translational approach used here, we extend Luke (2018) by discussing the application of translational neuroscience to Frank Parsons's (1909) foundational elements and the neuroscience of work-related stress as described by Luke and Gibbons (2022).

2024 CACREP Standards

This chapter addresses 2024 Council for Accreditation of Counseling and Related Educational Programs (CACREP) Standards pertinent to the Foundational Counseling Curriculum (Section 3) area of Career Development (Standard D):

- Approaches for conceptualizing the interrelationships among and between work, socioeconomic standing, wellness, disability, trauma, relationships, and other life roles and factors (Standard D.2.)
- Strategies for assessing abilities, interests, values, personality, and other factors that contribute to career development (Standard D.5.)
- Strategies for facilitating client skill development for career, educational, and life-work planning and management (Standard D.9.)

This chapter also addresses the following Entry-Level Specialized Practice Area (Section 5) standard for Career Counseling (Standard B):

- Factors that affect clients' attitudes toward work and their career decision-making processes (Standard B.1.)

Clinical Case Study: Larissa

Larissa is a woman in her 40s who meets with you for an intake to discuss phase-of-life issues. Your initial intake paperwork shows that she has been a stay-at-home mom rearing three children for the past 14 years. Larissa also recently separated from her husband of more than 15 years. She has a bachelor's degree in sociology but has never worked in a job that required this or any other degree. At this time, her children are in middle and high school and require a different type of active parenting, allowing her time to reflect on her role in the family, community, and world. During the first session, Larissa reports feeling some emptiness as she has entered her 40s and is no longer parenting small children. She thinks she wants to further explore occupational options (simple career advice, right?). As she has researched job options, she has been experiencing increased anxiety with disrupted sleep, including nightmares about getting fired from jobs. She has a history of anxiety, although it has not been treated or even diagnosed. She states that she feels the additional pressure of being a single mother and sole breadwinner, so she is looking for work that meets her financial needs as well as being work that she cares about. Neuro-informed career-focused counseling (Luke, 2018; Luke & Gibbons, 2022) provides a framework that can be used with all career counseling approaches. It describes the translational neuroscience of Parsons's (1909) tripartite model of career decision-making as well as stress and anxiety as they relate to these decisions.

> **Reflective Questions**
>
> Larissa may or may not meet the criteria for a mental health disorder, but she has psychosocial stressors impinging on her career decision-making.
> Is this a career or mental health issue?
> Why?

Principles of Neuro-Informed Career-Focused Counseling

Self-Knowledge in the Brain

Career development luminaries such as Donald Super have long understood the importance of a person's neurophysiology in career development. Super (1980) connected neurophysiology to self-concept. Super asserted that career development was essentially about the development of self-concept comprising inherited attitudes, neural and endocrine functioning, and role approval by peers and people in positions of authority. A person's vocational identity is essentially the self-at-work (Super, 1955, 1980), and it changes over time and across environments (i.e., life span, life space). Work performance and satisfaction are optimal when the self (intrapersonal) can be expressed in the work environment (interpersonal). The importance of outwardly expressing a self-concept has been emphasized not only by Parsons and Super but also by other important career theorists such as Edward Kellog Strong Jr., John Holland, and Mark L. Savickas.

This process of projecting the self into the future has been referred to as *mental time travel* (D'Argembeau, 2020). Where might one find the self in the brain, in order to project it into the future? One proposition is that the self is the totality of one's experiences and that these experiences are stored as memories—specifically, as *autobiographical memory* (AM). This type of memory refers to the memories related to oneself in the context of experiences, relationships, and environments. AM involves "memory systems that encode, consolidate, and retrieve personal events and facts" (Fossati, 2013, p. 487). It is "a semantic memory that involves representations of the self, frequently involving the precuneus" (Rolls, 2018, p. 578). The self is a (re)construction of memories drawn from these key components and integrated into a gestalt that is identity. Whereas memory is generally governed by the executive function of the hippocampus in connection with the emotional memory governance of the amygdala, AM involves the prefrontal cortex (PFC), medial temporal lobe, limbic system, and occipital lobe (Barry et al., 2018). Freton et al. (2014) noted that the precuneus and ventromedial PFC involved in AM have ramifications for "both emotion regulation and self-related processes" (p. 959). AM involves multiple neural structures required for memory storage and retrieval, such as the medial PFC, posterior cingulate cortex, and parietal lobe, which give rise to self-processing and self-representation (Fossati, 2013).

Application to the Case of Larissa

The situation in which Larissa finds herself is an example of how career issues and mental health issues are interrelated. Larissa's vocational identity is most certainly tied to her personal identity. Larissa has frequently made life decisions that served the purpose of mitigating her anxiety rather than pursuing her career goals. In investigating further, you find that her anxiety has been a constant companion throughout her life. Neuro-informed career-focused counseling might assist her in exploring how early memories of self and work have contributed to her anxiety and tendencies toward avoidance. Neuro-informed career-focused counseling will assist Larissa in exploring her memory function as well as her developmental perspective, as we show later in this chapter. The relationship between mental health (in this case, anxiety) and career issues results in and from work-related stress, which we consider next.

Occupational Knowledge in the Brain

Occupational knowledge involves information about the *world of work*, which includes job titles, skills, knowledge, and interests. To gain occupational knowledge, an individual must be in close proximity to work-related information, either through relationships with others or through direct contact with the information (Luke & Gibbons, 2022). Individuals can learn about work via relationships with family and friends and, especially, through social media content creators. They can also learn directly through web research and first-person interviews about the work an individual performs. A critical factor in obtaining occupational knowledge is motivation, which determines where an individual falls on the spectrum of passive to active engagement with the learning process. Perhaps the most significant consideration for assisting clients with accessing occupational information is that both the motivation to access and the type of information accessed are determined by the individual's self-knowledge (Luke & Gibbons, 2022). Writing about episodic prospection, Verfaellie et al. (2019) described the process of drawing on memory to create hypotheses about the self in the future. In other words, we use AM to construct possible futures. These possible futures influence our occupational interests and result in gathering, or not gathering, information about particular work.

> **Reflective Question**
>
> How would you describe Larissa's occupational knowledge and vocational identity at the point at which she enters counseling?

Application to the Case of Larissa

For Larissa, her past identity is failing to help her project herself into the future. Because she has limited the occupational information she seeks, she feels limited in her options. This leads to decreased motivation. For her to expand her search for and integration of new occupational information, she will need to understand who she is now in a work context and to future cast this identity into the world of work and occupational information. An early phase of career counseling would be directly discussing with Larissa her propensity toward avoidance and setting homework tasks that facilitate approaching/being exposed to anxiety-generating and feared tasks that are crucial to career exploration, such as completing career interest questionnaires, learning about occupations of interest, interviewing people who work in occupations of interest, and work shadowing.

Decision-Making in the Brain

Decision-making is more than a cognitive process or mere action. Purves et al. (2018) recognized that decision-making involves perception, attention, emotion, and memory—all of which require the functions of their respective brain systems. In fact, these authors asserted that "behavior depends not just on sensory input, but on remembered information, goals, and prediction about what might happen" (Purves et al., 2018, p. 724). In other words, decision-making involves multiple systems, like executive functions in the PFC, short-term memory, reward evaluation, conflict resolution, and response inhibition. Decision-making can involve more primitive brain regions driven more by the need to survive than to thrive. When working with individuals in career decision-making, consideration must be given to the fact that many decisions are made reflexively, even reactively, based on the individual's sense of self and projection of that self into future work-related simulations (see D'Argembeau, 2020; Verfaellie et al., 2019).

Application to the Case of Larissa
Sympathetic activation is associated with responding to stress in our environment: Sensing a threat initiates the release of adrenaline and cortisol in preparation to address the threat. In Larissa's sympathetically activated stress state, she is likely to have a preference or bias for surviving and avoiding threats when making decisions. In such a state, Larissa is likely to avoid tasks associated with threats and thus be less likely to take risks. This avoidance can keep Larissa stuck from moving forward with career change, out of fear that any change will place her survival at risk. Getting her back into parasympathetic functioning is a prerequisite for effective decision-making that facilitates taking risks and moving forward with career change.

Work and Stress in the Brain

Neuroscience can be readily applied to career-focused counseling practice when considering the self under stress. Stress is both an objective, observable phenomenon and a subjective, internally experienced one (Purves et al., 2018). Craig A. McEwen (2022) summarized his brother Bruce's groundbreaking work on the neurobiology of stress and, in particular, the notion of allostatic load (B. S. McEwen, 1998) and its sociological implications. Briefly, *allostatic load* is the effect of stress on the brain and body over time (B. S. McEwen, 2017). It involves the efforts of multiple systems in restoring neurobiological harmony and goes beyond earlier concepts of homeostasis (B. S. McEwen, 2017). It comes about in three ways: "(1) frequent activation of allostatic systems; (2) failure to shut off allostatic activity after stress; [and] (3) inadequate response of allostatic systems leading to elevated activity of other, normally counter-regulated allostatic systems after stress" (B. S. McEwen, 1998, p. 33). Luke and Schimmel (2023) likened allostatic load in these three ways to a leaky faucet. First, repeatedly turning the water on and off (B. S. McEwen's, 1998, first instance) will, over time, wear down the fittings, leading to a leaky faucet (second instance). Any leaks and drips are often uneven and therefore unpredictable, outside of our immediate control, and difficult to tune out (unavoidable). Eventually, the faucet may stop working altogether (third instance). The attention given to resolving all these drips and leaks and shutdowns is allostatic load.

In addition to external, world-of-work factors, there are individual factors (worker traits) that increase an individual's susceptibility and response to stressors (Zunker, 2016). In other words, work stresses can have an interpersonal and intrapersonal component. The hope for Larissa and her career-anxiety dilemma is realized in Fisher and Berkman's (2015) findings that both acute and chronic stress show *neural plastic responses* (i.e., changes in nerve cells and their related functioning) to counseling interventions. Larissa's work-related stressed brain can change, grow, and improve.

The limitations of the human stress response system have implications for addressing work stress in career-focused counseling. The general stress mechanism in the brain and autonomic nervous system is found in the hypothalamic-pituitary-adrenal (HPA) axis, identified early on in neuroscience research (Selye, 1955). The system works by interpreting sensory material (input through the five senses from an external source) as threatening or nonthreatening. It is important to note that the HPA axis does not discriminate between perception and reality; in terms of the stress response, it is all real. Sensory input from the five senses is routed through the thalamus and limbic region—namely, the amygdala—to determine whether what is being perceived is similar in any meaningful way to stored material with fear or anxiety salience. Once a threat (stress) is perceived, the hypothalamus activates the pituitary (the master gland), which triggers the release of adrenocorticotropic hormone, which stimulates the adrenal glands to release cortisol. During this time, epinephrine and norepinephrine are also released by way of the sympathetic-adrenal-medullary axis either to prepare the body for fight or flight or to prepare to freeze when the system is utterly overwhelmed (LeDoux, 2003).

It is important to note that this description represents a significant oversimplification of the process, but the process is described in this way for the sake of clarity and clinical application. The autonomic nervous system is an efficient system in terms of preparing the body to protect itself. The sympathetic nervous system has three features in that it prioritizes speed over accuracy, efficiency over effectiveness, and safety and survival over development. Over time, these priorities, in the context of allostatic load, wear away at an individual's ability to make and pursue goals, instead facilitating habitual, rote behaviors (Arnsten et al., 2021; Cordner & Tamashiro, 2016; Taylor et al., 2014). Regardless of the speed and accuracy of the sympathetic nervous system, it has clear physical effects on the body, and work-related stress is no exception. McDonald and Hite (2023) reviewed the literature on stress and strain in the workplace, noting physiological effects, such as headaches, irritable bowel syndrome, high blood pressure, and so forth; psychological effects, including anxiety and depression; and behavioral effects, such as substance use, being unpunctual, or missing work.

Application to the Case of Larissa

The implications of these stress-related neural phenomena are clear. Larissa may find it difficult to concentrate in response to anxiety and acute stress, a state to which she may have grown accustomed. Any challenges in following through on counselor recommendations for moving forward in exploring potential careers may be less about her resistance and more about resilience. Neuro-informed career-focused counselors may begin interventions focused on parasympathetic activation, such as breathwork, before fully engaging in more traditional career counseling interventions.

Neuro-Based Metaphors in Career-Focused Counseling

Research has demonstrated the clinical utility of metaphor use in counseling (Tay, 2012; Wickman et al., 1999), and counselors can take a metaphoric approach to the integration of neuroscience and counseling (Luke, 2015; Michael & Luke, 2016). Wickman et al. (1999) offered a rationale for the use of metaphors in counseling:

> Traditionally, counselors have developed metaphors to demonstrate empathy and to suggest alternative interpretations of presenting problems. This use of metaphor, created by the counselor, does not change a client's problems; rather, it changes perception of the problem and allows for solutions as yet unconsidered. In this manner, metaphor has provided both a linguistic tool to facilitate empathy and an intervention technique with a history of therapeutic value. (p. 389)

From this perspective, metaphor is just a mnemonic; thus, in the metaphoric model, we use metaphor somewhat interchangeably with other linguistic-empathic devices, such as similes, analogs, and examples. Brain and central nervous system structures, systems, and functions provide numerous analogs to human psychological and behavioral functioning. The metaphoric approach can be applied in two distinct ways: (a) using examples from everyday life as a metaphor for brain function to help clients understand that this is a "brain thing," not a character deficit; and (b) using brain function as a metaphor for client experiences.

Brain-Based Metaphors as Normalizing and Demystifying

Neuroscience offers ways for counselors to assist clients in accepting their career-related presenting problem as often being neurological, rather than character based. The field of translational neuroscience (Feldstein-Ewing & Chung, 2013; Fisher & Berkman, 2015) seeks to bridge the gap between neurobiological functions and psychosocial behavior and therapeutic interventions. For example, Fisher and Berkman (2015) identified multiple brain structures involved in the neural mechanisms of addiction. This translational approach has important implications for career-focused counseling. Clients can understand what happens in their brain and central nervous system when they are under acute or prolonged stress by examining the effects of the HPA axis as a function of the sympathetic branch of the autonomic nervous system. When stressed, the body goes into action via the following:

- Increasing the heart rate, which fuels major muscle groups
- Dilating the pupils, which take in more light to see a threat
- Stimulating the adrenal glands for sudden (brief) bursts of energy and clarity

- Slowing or stopping digestion to redirect the body's resources to life-saving functions
- Constricting blood vessels to minimize blood loss from injury

One of these components—stimulation of the adrenal glands—involves cortisol, and its impact on stress management can be applied to career-related issues. Cortisol functions as a glucocorticoid during stress responses, effectively speeding the metabolism for energy. As you can imagine, the life-saving utility of cortisol is balanced by its deleterious effects on the individual if the stress is unrelenting or if traumatic stress occurs. In other words, stress such as the kind Larissa experienced in the transition from stay-at-home mom to employment shows cortisol's darker side—that is, decreased neuroplasticity (adaptability of the cells and synapses), dendrite degeneration (resulting in diminished message transfer), deficits in cell remyelination (cell insulation that promotes neural efficiency), cell death, and inhibition of neurogenesis and neural growth (Kindsvatter & Geroski, 2014). In addition, and perhaps most concerning with career-focused counseling clients who have lived with chronic anxiety, high levels of cortisol correlate with compromised hippocampal function. The hippocampus functions as the executive regulator of memory, so individuals with diminished hippocampal function demonstrate deficits in short- and long-term memory as well as in capacity for new learning (LeDoux, 2003).

Using a Metaphoric Approach to Understanding Brain Function

Amblyopia is an eye-brain condition wherein the eye and brain do not communicate effectively, if at all. In reaction, the brain assumes the eye is not functional or even present, so it limits connections between the eye and the occipital lobe of the brain. Amblyopia is an example of *synaptic pruning*, whereby the brain trims back unused circuits and synapses. The result is the ironically termed "lazy eye" and even blindness. The treatment is to cover the working eye to compel reliance on the other eye and to stimulate the brain to maintain connections. The eye is not actually lazy; it is understimulated and unsupported. Larissa will most certainly benefit from this reframing of motivation and effort in terms of her career.

This concepts of synaptic pruning and synaptogenesis, both concepts related to neuroplasticity, can also be seen in maze completion. Many people have seen or completed a maze at some point in their lives, so the example has an element of universality. With Larissa, the counselor might produce a basic paper-based maze and ask her to complete it while the counselor times her. This may produce some anxiety in Larissa, so the counselor would reassure her that it is a low-stakes activity. In the session, the counselor would use three to five trials in succession, recording the time it took for Larissa to complete each trial. In reflecting on the results, Larissa would likely notice a common occurrence: Overall, her time decreases with each attempt.

This simple activity demonstrates how the brain learns from both successes and failures. Maze trials offer clients such as Larissa hope for learning, which represents growth and change. In effect, her stronger "mothering eye" has been covered, allowing the weaker "vocational eye" to get stronger. As expectations for this unused tool increase, so does its performance. Career-skills synaptogenesis takes place, wherein the necessary networks of career problem-solving are formed and reformed, reminding Larissa of previous successes in these areas. This progress confirms that Larissa has been understimulated rather than lazy or idle.

> **Reflective Question**
>
> How can a counselor help Larissa understand why she might have difficulty with concentration and task completion under stress during a session without Larissa fearing that she has a sort of brain damage?

Our Brain-Based Approach to the Case of Larissa

As we noted earlier in this chapter, neuroscience offers ways for counselors to assist clients in accepting their career-related presenting problem as often being neurological rather than character based. We might conceptualize Larissa's dilemma as an overactive sympathetic response system in which stress and anxiety are the dominant experience, leading to paralysis of action. Up until her separation from her husband, she had cultivated skills related to her role in the home, avoiding anxiety-ridden thoughts that she might need certain skills in the future. But suddenly this arrangement, her job, and her role have changed. Like many counseling clients, Larissa entered counseling because the tasks she is facing have overwhelmed her capacity to complete them (i.e., stress).

Through counseling, Larissa developed skills that facilitated a shift to parasympathetic states of relaxation and calm. Relaxation techniques—for example, diaphragmatic breathing; mantram repetition; and sensory-based coping, such as the use of soothing music and aromatherapy—gradually helped Larissa to reduce baseline stress. By the end of the seventh session, Larissa's presentation in counseling sessions was more grounded. She acknowledged some angst in not knowing which direction her career would take next, but she felt ready to take steps toward clarifying her career goals.

Although establishing better regulation of the stress response is helpful foundational work, Larissa will remain motionless in exploring career options unless her avoidance is addressed directly. Avoidance, in our experience, tends to be resilient despite other counseling gains. To help Larissa make progress, the counselor helped Larissa to identify and act on homework assignments that were designed to expose her to anxiety-provoking career exploration and thus work through her anxiety. By the 16th session, Larissa had explored several viable career paths, applied for a dozen jobs in the realm of her sociology degree, and had enrolled in a graduate course in nonprofit management. These homework assignments were not easy for Larissa, and she often reported feeling more anxious before completing the task (after which, the anxiety significantly reduced). Eventually though, Larissa felt enough sense of achievement and self-efficacy that she was able to make progress. This reinforced her sense of self-concept, and the counselor helped Larissa to reauthor her past narrative (Savickas, 2012) that it would be impossible for her to find a fulfilling career in mid-life.

In summary, neuro-informed career-focused counseling (a) aided Larissa in normalizing her experience of confusion, ambivalence, and generally feeling lost about what to do next; (b) normalized her avoidance as a stress response; (c) helped her to take action to explore her occupational options and to move ahead with her career transition; and (d) helped her to reauthor and reconstruct her life and career narratives.

> **Reflective Question**
>
> How might Larissa's understanding of the brain help her to be successful in her studies and next career?

Quiz

1. According to Donald Super (1980), vocational identity is:
   a. A projection of the self into the world of work.
   b. The status one attains as a result of one's job.
   c. A person's sense that they have selected the right career.
   d. An unnecessary component of career counseling.
2. Which system is activated during times of stress, including work stress?
   a. Parasympathetic nervous system.
   b. Autonomic nervous system.
   c. Mesolimbocortical dopamine system.
   d. HPA axis.

References

Arnsten, A. F. T., Condon, E. M., Dettmer, A. M., Gee, D. G., Ka Shu Lee, Mayes, L. C., Stover, C. S., & Tseng, W.-L. (2021). The prefrontal cortex in a pandemic: Restoring functions with system-, family-, and individual-focused interventions. *American Psychologist, 76*(5), 729–743. https://doi.org/10.1037/amp0000823

Barry, T. J., Chiu, C. P. Y., Raes, F., Ricarte, J., & Lau, H. (2018). The neurobiology of reduced autobiographical memory specificity. *Trends in Cognitive Sciences, 22*(11), 1038–1049. https://doi.org/10.1016/j.tics.2018.09.001

Cordner, Z. A., & Tamashiro, K. L. K. (2016). Effects of chronic variable stress on cognition and Bace1 expression among wild-type mice. *Translational Psychiatry, 6*(7), Article e854. https://doi.org/10.1038/tp.2016.127

Council for Accreditation of Counseling and Related Educational Programs. (2024). *2024 CACREP standards.* https://www.cacrep.org/wp-content/uploads/2023/06/2024-Standards-Combined-Version-6.27.23.pdf

D'Argembeau, A. (2020). Zooming in and out on one's life: Autobiographical representations at multiple time scales. *Journal of Cognitive Neuroscience, 32*(11), 2037–2055. https://doi.org/10.1162/jocn_a_01556

Feldstein-Ewing, S. W., & Chung, T. (2013). Neuroimaging mechanisms of change in psychotherapy for addictive behaviors: Emerging translational approaches that bridge biology and behavior. *Psychology of Addictive Behaviors, 27*(2), 329–335.

Fisher, P. A., & Berkman, E. T. (2015). Designing interventions informed by scientific knowledge about effects of early adversity: A translational neuroscience agenda for next-generation addictions research. *Current Addiction Reports, 2*, 347–353.

Fossati, P. (2013). Imaging autobiographical memory. *Dialogues in Clinical Neuroscience, 15*(4), 487–490. https://doi.org/10.31887/DCNS.2013.15.4/pfossati

Freton, M., Lemogne, C., Bergouignan, L., Delaveau, P., Lehéricy, S., & Fossati, P. (2014). The eye of the self: Precuneus volume and visual perspective during autobiographical memory retrieval. *Brain Structure and Function, 219*, 959–968.

Kindsvatter, A., & Geroski, A. (2014). The impact of early life stress on the neurodevelopment of the stress response system. *Journal of Counseling & Development, 92*(4), 472–480. https://doi.org/10.1002/j.1556-6676.2014.00173.x

LeDoux, J. (2003). *Synaptic self: How our brains become who we are.* Penguin Books.

Luke, C. (2015). *Neuroscience for counselors and therapists: Integrating the sciences of mind and brain.* Sage.

Luke, C. (2018). *Career focused counseling: Integrating theory, research, and neuroscience.* Cognella.

Luke, C., & Gibbons, M. M. (2022). *Career-focused counseling: Integrating culture, development, and neuroscience.* Cognella.

Luke, C., & Redekop, C. (2016). Supervision of co-occurring career and mental health concerns: Application of an integrated approach. *Career Planning and Adult Development Journal, 32*(1), 130–140.

Luke, C., & Schimmel, C. J. (2023). Using neuroscience-informed group work with children and adolescents affected by the pandemic. *The Journal for Specialists in Group Work, 48*(1), 20–31. https://doi.org/10.1080/01933922.2022.2158972

McDonald, K. S., & Hite, L. M. (2023). *Career development: A human resource development perspective* (2nd ed.). Taylor & Francis.

McEwen, B. S. (1998). Stress, adaptation, and disease: Allostasis and allostatic load. *Annals of the New York Academy of Sciences, 840*(1), 33–44. https://doi.org/10.1111/j.1749-6632.1998.tb09546.x

McEwen, B. S. (2017). Neurobiological and systemic effects of chronic stress. *Chronic Stress, 1.* https://doi.org/10.1177/2470547017692328

McEwen, C. A. (2022). Connecting the biology of stress, allostatic load and epigenetics to social structures and processes. *Neurobiology of Stress, 17,* Article 100426. https://doi.org/10.1016/j.ynstr.2022.100426

Michael, T., & Luke, C. (2016). Utilizing a metaphoric approach to teach the neuroscience of play therapy: A pilot study. *International Journal of Play Therapy, 25*(1), 45–52. https://doi.org/10.1037/pla0000015

Parsons, F. (1909). *Choosing a vocation.* Houghton Mifflin.

Purves, D., Augustine, G. J., Fitzpatrick, D., Hall, W. C., Lamantia, A. S., Mooney, R. D., Platt, M. L., & White, L. E. (2018). *Neuroscience* (6th ed.). Sinauer Associates.

Rolls, E. T. (2018). The storage and recall of memories in the hippocampo-cortical system. *Cell and Tissue Research, 373*(3), 577–604.

Savickas, M. L. (2012). Life design: A paradigm for career intervention in the 21st century. *Journal of Counseling & Development, 90*(1), 13–19. https://doi.org/10.1111/j.1556-6676.2012.00002.x

Selye, H. (1955). Stress and disease. *Science, 122*(3171), 625–631.

Super, D. E. (1955). Transition: From vocational guidance to counseling psychology. *Journal of Counseling Psychology, 2*(1), 3–9.

Super, D. E. (1980). A life-span, life-space approach to career development. *Journal of Vocational Behavior, 16*(3), 282–298. https://doi.org/10.1016/0001-8791(80)90056-1

Tay, D. (2012). Applying the notion of metaphor types to enhance counseling protocols. *Journal of Counseling & Development, 90*(2), 142–149. https://doi.org/10.1111/j.1556-6676.2012.00019.x

Taylor, S. B., Anglin, J. M., Paode, P. R., Riggert, A. G., Olive, M. F., & Conrad, C. D. (2014). Chronic stress may facilitate the recruitment of habit- and addiction-related neurocircuitries through neuronal restructuring of the striatum. *Neuroscience, 280,* 231–242. https://doi.org/10.1016/j.neuroscience.2014.09.029

Verfaellie, M., Wank, A. A., Reid, A. G., Race, E., & Keane, M. M. (2019). Self-related processing and future thinking: Distinct contributions of ventromedial prefrontal cortex and the medial temporal lobes. *Cortex, 115,* 159–171. https://doi.org/10.1016%2Fj.cortex.2019.01.028

Wickman, S. A., Daniels, M. H., White, L. J., & Fesmire, S. A. (1999). A "primer" in conceptual metaphor for counselors. *Journal of Counseling & Development, 77*(4), 389–394. https://doi.org/10.1002/j.1556-6676.1999.tb02464.x

Woolf, S. H. (2008). The meaning of translational research and why it matters. *JAMA, 299*(2), 211–213.

Zunker, V. (2016). *Career counseling: A holistic approach* (6th ed.). Cengage Learning.

Chapter 13

Neuro-Informed Group Work

*Chad Luke, Joel F. Diambra,
and Christine J. Schimmel*

Neuroscience has the potential to explain, inform, and guide group process. The integration of neuroscience with group counseling need not create a whole new paradigm of group counseling (Beeson & Luke, 2023; Luke & Schimmel, 2022). Rather, neuroscience both illuminates group processes and adds illustrative imagery for interventions (Luke & Diambra, 2017). This chapter describes a neuroscience-informed approach to group counseling that is organized by group facilitation fundamentals, with special consideration to applying therapeutic factors of group work (Yalom, 1995; Yalom & Leszcz, 2020) through the lens of neuroscience. We consider how neuroscience can be used to engage participants, establish group norms, and get to the working stage so that members experience the therapeutic factors of group work. This chapter describes the case of Ricky to help readers facilitate the group process by understanding the neuroscience of group interaction. Reflective questions are provided at the end of each section to assist you in reflecting upon the material and taking a thoughtful approach to facilitating the group process.

Luke and Schimmel (2023) reviewed the literature surrounding group counseling, the neuroscience of group counseling, and the neuroscience of stress and group counseling interventions for stress. We summarize the highlights of their review here. First, group counseling is an empirically supported treatment modality for working with children, adolescents, and adults (Luke & Schimmel, 2023; Schimmel & Jacobs, 2019). Groups are often used to great effect in addressing a range of needs, from social skills to anxiety and mood disorders to more severe and persistent mental health issues (see Kul & Hamamci, 2021). One component of group work that has demonstrated effectiveness is psychoeducation, an approach that works well with neuroscience education (i.e., neuroeducation; Baourda et al., 2022). Second, group counseling has been shown to be effective with members of minoritized populations, such as African American youth (Havlik et al., 2020). In addition, Masten et al. (2021) identified a host of therapeutic and resilience factors in group counseling. The list includes belonging and cohesion, self-regulation, agency, problem-solving and planning, purpose and a sense of meaning, and positive habits, to name a few. Wei et al. (2021) described the upward spiral of positive outcomes as a benefit of group counseling.

2024 CACREP Standards

This chapter addresses 2024 Council for Accreditation of Counseling and Related Educational Programs (CACREP) Standards pertinent to the Foundational Counseling Curriculum (Section 3) area of Group Counseling and Group Work (Standard F):

- Therapeutic factors of group work and how they contribute to group effectiveness (Standard F.3.)
- Characteristics and functions of effective group leaders (Standard F.4.)
- Types of groups, settings, and other considerations that affect conducting groups (Standard F.7.)

Reflective Questions

Would a process group or psychoeducational group be most helpful for Ricky? Why?
Is there something specific Ricky needs to learn, and if so, is this best accomplished through cognitive assent or experiential learning?

Clinical Case Study: Ricky

Ricky is a 17-year-old male who was referred by the court to an anger management group following a road rage incident. He had received his learner's permit and was driving a friend's car unsupervised when he got into an accident. The fault was unclear, but Ricky got out of his car and kicked in the taillights of the other driver's car. In addition, he was threatening and intimidating toward the other driver, whom he even shoved out of the way. This was not Ricky's first physical altercation; many previous incidents had occurred in middle and high school. As a result, the judge revoked Ricky's learner's permit and required him to complete traffic school and anger management treatment as a condition of having his permit reinstated. Because of the limited availability of group counseling options in Ricky's area and the severity of his history of fighting, Ricky must participate in an adult, male-only stress management group. In reviewing Ricky's intake paperwork, the counselor learns that Ricky is the second of four children living in a single-parent home in a two-bedroom apartment in a rural community 40 miles outside of a large metropolitan area. Ricky's mother assisted him in completing the paperwork, reporting that he has struggled with attention, emotion regulation, and impulsivity since he was 3 years old. He has a few friends between the ages of 17 and 21, some of whom dropped out of high school, and together they drink alcohol and smoke marijuana. Ricky refuses to elaborate on the frequency or amount of his substance misuse. During intake, Ricky acknowledges that his anger gets the best of him at times, but he does not want to work on it. He says he just wants to "get this s— done" so he can obtain a driver's license.

Group Facilitation Fundamentals

Professional counselors use several foundational facilitation skills when leading groups. In cases such as Ricky's, when the client is somewhat reluctant to attend group or is ordered by some other person or institution to attend group, it is common for clients to use silence, abrasiveness, circumlocution, or other behaviors to avoid *making contact* with other group members and themselves. Whether done intentionally or not, these patterns of communication have likely contributed to the struggles these clients now face. When planning a group, the leader might consider whether a process group or psychoeducational group is indicated. In neuro-informed group counseling, however, process groups can and should be combined with psychoeducation or *process-based neuroeducation* (Luke & Schimmel, 2022). This approach uses neuroscience to illuminate client concerns—in Ricky's case, stress mismanagement leading to aggression. Neuroscience aids clients in understanding what is happening inside them, when, and how, and it offers insight and tools for managing these internal states and external behaviors (Beeson & Luke, 2023). Focusing on process-based education assists clients in identifying their blind spots by experiencing and elucidating dynamics happening in the here and now and offering them a chance to reconsider their interactive patterns and try a different approach to handling strong emotions.

Establishing Group Norms

When setting up a group, it is essential to talk to members about *group norms* or guidelines for interaction (Yalom & Leszcz, 2020). Neuroscience can inform our approach to establishing norms for member storytelling, safety, and respect.

Storytelling and the Neuroscience of Memory

One of the truisms of counseling is that it is the counselor's responsibility to teach clients how to be group members (Woodside & Luke, 2018). This is critical both to beginning and to finishing well in counseling, because group members are not likely to know either the group facilitator's expectations of them in this environment or what they can or should expect from counseling. For example, in our combined clinical experience, group members often believe that for counseling to be effective, they must tell the full story of their experience, repeatedly and in great detail. Members are correct to believe this in that they are attached to their stories and understanding clients' narratives is instructive. However, some clients have been misled into believing that they need to tell their stories with all the details because doing so is crucial for success. Given the current state of neuroscience and its integration with counseling, this may be erroneous reasoning. A group member's need to share their story may be linked to the memory's role in constructing a personal narrative when influenced by fluctuating perceptions and emotions.

Luke (2020) described the neuroscience of memory using the analogy of orange juice from concentrate. Orange juice from concentrate begins as freshly squeezed juice, which is then dehydrated and frozen. Later, the juice is thawed, and water is added. Technically, the result is orange juice, but there are differences in taste and quality between fresh-squeezed and reconstituted juice. Likewise, initial memory formation and retrieval, as occurs via the hippocampus, changes memory. Experiences go in one way and come out another, based on the emotional state and circumstances under which they are retrieved. These reconstituted memories resemble the original memory but are different. In addition, as memories are retrieved and filtered through the amygdala for emotional valence, details of the experience may shift as the brain attempts to reconstruct the emotional intensity. This is how the length of a fish in the 10th retelling of a "fish tale" may be longer than it was originally. This change is not necessarily about lying, but it is about reconstructing (reconstituting) the details of a story to match the emotion from the original experience. Group counselors may unintentionally do clients a disservice by encouraging them to relive events and experiences.

Furthermore, an individual's self-debasing or self-aggrandizing storytelling, in an attempt to establish reasons for their state, can have the effect of manipulating individual group members into agreeing with the client and enabling the client's helplessness. In turn, this type of unhealthy collusion can negatively affect the group dynamics in general. As an alternative, we invite group members to describe how a particular story is affecting them in the present time and place. For example, as a member begins to tell a story, we might ask the member to pause in the telling of it and focus on their breathing as the member recalls the situation. Reconsidering the memory in the here and now of emotions and context not only is much more informative than sharing the content of the story but also seems to have more positive clinical impact on the individual and the group as a whole.

> **Reflective Questions**
>
> What benefits and drawbacks might Ricky's storytelling serve him in the world outside of group?
> How might you use the group process to help Ricky develop greater self-reflective intentionality about what he shares and why he shares it?

If Ricky is allowed to reiterate his displeasure at being required to attend the group and retells the reasons he should not have to participate in group (e.g., because he was treated unfairly), the group leader may actually be reinforcing Ricky's displeasure (Hayes, 2004) by creating the perception that the past is present in the present. To establish a limited storytelling norm, Ricky's counselor describes the phenomenon of *contagion*, wherein the mirror neuron system in the brains of the group members simulates Ricky's described experience. This can lead to the emotional experiences of one person encroaching on the emotions of the observers (Iacoboni, 2009), particularly in vulnerable individuals.

Safety and the Neuroscience of Relationships

Porges's (2011) polyvagal theory describes regulation from the perspective of the 10th cranial nerve, the vagus nerve, and the ability to perceive threats and safety unconsciously. Porges referred to this as *neuroception.* The sensitivity of group members to threat and safety cues means that even the group meeting room must be selected and set up with intention (Porges & Flores, 2017). Seating members in a circle where every member sees every other person easily reduces implicit defense responses while initiating social networks in the brain (Porges & Flores, 2017). Porges's (2011) work supports the idea that room setup, a factor long recognized by group counselors, is an important component of the group process (Yalom, 1995). The room setup can activate what Cozolino (2014) referred to as the *social synapse,* the semipermanent connections between individuals that are analogous to neural synapses.

Respect and the Neuroscience of Communication

One of the important ingredients of the group process is the extending and receiving of respect. *Respect* means extending dignity to others, even when they may not present themselves in dignified ways. Respect is both dispositional and behavioral, and group provides an ideal environment in which to practice sending and receiving respect. In our groups, we emphasize that respect is defined by the person receiving messages, and the sending member cannot determine for the receiving member what is and is not respectful (Luke & Diambra, 2017). Respect is important for "quieting" the sympathetic nervous system and its threat response. Expressing boundaries and needs sets an expectation for respect and, by extension, safety. This can support the activation of the parasympathetic nervous system, leading to more regulated responses in the group context.

Establishing neural safety in group requires an acknowledgment that there are many nervous systems interacting in one place, so concepts such as respect must be discussed openly. In fact, the nervous system of any given individual is made up of their inheritance: sociocultural, genetic and biologic, relational, environmental, experiential, cognitive, affective, and behavioral (Luke, 2020; Luke & Schimmel, 2022). Individuals develop and make choices based on this multidimensional inheritance, and they bring this history with them into group. Therefore, group members must be able to communicate their expectations regarding respect, as well as learn to accept the experiences of respect and disrespect from others in nonthreatening ways. In essence, group is a place where members can begin to discern the functioning of their nervous system in a safe place.

We encourage participants to refrain from discussing in detail the behaviors of individuals who are not group members. Clients seem quite attentive to the actions of other people in their lives and will frequently use group time to discuss the actions/antics of people who are outside the group. As an alternative, we have a therapeutic phrase that we often use in gossip situations: "Don't tell us; show us. Show us how thinking about this experience is affecting you." As we have learned from Gestalt therapy and other experiential approaches, words can deceive, but the body rarely lies. For example, Ricky may initially deny feeling angry, but the group leader or another seasoned member may be attuned to Ricky bouncing his leg up and down quickly. Reflecting this back to Ricky provides him, along with other members, the opportunity to observe and reflect on the dissonance. Other members can be invited to tune in to their own body as they see Ricky's anger. Instead of allowing or encouraging Ricky to focus on details that distract from his actual feelings, the leader could encourage Ricky to redirect his focus on his physical movement and share more deeply about what he is feeling in the moment.

> **Reflective Question**
>
> What purpose might the concept of gossip manifest in a men's group on stress and anger?

Herein is the junction between group facilitation and neuroscience integration. In teaching and writing about the integration of neuroscience and counseling, we often ask clients and students to think about brain function in terms of the internet rather than as a telephone line. Despite descriptions of neurobiological structures, systems, and functions, the brain is best understood as a system of systems, constantly and inextricably connected. For the sake of simplicity, descriptions of neuronal communication are reduced to neuron-to-neuron communication. In the same way, and in terms of Gestalt therapy, accurate communication requires *holism*, the sense that words and behaviors are united (Perls, 1976). Counseling is more effective when clients can move out from behind their words (and words about others) to demonstrate their experience of relationships within the group environment.

Ricky may have learned that indirect communication is the least risky way of getting his needs met. In group, the counselor will assist him in bringing his communication into direct contact with others, both by speaking directly to group members when talking to and about them and via role plays in which he can practice direct communication. In the group, the counselor and members can supply Ricky with cues that remind him to regulate his sympathetic response to this new type of communication.

Focusing on Process: The Neuroscience of Group Therapeutic Factors

One of the ways that novice group leaders can get into trouble is by getting the "math" wrong. Counseling a group of five members is not like counseling five group members times one individual session. Group interactions are much more complicated than this simple math would indicate. Rather, it is more accurate to think of it as five group members times the number of individuals times the dyads and triads in the group, all multiplied by the group facilitator's interactions! This math exponentially increases the number of interactions and relationship dynamics occurring in the group counseling setting. Group work requires continual attention to each individual as well as to the interactions between any two or more group members, plus the potential responses of all group members to any comment or issue.

Managing these types of interactional dynamics requires a focused counselor. Information in Chapter 9 regarding how counselors can best attend is especially important to group work. One way that we might leverage these dynamics in a group with Ricky is to have each member of the group represent Ricky's self-talk. For example, we have asked clients like Ricky to identify self-defeating statements and then assign each group member one of those statements to repeat. Then, Ricky might be asked to stand in the center of the group and be asked to perform a Mini-Mental State Examination (Vertesi et al., 2001) item. For instance, the counselor may ask Ricky to count backward from 100 by sevens (reverse serial sevens). As Ricky prepares for the task, we also ask each group member to say the self-statement they are portraying aloud, increasing their volume each time. Rarely is a client able to complete this task without getting flustered, and for good reason. Through this exercise, Ricky can directly experience his negative self-talk during a basic task, highlighting his struggle to perform more complicated daily social tasks. He was just unaware of this clamor before this intervention.

> **Reflective Question**
>
> What might have been the consequence if the facilitator had focused solely on Ricky's story instead of asking other members to give input?

Groups that reach the working stage experience numerous benefits, including the ability of members to experience therapeutic factors of group work across all stages of group development. Yalom (1995) identified 11 factors that make group therapy effective and lead to positive outcomes, and for decades, these factors have driven the field of group counseling. The neuroscience implications of these factors (Yalom & Leszcz, 2020) are presented in Table 13.1.

TABLE 13.1

Neuroscience of Therapeutic Factors

Therapeutic Factor	Neuroscience Foundations and Implications for Group Counseling
Altruism	Altruism has a genetic and a developmental basis that must be fostered to prepare group members for social connection (Andreoni & Rao, 2011; Brouzos et al., 2021; De Dreu & Kret, 2016; Insel, 2010; Sonne & Gash, 2018)
Cohesion	Humans are wired for connection and belonging at the cellular level, but this must be channeled in a safe environment (Burlingame et al., 2018; Cikara & Van Bavel, 2014; Inzlicht et al., 2012; Norcross & Prochaska, 2018; Schermer, 2010; Schore, 2020)
Hope	Believing is seeing, so group leaders build hope in group members by stimulating positive affect and a sense of agency (Brouzos et al., 2021; Davis & Montag, 2019; Byrd & Luke, 2021)
Universality	Group members' egocentrism is a developmental process that can be challenged systematically in group (Brouzos et al., 2021; Denninger, 2010; Porges & Flores, 2017; Preckel et al., 2018)
Imparting information	Neuroeducation is best delivered through play and experiential activities, especially for stressed and anxious group members who may struggle to find words for their experience (Luke, 2020; Miller & Beeson, 2021)
Family dynamics	Involving parents in both play and psycho- or neuroeducation is vital, as is providing space for children and adolescents to relearn new perspectives and behaviors related to their family by practicing in group (Glynn et al., 2021; Lamar et al., 2021)
Imitative behaviors	Mirroring plays a critical role in learning through modeling and imitation, and group is a great place to harness this biological design (Ivey & Daniels, 2016; Purves et al., 2018; Ramachandran, 2000)
Socialization techniques	Group provides a laboratory of sorts that encourages healthy and effective expression of feelings and responses to the difficult expressions of feelings from others (Cozolino, 2017; Flores, 2010; Luke & Diambra, 2017; Schore, 2020; Wei et al., 2021)
Interpersonal learning	Group members are exposed to new and different ways of relating as people, which challenges their own communication and coping strategies (Brouzos et al., 2021; Palumbo et al., 2017; Siegel, 2006; Tan, 2017) Since
Catharsis	group members may not be forthcoming in disclosing their stress and anxiety levels, group provides a place for them to relieve tension through alternative forms of expression (Brouzos et al., 2021; Panksepp & Biven, 2012)
Existential factors	Adolescents interpret and make meaning from experiences differently from adults, so uncovering these meaning-making processes is important for recalibrating beliefs, feelings, and behaviors (Siegel, 2020)

Note. Adapted from *Applying Neuroscience to Counseling Children and Adolescents: A Guide to Brain-Based Interventions*, by C. Luke and C. J. Schimmel, pp. 116–120. Copyright © 2022 by Cognella. Adapted with permission.

> **Reflective Question**
>
> How might you draw on the power of group therapeutic factors to help Ricky?

Luke and Schimmel (2022) discussed three ways in which neuroscience informs group work. First, neuroscience validates and underscores the deep therapeutic processes involved in group, including interpersonal autonomic synchrony (Palumbo et al., 2017). Second, the neuroscience underlying therapeutic factors aids in guiding interventions. Third, neuroscience offers metaphors for human functioning (Luke, 2020) that facilitate engaging group activities (Luke & Schimmel, 2023). Therefore, our brain-based approach to group involves leveraging the neuroscience of group therapeutic factors through metaphors and demonstrations.

For example, we may use neuroeducation as *imparting information* to build group *cohesion*. There is a simple illustration that group members tend to understand regarding the blind spot in the eye. Luke (2020) described how counselors can use this analogy, and we apply that to Ricky's situation here. The optic nerve attaches to the back of the eye via the retina. At the point of this connection, there are no light receptors, which creates a blind spot in the field of vision. This blind spot is largely undetectable because the two eyes have overlapping fields of vision, which compensates for the missing information. The occipital lobe stitches together the two images to provide one complete picture.

In a group session, members' blind spot can be revealed through a simple experiment, wherein a plus (+) sign is drawn on the far-right side of an 8 × 11-inch piece of paper. On the far-left side of the paper, an O symbol is drawn (see Figure 13.1). Now, group members are instructed to close their right eye and look at the + sign with their left eye and then to move the paper closer to and further away from their face (up to 12 inches away). At a certain point, the O symbol will disappear. The application of this activity highlights that we all have blind spots, whether we acknowledge them or not, and that the ways in which we fill in missing information will affect how we function in relationships. Furthermore, the group has the effect of being that second eye that helps us see more fully and improve the functioning of our nervous system. Using this activity can be a fun and "eye-opening" way to engage group members.

FIGURE 13.1

Blind Spot Experiment

> **Reflective Question**
>
> How could a brain-based understanding of humans' interconnectedness help the group leader engage with Ricky?

Our Brain-Based Approach to the Case of Ricky

In our brain-based approach to group counseling, we prepared Ricky for the group during the planning and prescreening processes. We facilitated positive expectations for the group experience, helping Ricky tolerate the ambiguity of the group experience. Once in group, we used experiential activities to engage Ricky. For instance, when Ricky experienced the intolerable state of not storytelling, we encouraged him to demonstrate his feelings related to holding back. When he did, his behaviors signaled memories, cognitions, and emotions in other group members as their brains and bodies connected to his. Through this process, Ricky came to develop a new skill set, that of bringing his behavior into harmony, or *congruence*, with his verbalizations. Ricky's body acted out the feelings in his limbic system, which up to this point had been circumvented by his cerebral cortex. This led him to share his feelings verbally, rather than simply letting his feelings "leak" into his behaviors. This could work for Ricky in reverse as well. Ricky could address his anger via his body, through the so-called *bottom-up* process. Rather than focusing on his self-talk, which is a top-down, forebrain-based intervention, Ricky could be guided to use his body to calm his mind. Exercises such as breathing, yoga, and progressive muscle relaxation are examples of such activities. A renewed sense of wholeness played out in Ricky's relationships with his fellow group members. Eventually, Ricky became more congruent in his relationships outside of group in his personal life. By the close of his group experience, Ricky expressed fewer mixed messages, reduced his passive-aggressive comments, and reported more relational satisfaction.

Conclusion

This chapter offered a method for facilitating group work that is experiential and process based using neuroscientific discoveries to enlighten our understanding of how such an approach can be advantageous. Group facilitators do not need extensive neuroscientific knowledge to use the principles of neuro-informed counseling. In fact, Yalom and Leszcz's (2020) admonition to stay in the here and now of group, and for facilitators to illuminate the group process, rings true in our brain-based approach to group counseling.

Quiz

1. What factors make group so complex?
    a. The numerous points of interactions (dyads and triads) between and among members to which a facilitator must attend.
    b. The sheer number of people in a group at a given time.
    c. Therapeutic factors that make it difficult for the group to come together to work on their goals.
    d. The need for group leaders to maintain control of the group at all times and members' unpredictability.
2. According to the authors of this chapter, what is the problem with storytelling from a neuroscience perspective?
    a. Nothing is wrong with storytelling because catharsis is one of the therapeutic factors.
    b. Group members are often disingenuous and should not be trusted to tell their story.
    c. Storytelling can activate group members' fight-or-flight response; a "no storytelling" guideline is for their protection.
    d. Memory is not recall as much as it is reconstitution, so storytelling in the wrong context can bias the story and distort perception.

References

Andreoni, J., & Rao, J. M. (2011). The power of asking: How communication affects selfishness, empathy, and altruism. *Journal of Public Economics, 95*(7–8), 513–520. https://doi.org/10.1016/j.jpubeco.2010.12.008

Baourda, V. C., Brouzos, A., Mavridis, D., Vassilopoulos, S. P., Vatkali, E., & Boumpouli, C. (2022). Group psychoeducation for anxiety symptoms in youth: Systematic review and meta-analysis. *The Journal for Specialists in Group Work, 47*(1), 22–42. https://doi.org/10.1080/01933922.2021.1950881

Beeson, E. T., & Luke, C. (2023). Neuroscience-informed counseling: Another lens to view the human experience. In R. Fulmer (Ed.), *Counseling and psychotherapy: Theory and beyond* (pp. 254–310). Cognella.

Brouzos, A., Vassilopoulos, S. P., Stavrou, V., Baourda, V. C., Tassi, C., & Brouzou, K. O. (2021). Therapeutic factors and member satisfaction in an online group intervention during the COVID-19 pandemic. *Journal of Technology in Behavioral Science, 6*(4), 609–619.

Burlingame, G. M., McLendon, D. T., & Yang, C. (2018). Cohesion in group therapy: A meta-analysis. *Psychotherapy, 55*(4), 384–398. https://doi.org/10.1037/pst0000173

Byrd, R., & Luke, C. (2021). *Counseling children and adolescents: Cultivating empathic connection*. Routledge.

Cikara, M., & Van Bavel, J. J. (2014). The neuroscience of intergroup relations: An integrative review. *Perspectives on Psychological Science, 9*(3), 245–274. https://doi.org/10.1177/1745691614527464

Council for Accreditation of Counseling and Related Educational Programs. (2024). *2024 CACREP standards*. https://www.cacrep.org/wp-content/uploads/2023/06/2024-Standards-Combined-Version-6.27.23.pdf

Cozolino, L. (2017). *The neuroscience of psychotherapy: Healing the social brain* (3rd ed.). W. W. Norton & Co.

Davis, K. L., & Montag, C. (2019). Selected principles of Pankseppian affective neuroscience. *Frontiers in Neuroscience, 12*, Article 1025. https://doi.org/10.3389/fnins.2018.01025

De Dreu, C. K., & Kret, M. E. (2016). Oxytocin conditions intergroup relations through upregulated in-group empathy, cooperation, conformity, and defense. *Biological Psychiatry, 79*(3), 165–173. https://doi.org/10.1016/j.biopsych.2015.03.020

Denninger, J. W. (2010). Commentary on the neurobiology of group psychotherapy: Group and the social brain: Speeding toward a neurobiological understanding of group psychotherapy. *International Journal of Group Psychotherapy, 60*(4), 595–604. https://doi.org/10.1521/ijgp.2010.60.4.595

Flores, P. J. (2010). Group psychotherapy and neuro-plasticity: An attachment theory perspective. *International Journal of Group Psychotherapy, 60*, 546–570. https://doi.org/10.1521/ijgp.2010.60.4.546

Glynn, L. M., Davis, E. P., Luby, J. L., Baram, T. Z., & Sandman, C. A. (2021). A predictable home environment may protect child mental health during the COVID-19 pandemic. *Neurobiology of Stress, 14*, Article 100291. https://doi.org/10.1016/j.ynstr.2020.100291

Havlik, S., Malott, K., Davila, J. D., Stanislaus, D., & Stiglianese, S. (2020). Small groups and first-generation college goers: An intervention with African American high school seniors. *The Journal for Specialists in Group Work, 45*(1), 22–39. https://doi.org/10.1080/01933922.2019.1699618

Hayes, S. C. (2004). Acceptance and commitment therapy, relational frame theory, and the third wave of behavioral and cognitive therapies. *Behavior Therapy, 35*(4), 639–665. https://doi.org/10.1016/S0005-7894(04)80013-3

Iacoboni, M. (2009). *Mirroring people: The new science of how we connect with others*. Picador.

Insel, T. R. (2010). The challenge of translation in social neuroscience: A review of oxytocin, vasopressin, and affiliative behavior. *Neuron, 65*(6), 768–779. https://doi.org/10.1016/j.neuron.2010.03.005

Inzlicht, M., Gutsell, J. N., & Legault, L. (2012). Mimicry reduces racial prejudice. *Journal of Experimental Social Psychology, 48*(1), 361–365. https://doi.org/10.1016/j.jesp.2011.06.007

Ivey, A. E., & Daniels, T. (2016). Systematic interviewing microskills and neuroscience: Developing bridges between the fields of communication and counseling psychology. *International Journal of Listening, 30*(3), 99–119. https://doi.org/10.1080/10904018.2016.1173815

Kul, A., & Hamamci, Z. (2021). The effect of an anxiety-coping program for children based on cognitive behavioral therapy on 4th graders' anxiety levels. *Education Quarterly Reviews, 4*(2), 287–300. https://doi.org/10.31014/aior.1993.04.02.280

Lamar, M. R., Speciale, M., Forbes, L. K., & Donovan, C. (2021). The mental health of US parents during the COVID-19 pandemic. *Journal of Mental Health Counseling, 43*(4), 319–335. https://doi.org/10.17744/mehc.43.4.03

Luke, C. (2020). *Neuroscience for counselors and therapists: Integrating the sciences of mind and brain* (2nd ed.). Cognella.

Luke, C., & Diambra, J. (2017). Neuro-informed group work. In T. Field, L. Jones, & L. Russell-Chapin (Eds.), *Neurocounseling: Brain-based clinical approaches* (pp. 179–194). American Counseling Association. https://doi.org/10.1002/9781119375487.ch12

Luke, C., & Schimmel, C. J. (2022). *Applying neuroscience to counseling children and adolescents: A guide to brain-based interventions*. Cognella.

Luke, C., & Schimmel, C. J. (2023). Using neuroscience-based group work with children and adolescents affected by the pandemic. *The Journal for Specialists in Group Work, 48*(1), 20–31. https://doi.org/10.1080/01933922.2022.2158972

Masten, A. S., Lucke, C. M., Nelson, K. M., & Stallworthy, I. C. (2021). Resilience in development and psychopathology: Multisystem perspectives. *Annual Review of Clinical Psychology, 17*, 521–549. https://doi.org/10.1146/annurev-clinpsy-081219-120307

Miller, R., & Beeson, E. T. (2021). *The neuroeducation toolbox: Practical translations of neuroscience in counseling and psychotherapy*. Cognella.

Palumbo, R. V., Marraccini, M. E., Weyandt, L. L., WilderSmith, O., McGee, H. A., Liu, S., & Goodwin, M. S. (2017). Interpersonal autonomic physiology: A systematic review of the literature. *Personality and Social Psychology Review, 21*(2), 99–141. https://doi.org/10.1177/1088868316628405

Panksepp, J., & Biven, L. (2012). *The archaeology of mind: Neuroevolutionary origins of human emotions*. Norton.

Perls, F. (1976). *The Gestalt approach and eyewitness to therapy*. Bantam Books.

Porges, S. W. (2011). *The polyvagal theory: Neurophysiological foundations of emotions, attachment, communication, and self-regulation*. Norton.

Porges, S. W., & Flores, P. J. (2017). Group psychotherapy as a neural exercise: Bridging polyvagal theory, and attachment theory. *International Journal of Group Psychotherapy, 67*, 202–222.

Preckel, K., Kanske, P., & Singer, T. (2018). On the interaction of social affect and cognition: Empathy, compassion and theory of mind. *Current Opinion in Behavioral Sciences, 19*, 1–6. https://doi.org/10.1016/j.cobeha.2017.07.010

Prochaska, J. O., & Norcross, J. C. (2018). *Systems of psychotherapy: A transtheoretical analysis.* Oxford University Press.

Purves, D., Augustine, G. J., Fitzpatrick, D., Hall, W. C., Lamantia, A. S., Mooney, R. D., Platt, M. L., & White, L. E. (2018). *Neuroscience* (6th ed.). Sinauer Associates.

Ramachandran, V. S. (2000, May 31). *Mirror neurons and imitation learning as the driving force behind "the great leap forward" in human evolution.* Edge. http://www.edge.org/3rd_culture/ramachandran/ramachandran_p1.html

Schermer, V. L. (2010). Mirror neurons: Their implications for group psychotherapy. *International Journal of Group Psychotherapy, 60*(4), 486–513. https://doi.org/10.1521/ijgp.2010.60.4.486

Schimmel, C. J., & Jacobs, E. (2019). Small group counseling. In A. Vernon & C. J. Schimmel (Eds.), *Counseling children and adolescents* (5th ed., pp. 418–456). Cognella.

Schore, A. N. (2020). Forging connections in group psychotherapy through right brain-to-right brain emotional communications. Part 1: Theoretical models of right brain therapeutic action. Part 2: Clinical case analyses of group right brain regressive enactments. *International Journal of Group Psychotherapy, 70*(1), 29–88. https://doi.org/10.1080/00207284.2019.1682460

Siegel, D. J. (2006). An interpersonal neurobiology approach to psychotherapy: Awareness, mirror neurons, and neural plasticity in the development of well-being. *Psychiatric Annals, 36*(4), 248–256.

Siegel, D. J. (2020). *The developing mind: How relationships and the brain interact to shape who we are.* Guilford Press.

Sonne, J. W., & Gash, D. M. (2018). Psychopathy to altruism: Neurobiology of the selfish-selfless spectrum. *Frontiers in Psychology, 9,* Article 575. https://doi.org/10.3389/fpsyg.2018.00575

Tan, L. (2017). When neurobiology meets psychotherapy: Mirror neurons, the social brain and group work with the addicted population. *Addiction Research, 1*(1), 1–4.

Vertesi, A., Lever, J. A., Molloy, D. W., Sanderson, B., Tuttle, I., Pokoradi, L., & Principi, E. (2001). Standardized Mini-Mental State Examination. Use and interpretation. *Canadian Family Physician, 47*(10), 2018–2023.

Wei, M., Wang, L., & Kivlighan, D. M., Jr. (2021). Group counseling change process: An adaptive spiral among positive emotions, positive relations, and emotional cultivation/regulation. *Journal of Counseling Psychology, 68*(6), 730–745. https://doi.org/10.1037/cou0000550

Woodside, M., & Luke, C. (2018). *Empowering the practicum students: A developmental guide.* Cognella.

Yalom, I. D. (1995). *Theory and practice of group psychotherapy* (5th ed.). Basic Books.

Yalom, I. D., & Leszcz, M. (2020). *Theory and practice of group psychotherapy* (6th ed.). Basic Books.

Part III
ADVANCED APPLICATIONS

The first two sections of the text provided foundational knowledge and described how neuroscience can be integrated into every aspect of counseling practice. This third section presents advanced applications of a brain-, mind-, and body-based approach. The foci of the chapters in this section are applications to research and counselor supervision, as well as basic principles for integrating neuroscience into your own counseling practice.

Chapter 14

Enhancing Counseling Practice With Neuroscience-Informed Research

G. Michael Russo, Eric T. Beeson,
and Isaac Burt

This chapter proposes foundational steps that counselors can take to enhance their understanding of neuroscience-informed counseling research, an understanding that could be beneficial for supporting client success and wellness. In addition, this chapter aims to empower counselors to reconsider the role of research as an incredibly powerful form of client and systemic advocacy.

2024 CACREP Standards

This chapter addresses 2024 Council for Accreditation of Counseling and Related Educational Programs (CACREP) Standards pertinent to the Foundational Counseling Curriculum (Section 3) area of Research and Program Evaluation (Standard H):

- The importance of research in advancing the counseling profession, including the use of research to inform counseling practice (Standard H.1.)
- Identification and evaluation of the evidence base for counseling theories, interventions, and practices (Standard H.2.)
- Analysis and use of data in research (Standard H.6.)

This chapter also addresses the following Doctoral Standards for Counselor Education and Supervision (Section 6) in the area of Doctoral Curriculum (Standard B):

- Methods for evaluating counseling effectiveness (Standard B.1.e.)

Case Study: Staff Sergeant Ryan

Susan is an independent licensed professional counselor and is working with Ryan, an active-duty staff sergeant in the U.S. Army who has been experiencing posttraumatic stress disorder (PTSD) symptoms since suffering a blast-related injury. Ryan identifies as a cisgender male who uses he/him pronouns. He is 26 years of age and married, and his social and emotional development falls within the normal range for his age. He holds a high school diploma, which is his highest level of education. Ryan identifies as Christian and reports a strong sense of support and strength through his faith. He identifies as biracial and maintains close relationships with his White mother and Black father. He proudly reports that he is the fourth generation of his family to serve in the U.S. Army. Ryan has a close network of friends whom he identifies as his brothers and sisters. In addition, he draws a strong sense of support from his spouse. Ryan identifies that to his knowledge, he is the first person in his family to seek professional support for mental health treatment. He denies current alcohol or substance use and denies any history of substance use but consumes an average of 32 ounces of coffee per day. Ryan reports experiencing flashbacks to the moments before his convoy hit a roadside improvised explosive device (IED). Ryan reports that following his blast injury, he felt guilt associated with surviving and experienced passive suicidal ideation without intent. Other than this instance, Ryan says, he has never experienced suicidal ideation in any form. He denies concerns relating to his food consumption, weight, muscle mass, and/or physical appearance. In terms of sleep, he reports difficulty staying asleep, averaging only 6 hours of sleep per night because of night terrors. Ryan pays for services out of pocket to avoid insurance reimbursement claims and is reluctant to accept medical treatments such as psychotropic medications, which he believes might have negative repercussions for his Army career.

Susan, having learned that an IED explosion can result in traumatic brain injury (TBI), encourages Ryan to seek further medical assessment to explore potential impacts and/or treatments regarding TBI-related injuries. Because attempts to utilize prolonged exposure and other trauma-focused approaches have resulted in above-average levels of distress for Ryan, Susan thinks a neuroscience-informed counseling approach could be beneficial. However, she is not entirely sure how to begin integrating neuroscience into her counseling practice with Ryan.

Upon consulting with her former professors, Susan begins taking steps to enhance her understanding of neuroscience-informed counseling literature. As she begins this journey, Susan quickly realizes that she has limited understanding of basic neuroanatomy and neurophysiology. This lack of knowledge prevents her from adequately evaluating the extent to which theories of neuroscience are grounded in evidence-based studies. The remainder of the chapter is focused on steps that counselors like Susan can take to better understand, and to ethically implement, counseling interventions that have a sound basis in neuroscience.

Understanding Neuroscience-Informed Counseling Research

To begin this journey of understanding neuroscience-informed counseling research, Susan spoke with her former professor, who suggested that she undertake a five-step process:

1. Identify neuroscience-informed counseling research that aligns with her areas of interest
2. Explore sources and find client voices in quantitative research
3. Read the research with a sense of curiosity
4. Understand the concept of generalizability and related statistical concepts such as significance testing and effect size
5. Consult with colleagues who have knowledge in the areas of neuroscience and neuroscience-informed counseling and integrate research findings into practice

Step 1: Identify Research in Your Areas of Interest

Starting from a strength-based perspective can be a powerful approach when embarking on a voyage of new counseling practices. Simply reflecting on areas that you are passionate about and/or skilled in can be a valuable foundation as you begin your neuroscience-informed counseling journey. In the case of Susan, she had experience and training in the treatment of people with traumatic experiences. However, she did not have the same level of expertise in approaching trauma from a neuroscience-informed counseling perspective. One of the first actions she had to take was to understand basic neuroanatomy and physiology of trauma in order to later understand how neuroscience can inform trauma treatment. A neuroscience taskforce commissioned by the American Mental Health Counselors Association determined that counselors must first learn foundational neuroanatomy and physiology before learning about neurophysiological aspects of clinical presentations and interventions (Field et al., 2022). Susan learned about the basics of the stress response system and hypothalamic-pituitary-adrenal axis.

As a member of the American Counseling Association (ACA), Susan had online access to articles published in ACA's *Journal of Counseling & Development*. Searching for those that combine neuroscience with trauma treatment, she found a wide range of articles, including literature reviews and outcome studies.

Reflective Question

Where else would you encourage Susan to look for neuroscience-informed counseling literature, and what specific resources would you recommend?

When exploring neuroscience-based research, Susan came across research that involves functional MRI (fMRI) and/or electroencephalography (EEG). Both methods have benefits as well as weaknesses that are important to understand to assist in evaluating research that has utilized these methods. However, one of them, fMRI, is an incredibly powerful tool for locating very precise areas in the brain that are activated by a particular event or phenomenon. It operates by identifying differences in oxygenation levels in the brain, which represent brain activity (Dewiputri & Auer, 2013). Data from fMRIs can provide information about a precise region of interest (ROI) and/or neuronal circuits in the brain that are associated with specific events or tasks. When exploring the regions in the brain, fMRI researchers will often describe voxels. A *voxel* is a series of computer-generated cubes that are 3 millimeters wide and are used to conceptualize activity within the brain through the analysis of oxygenation levels (Kriegeskorte et al., 2010). To obtain a degree of precision, multiple imaging samples are required, with each taking 100 milliseconds (ms) to complete; some images take 800–900 ms to complete (Kriegeskorte et al., 2010). Thus, the sampling rate of fMRI can be understood as 10 samples per second, with complete imaging times taking up to nearly a full second.

EEG sacrifices spatial orientation for speed. For example, one vendor (see www.brainmaster.com) has asserted that its EEG amplifiers can capture up to 1,024 samples per second and provide feedback to participants at the rate of 256 samples per second. Ortinski and Meador (2004) found that humans take an average of 300–500 ms to become conscious of stimuli, which means that neurofeedback using EEG can record information and feedback at a preconscious rate. EEG also differs from fMRI in that its data represent electrical activity created when nerves fire in the brain as opposed to oxygenation levels in the blood. This electrical activity is represented visually in the form of brain waves. Brain waves are often categorized and named based on the frequency with which they fire. Brain waves range in speed, with the lowest being delta and the most rapid being gamma (see the glossary for additional information about brain waves).

Step 2: Explore Sources and Find Client Voices in Quantitative Research

There are several sources of information available to counselors, and the origin of the source can greatly influence the way that counselors interact with the work. These sources are often categorized as primary or secondary/tertiary sources of information. Primary sources of information include the reporting of direct participant experiences via quantitative or qualitative means. In qualitative research, the participants' spoken or written responses to questions may be reported verbatim. In terms of quantitative research, participant voices are present in each data point. A person's endorsement of an item on a psychological test represents their voice. The person might think to themselves when taking this assessment, "Yes, this item sounds a lot like me," "No, this item does not sound like me," or "I really don't understand or care about this assessment, so I'm going to arbitrarily select this item." Regardless of the decisions that the person makes, their voice is omnipresent in all forms of psychometric assessment and quantitative research.

In comparison with primary sources, secondary/tertiary sources of information often involve the researcher creating a theoretically based argument through the synthesis of previously conducted primary research. Both categories of research (primary and secondary/tertiary) have their merits, but the implications associated with using this information can greatly impact the quality of counseling services that we provide.

Let us take, for example, Rock et al.'s (2012) model of the *healthy mind platter*, which illustrates the importance of balancing time among wellness-promoting activities in seven domains, such as time-in, connecting time, and downtime (see Figure 8.1). Rock et al. established the foundation of the model by synthesizing and categorizing neuroscience literature that aligned with the seven domains. The development of the domain of *playtime* arose when Rock et al. synthesized research completed by several authors. For example, Panksepp and Burgdorf (2003) were a secondary source in that they synthesized a subset of the literature that explored ultrasonic vocalization patterns in rat studies and research findings associated with human studies of joy. So, why does it matter that previous research was synthesized to create this model? Every time that research is conducted, there is room for error, particularly in social science research where there are often large degrees of variation between participants.

Does this mean that the healthy mind platter should not be used? No, of course not. We know there are many clinical benefits associated with encouraging our clients to be mindful of ways that promote their own wellness. However, the identification of primary, secondary, tertiary, and other such sources is important to consider, because the further one gets from primary sources, the more likely one is to make overgeneralizations or attribution-based errors.

Step 3: Read With a Sense of Curiosity

Once able to identify the differences between sources, as well as the strengths and limitations of various source types, Susan was better equipped to approach the literature with a sense of curiosity. When learning about a new body of knowledge, a meta-analysis or systematic review is a great place to start because it analyzes or summarizes findings from multiple studies conducted within that topic area. Susan identified a meta-analysis by Russo et al. (2022) that explored the use of neurofeedback to treat anxiety and PTSD. The meta-analysis consisted of studies in which all participants received neurofeedback (i.e., pre-experimental) as well as studies that assigned participants to a neurofeedback or alternatively comparable type of treatment, such as a treatment-as-usual condition (experimental research). Russo et al. found that neurofeedback is efficacious in reducing anxiety as well as symptoms of PTSD. In contrast, Susan found another recent meta-analysis by Balkin et al. (2022) that examined the use of eye movement desensitization and reprocessing (EMDR) for trauma and found less compelling evidence.

To make their claims, both Russo et al. (2022) and Balkin et al. (2022) relied on effect size metrics, one of which was Hedges's g, which is considered an unbiased effect size measure because sample size is considered in the computation. These study findings support the efficacy of neurofeedback in the reduction of symptoms associated with anxiety and PTSD among adults. This aspect was interesting to Susan because, based on this research, neurofeedback could be a viable avenue for treatment/referral.

Prior to presenting these findings to her client or continuing on with the steps suggested by her former professor, Susan decided to synthesize major points made by Russo et al. (2022). First, Susan noted that the authors collected information on every article that had been published on the topic and devoted significant discussion within the article to removing and accounting for bias. In terms of the failsafe *n*, Russo et al. reported that for each study included in the single-group study, a total of 720 missing or unpublished studies would need to be found to offset the findings. This aspect helped Susan to feel comfortable in incorporating the findings into her practice with Ryan, because the likelihood of finding hundreds of unpublished studies that indicated that neurofeedback was not effective for anxiety-spectrum disorders seemed highly unlikely.

Second, Susan noted that in both arms of the meta-analysis (single-group and between-group analyses), PTSD symptoms were reduced among participants who received neurofeedback (Russo et al., 2022). Furthermore, in future studies, 9 out of 10 participants would be likely to experience a reduction in anxiety-spectrum disorders in single-group treatment conditions and 8 out of 10 participants would be likely to experience a reduction in anxiety-spectrum disorders in between-group treatment conditions (Russo et al., 2022).

After exploring the literature, Susan decided to seek out additional training in neurofeedback. She also began to look for counselors who already utilize neurofeedback to incorporate into her network of client referrals.

Step 4: Understand Generalizability and Types of Statistical Significance

Reviewing her notes from the meta-analysis, Susan wondered whether the findings would be applicable to Ryan. She also found herself perplexed about the differences between statistical, practical, and clinical significance in the research findings. Thus, she decided to review her statistics textbook and notes from a research methodologies course she had taken in her university counseling program.

Generalizability

When exploring primary sources of literature, counselors should keep in mind the degree of similarities or differences between the reported experiences of study participants and the experiences of the client whom the counselor aims to serve. This aspect is especially salient when exploring primary sources of neuroscience research, which often include animal studies. Rarely, if ever, can counselors generalize findings that are based on animal studies to their human clients. However, animal studies can be important for informing counseling practices or future counseling research, as shown by the work of psychologist B. F. Skinner. Skinner's theory of *operant conditioning* posits that animals are more likely to continue a behavior if they receive reinforcement as a consequence for performing it. This theory was highly influential to modern forms of behaviorist intervention, such as applied behavior analysis, behavior modification, and token economy systems, and even to early models of cognitive behavior therapy.

An equally troublesome issue is the lack of diversity when using human subjects (De Wolfe et al., 2021). Many rigorous experiments use large population samples but neglect to include a range of participants that reflects racial and ethnic diversity (Burt & Pankey, 2023). Although it is true that humans share 99.9% of genes, there are societal, cultural, and institutional factors that can affect the brain in various adaptive and negative ways (Aroke et al., 2019; National Human Genome Research Institute, 2018). As indicated by *epigenetics* (i.e., the study of how environments can enhance or suppress gene expression), culture matters at a neurobiological level (Jones et al., 2021). Thus, researchers cannot simply ignore how society's norms influence populations, especially those that have been historically disenfranchised or marginalized (Myers et al., 2022). In the United States, Black, Indigenous, and people of color (BIPOC) are particularly underrepresented in research (Burt & Pankey, 2023; Day-Vines et al., 2022). Counselors applying techniques from studies that lack diverse participants need to be cognizant of this oversight and recognize that generalizability may not exist.

A salient example in neuroscience concerns studies that utilize EEGs. Because EEG probes must adhere to the scalp, some populations (e.g., Blacks/African Americans) have not been included in some research samples (Wheaton, 2021). Much of the research using EEG has relied on apparatuses that produce acceptable-quality data only if the study participant has short or thin hair, thereby excluding participants who have another hair type or style (e.g., curly hair, braids, cornrows; Choy et al., 2021). Yet many researchers and practitioners still generalize the results to all populations, regardless of race or culture (Chatters et al., 2022). Although this tendency may be acceptable in the physical sciences, counselors and other professionals working in the social sciences need to operate at a different level (Johnson et al., 2022).

Furthermore, researchers should address the possible effects of racism and social inequalities when presenting their study results. The casting of blame on minority populations for environmental conditions is a current and historical concern that has implicitly reinforced scientific racism in the social sciences (Burt & Pankey, 2023). *Scientific racism*, operationally defined, is a pattern of antediluvian and archaic ideologies and ideas that yields dubious prejudiced or biased empirical conclusions (National Human Genome Research Institute, 2023). For example, Opara et al. (2022) pointed to a medical research study whose authors attributed disproportionate rates of COVID-19 infection in Black communities to potential genetic differences rather than to the risk vectors that result from environmental conditions linked to structural inequality and poverty. Opara et al. equated these researcher inferences with *drapetomania*, a diagnosis once used to label Black slaves who intended or attempted to escape slavery as suffering from a mental illness. In short, overlooking the social inequities that can lead to disproportionate negative mental health outcomes in favor of purported biogenic causes can reinforce racist attitudes that were on display during the era of slavery (Opara et al., 2022). The American Psychological Association (2021) acknowledged its history of supporting scientific racism and issued an apology. Specifically, they stated the following:

> The American Psychological Association . . . was complicit in contributing to systemic inequities, and hurt many through racism, racial discrimination, and denigration of people of color, thereby falling short on its mission to benefit society and improve lives. (para. 1)

Many scholars and practitioners alike unfortunately have replicated and promulgated research that blames social inequalities on BIPOC peoples and communities. In many neuroscience studies, the researcher's theory drives the interpretation of the data, not the other way around (Choy et al., 2021). To be wise consumers of neuroscience research, counselors need to recognize these shortcomings, avoid misattribution of biogenic cause for the effects of environmental conditions, and refrain from automatically generalizing results without considering cultural and societal factors. Counselors are encouraged to locate and appraise research that is informed by studies whose scope and participants closely align with the unique multicultural identities of the clients whom they aim to serve. Counselors can determine whether a particular article's findings can be generalized to address the needs of their client by determining the degree to which the participant identities reported in the study align with the identity of their client.

Statistical Significance

The first form of significance that we will discuss is statistical significance, a form that many will find familiar. *Statistical significance* quantifies the likelihood, or probability, that a researcher has incorrectly rejected the null hypothesis in favor of their proposed alternative hypothesis. Statistical significance reflects this probability in terms of a p value, usually presented as $p < .05$, $p < .01$, or $p < .001$. Respectively, these values reflect a 5%, 1%, or 0.1% chance that the researcher has incorrectly rejected the null hypothesis in favor of the alternative. For example, researchers recognize that $p < .05$ means that 5 times out of 100 the implications of the findings are completely wrong, and the null hypothesis should have been accepted over the alternative hypothesis.

This form of significance testing dates back to the work of Sir Ronald Fisher, who was a member of the Society for Psychical

Research. The Society was formed in 1882 to explore paranormal claims using scientific methodologies and experimentation (Weaver, n.d.). Fisher's concept of null hypothesis testing allowed that researchers could determine only one of two truths: Paranormal activity was real, or paranormal activity was not real. Evidence of the application of null hypothesis testing is found in Fisher's 1929 article, wherein he described the statistical likelihood that a person with clairvoyant powers could correctly guess a specific card drawn from a standard deck of 52 playing cards. Fisher (1929) proposed that it was possible to scientifically test whether the purported clairvoyant, Miss Josephine, possessed clairvoyant powers by "scor[ing] complete success once in 26 trials, whereas one with no clairvoyant powers whatever would score complete success once in 52 trials, or just half as often" (p. 192).

Hypothesis testing is not without flaws. Most notable of the flaws is the potential for sample size to skew results. To illustrate this point, let us consider the following scenario of comparing mean differences between two samples: A school counselor wants to see if students who completed their SAT mathematics preparation course scored statistically higher than an average person tested on the math section of the SAT. The school counselor convenes a sample of 25 students with an average score of 520 on the math section. The school counselor knows the average SAT math score for the prior year was 500, with a standard deviation of 100. Based on this knowledge, the school counselor uses the standard z-score formula:

$$z = \frac{\bar{x} - \mu}{\frac{\sigma}{\sqrt{n}}}; \; z = \frac{520 - 500}{\frac{100}{\sqrt{25}}}; \; z = \frac{520 - 500}{\frac{100}{5}}; \; z = \frac{20}{20}; \; z = 1$$

By applying and reducing the values, the school counselor obtains a z score of 1, which is not statistically significant at the .05 probability level based on the critical z value of 1.96. However, if the school counselor increases the sample size to 100 students, also with an average SAT math score of 520, the outcome is different: The z score is 2, which is statistically significant at the .05 level. As this scenario highlights, one fundamental flaw in statistical testing is that sample size alone can impact results. By introducing additional forms of significance testing, we become better equipped to recognize the strengths and limitations of our intervention.

Practical Significance

Another form of significance, practical significance, is determined based on effect size, or correlational analyses. *Practical significance* differs from statistical significance in that it explores the strength of the relationship between variables. In other words, we can determine if there is a small, medium, or large relationship between our therapeutic intervention and the reduction of client symptoms. Table 14.1 shows the conventions that are used to determine whether effect sizes are small, medium, or large.

TABLE 14.1

Conventions Used With Effect Size (ES) Measures

Common ES Measure	Small ES	Medium ES	Large ES
Cohen's f	0.10–0.24	0.25–0.39	0.40 and up
Pearson's r	0.10–0.23	0.24–0.36	0.37 and up
Eta squared (η^2)	0.01–0.058	0.059–0.137	0.138 and up
Cohen's d	0.20–0.49	0.50–0.79	0.8 and up
Hedges's g	0.30–0.49	0.50–0.66	0.67 and up
Standard mean difference (SMD)	0.30–0.49	0.50–0.66	0.67 and up

Many effect size measures can be viewed interchangeably. However, Cohen's *d* (Cohen, 1988) does not account for the effect that sample size has on results (Borenstein et al., 2009; Lipsey & Wilson, 2001). By contrast, Hedges's *g* includes a modification to account for sample size, which gives it the distinction of being an unbiased measure (Borenstein et al., 2009; Erford et al., 2010; Hedges, 1981; Lipsey & Wilson, 2001; Watson et al., 2016).

This interpretation promotes transparency when determining if a small, moderate, or large effect was noted in the study. In other words, effect size tells us the strength of the effect that the intervention (e.g., neurofeedback) had on reducing the dependent variable (e.g., scores on PTSD assessments). Practical significance allows us to answer the question, How efficacious is this treatment for this individual? The stronger the effect, the more beneficial the treatment is to empower client success and growth in therapy.

We can also ask a more advanced form of the aforementioned question: How efficacious is this treatment for this population? This line of questioning is what is considered in quantitative meta-analyses. In quantitative meta-analyses, one of the main goals is to understand the overall *efficacy* or results of a study in relation to all the other studies that have been published on the same topic (Borenstein et al., 2009). Meta-analyses are an excellent place to start when engaging with a new body of literature, because the author of the meta-analysis has collected and analyzed all available research within an entire body of literature to explore the efficacy, or practical significance, associated with the ability of an intervention or course of treatment to positively benefit an entire population of people. This collection of all published research creates the population, which is then analyzed to determine the overall degree of efficacy or inefficacy of the intervention. However, it is important to note that articles published during or after the publication of a meta-analysis are not included in the meta-analysis, which means that it is important to also consider any later articles when making clinical decisions.

Clinical Significance

Another form of significance testing concerns a question that many counselors ask themselves each and every day: Is my treatment clinically benefiting my client? *Clinical significance* does not include any specific type of mathematical formulae or test; instead, it requires the counselor to consider the whole person when determining significant lifestyle growth or changes (Barrio-Minton & Lenz, 2019). When addressing clinical significance, counselors ask themselves if this form or approach to treatment is helpful. This question empowers the counselor to consider all of the known factors within the client's life. These factors can include strengths and areas of growth, wellness, areas of oppression and privilege, multicultural identities, history, goals, and overall worldview, among many others that contribute to or detract from successful counseling. Each of these factors and more also contribute to brain-based structural and functional changes. Evidence of these changes can be found in human EEGs. Experiences such as exercise (Panksepp, 2007), sleep (Williamson & Feyer, 2000), languages spoken (Carrasco-Ortíz et al., 2017), nutrition and diet (Gutierrez et al., 2021), and subtle changes in mental state (Russo & Stevens, 2016) as well as issues of poverty, abuse, neglect, traumatic experiences, racism, sexism, and other social injustices (Zalaquett & Ivey, 2018) uniquely influence human EEG activity. Recognizing such factors as contributing to or inhibiting client success allows us to fairly evaluate the level of clinical significance that our treatment has on the therapeutic goals that our clients create for themselves.

From a practical standpoint, counselors can use information pertaining to practical and statistical significance in concert with clinical significance to evaluate whether counseling is adequately addressing a client's needs. One way to do this is to track client progress every odd number of sessions (e.g., 1st, 3rd, 5th) by administering formal assessments that are relevant to the client's goals (e.g., Beck Depression Inventory, Second Edition [see Chapter 7]; Outcome Rating Scale [Miller et al., 2003]; Quality of Life Inventory [Frisch et al., 2005]; Session Rating Scale [Duncan et al., 2003]). These data could be analyzed to explore change from the statistical and practical perspectives as well as from the clinical perspective to create a holistic evaluation of client progress. Not only could this information help counselors to remain in touch with client needs, but also the data obtained from formal assessments could be used to advocate on behalf of clients whose scores are still within clinical ranges but who have reached the limit of sessions covered by insurance. This process of combining formal and informal assessments of client progress toward their goals for therapy can be viewed as a form of counseling treatment evaluation.

After reviewing the concepts she had first learned in her research methodologies course, Susan identified several articles relevant to Ryan's concerns. Once Susan reflected on demographic representation, generalizability, and multiple types of significance for these sources, she determined whether these articles were applicable and useful to how she might best help Ryan. Then she felt empowered to move on to the next step, consulting with additional trusted sources and integrating her research results into practice.

Step 5: Consult With Trusted Sources; Integrate Research Into Practice

To further her practice, Susan decided to pursue additional training and continuing education credits by first investigating those organizations recommended by colleagues or in professional counseling publications (e.g., Russo et al., 2022). She reached out to connect with neuroscience interest networks and committees within professional counseling associations and also began attending monthly meetings and webinars on neuroscience topics.

Susan also explored the training and certification requirements of the Biofeedback Certification International Alliance (BCIA; n.d.-a) and reviewed an online database of local certified practitioners (BCIA, n.d.-b) whose expertise she thought could be beneficial to Ryan's course of treatment.

After searching the literature to identify possible interventions that would address Ryan's needs, Susan identified two interventions using neurofeedback and EMDR. Understanding that research on the efficacy of neurofeedback with African Americans is limited, Susan was encouraged to find studies that found efficacy for EMDR with veterans in samples that included African American participants. After further consulting with trusted colleagues, Susan decided to seek training that would allow her to provide EMDR to Ryan in future sessions. In the meantime, she offered to refer Ryan to a local counselor who was already certified in neurofeedback. Susan also decided to begin using individualized formal assessments to evaluate the effects of her counseling interventions on her clients. She would explore statistical and practical significance while also considering the clinical significance that she was observing in session. Susan felt empowered to continue to evaluate her counseling practice while also being of service to the people who were seeking support.

Conclusion

In this chapter, we proposed a five-step approach to the exploration of neuroscience-informed counseling literature. As part of this five-step process, we suggested a range of resources that can be beneficial to counselors who want to better understand how neuroscience can promote client growth. Such resources include professional organizations and interest networks, counseling journals and publications that integrate neuroscience into counseling practice, and peer-based groups that empower counselors to collaborate and to become informed about the role that neuroscience can have in counseling practice.

In addition, this chapter addressed the issue of the generalizability of research results, the importance of including diverse population groups among neuroscience study participants, and the role of significance testing in research and practice. It is hoped that this chapter will leave the reader with a sense of curiosity for further exploration of neuroscience-informed research.

Quiz

1. You find an article that reports mean differences between two samples. What kind of significance is being discussed with this metric?
    a. Statistical significance.
    b. Practical significance.
    c. Clinical significance.
    d. Super significance.
2. Client outcome research can be viewed as a form of _____.
    a. Systemic advocacy.
    b. Client advocacy.
    c. Self-promotion.
    d. Both a and b.

References

American Psychological Association. (2021, December). *Apology to people of color for APA's role in promoting, perpetuating, and failing to challenge racism, racial discrimination, and human hierarchy in the U.S.: Resolution adopted by the APA Council of Representatives on October 29, 2021.* https://www.apa.org/about/policy/racism-apology

Aroke, E. N., Joseph, P. V., Roy, A., Overstreet, D. S., Tollefsbol, T. O., Vance, D. E., & Goodin, B. R. (2019). Could epigenetics help explain racial disparities in chronic pain? *Journal of Pain Research, 12*, 701–710. https://doi.org/10.2147%2FJPR.S191848

Balkin, R. S., Lenz, A. S., Russo, G. M., Powell, B. W., & Gregory, H. M. (2022). Effectiveness of EMDR for decreasing symptoms of overarousal: A meta-analysis. *Journal of Counseling & Development, 100*(2), 115–122. https://doi.org/10.1002/jcad.12418

Barrio-Minton, C. A., & Lenz, A. S. (2019). *Practical approaches to applied research and program evaluation for helping professionals.* Routledge.

Biofeedback Certification International Alliance. (n.d.-a). *BCIA-accredited neurofeedback didactic programs US & Canada.* https://bcia.memberclicks.net/assets/NFCommonDocs/NF%20Didactic%20Training%20Programs.pdf

Biofeedback Certification International Alliance. (n.d.-b). *Find a practitioner - Consumers.* https://www.bcia.org/consumers-find-a-practitioner

Borenstein, M., Hedges, L. V., Higgins, J. P. T., & Rothstein, H. R. (2009). *Introduction to meta-analysis.* Wiley.

Burt, I., & Pankey, B. (2023). Critically analyzing the field of neuroscience and its therapeutic application with Black populations. *Journal of Multicultural Counseling and Development, 51*(3), 174–182. https://doi.org/10.1002/jmcd.12274

Carrasco-Ortíz, H., Velázquez Herrera, A., Jackson-Maldonado, D., Avecilla Ramírez, G. N., Silva Pereyra, J., & Wicha, N. Y. Y. (2017). The role of language similarity in processing second language morphosyntax: Evidence from ERPs. *International Journal of Psychophysiology, 117*(1), 91–110. https://doi.org/10.1016/j.ijpsycho.2017.04.008

Chatters, L. M., Taylor, R. J., & Schulz, A. J. (2022). The return of race science and why it matters for family science. *Journal of Family Theory & Review, 14*(3), 442–462. https://doi.org/10.1111/jftr.12472

Choy, T., Baker, E., & Stavropoulos, K. (2021). Systemic racism in EEG research: Considerations and potential solutions. *Affective Science, 3*(1), 14–20. https://doi.org/10.1007/s42761-021-00050-0

Cohen, J. (1988). *Statistical power analysis for the behavioral sciences.* Routledge.

Council for Accreditation of Counseling and Related Educational Programs. (2024). *2024 CACREP standards.* https://www.cacrep.org/wp-content/uploads/2023/06/2024-Standards-Combined-Version-6.27.23.pdf

Day-Vines, N. L., Bryan, J., Brodar, J. R., & Griffin, D. (2022). Grappling with race: A national study of the broaching attitudes and behavior of school counselors, clinical mental health counselors, and counselor trainees. *Journal of Multicultural Counseling and Development, 50*(1), 25–34. https://doi.org/10.1002/jmcd.12231

De Wolfe, T. J., Arefin, M. R., Benezra, A., & Rebolleda Gómez, M. (2021). Chasing ghosts: Race, racism, and the future of microbiome research. *mSystems, 6*(5), Article e00604-21. https://doi.org/10.1128/mSystems.00604-21

Dewiputri, W. I., & Auer, T. (2013). Functional magnetic resonance imaging (fMRI) neurofeedback: Implementations and applications. *Malaysian Journal of Medical Sciences, 20*(5), 5–15.

Duncan, B. L., Miller, S. D., Sparks, J. A., Claud, D. A., Reynolds, L. R., Brown, J., & Johnson, L. D. (2003). The Session Rating Scale: Preliminary psychometric properties of a "working" alliance measure. *Journal of Brief Therapy, 3*(1), 3–12.

Erford, B. T., Savin-Murphy, J. A., & Butler, C. (2010). Conducting a meta-analysis of counseling outcome research: Twelve steps and practical procedures. *Counseling Outcome Research and Evaluation, 1*(1), 19–43. https://doi.org/10.1177/2150137809356682

Field, T. A., Moh, Y. S., Luke, C., Gracefire, P., Beeson, E. T., & Russo, G. M. (2022). A training model for the development of neuroscience-informed counseling competencies. *Journal of Mental Health Counseling, 44*(3), 266–281. https://doi.org/10.17744/mehc.44.3.05

Fisher, R. A. (1929). The statistical method in psychical research. *Proceedings of the Society for Psychical Research, 39*(1), 182–192.

Frisch, M. B., Clark, M. P., Rouse, S. V., Rudd, M. D., Paweleck, J. K., Greenstone, A., & Kopplin, D. A. (2005). Predictive and treatment validity of life satisfaction and the Quality of Life Inventory. *Assessment, 12*(1), 66–78. https://doi.org/10.1177/1073191104268006

Gutierrez, L., Folch, A., Rojas, M., Cantero, J. L., Atienza, M., Folch, J., Camins, A., Ruiz, A., Papandreou, C., & Bulló, M. (2021). Effects of nutrition on cognitive function in adults with or without cognitive impairment: A systematic review of randomized controlled clinical trials. *Nutrients, 13*(11), 1–40. https://doi.org/10.3390/nu13113728

Hedges, L. V. (1981). Distribution theory for Glass's estimator of effect size and related estimators. *Journal of Educational Statistics, 6*(2), 107–128. https://doi.org/10.3102/10769986006002107

Johnson, K. F., Ieva, K., & Byrd, J. (2022). Introduction to the special issue on school counselors addressing education, health, wellness, and trauma disparities. *Professional School Counseling, 26*(1b). https://doi.org/10.1177/2156759X221105451

Jones, D. E., Park, J. S., Gamby, K., Bigelow, T. M., Mersha, T. B., & Folger, A. T. (2021). Mental health epigenetics: A primer with implications for counselors. *The Professional Counselor, 11*(1), 102–121. https://doi.org/10.15241/dej.11.1.102

Kriegeskorte, N., Cusack, R., & Bandettini, P. (2010). How does an fMRI voxel sample the neuronal activity pattern: Compact-kernel or complex spatiotemporal filter? *NeuroImage, 49*(3), 1965–1976. https://doi.org/10.1016/j.neuroimage.2009.09.059

Lipsey, M. W., & Wilson, D. B. (2001). *Practical meta-analysis (L. Bickman & D. J. Rog, Eds.).* Sage.

Miller, S. D., Duncan, B. L., Brown, J., Sparks, J. A., & Claud, D. A. (2003). The Outcome Rating Scale: A preliminary study of the reliability, validity, and feasibility of a brief visual analog measure. *Journal of Brief Therapy, 2*(2), 91–100.

Myers, L. J., Lodge, T., Speight, S. L., & Haggins, K. (2022). The necessity of an emic paradigm in psychology. *Journal of Humanistic Psychology, 62*(4), 488–515. https://doi.org/10.1177/00221678211048568

National Human Genome Research Institute. (2018, September 7). *Genetics vs. genomics fact sheet.* National Institutes of Health. https://www.genome.gov/about-genomics/fact-sheets/Genetics-vs-Genomics

National Human Genome Research Institute. (2023, August 18). *Scientific racism*. National Institutes of Health. https://www.genome.gov/genetics-glossary/Scientific-Racism

Opara, I. N., Riddle-Jones, L., & Allen, N. (2022). Modern day drapetomania: Calling out scientific racism. *Journal of General Internal Medicine, 37*(1), 225–226. https://doi.org/10.1007/s11606-021-07163-z

Ortinski, P., & Meador, K. J. (2004). Neuronal mechanisms of conscious awareness. *Archives of Neurology, 61*(7), 1017–1020. https://doi.org/10.1001/archneur.61.7.1017

Panksepp, J. (2007). Can play diminish ADHD and facilitate the construction of the social brain? *Journal of the Canadian Academy of Child and Adolescent Psychiatry, 16*(2), 57–66.

Panksepp, J., & Burgdorf, J. (2003). "Laughing" rats and the evolutionary antecedents of human joy? *Physiology & Behavior, 79*(3), 533–547. https://doi.org/10.1016/S0031-9384(03)00159-8

Rock, D., Siegel, D. J., Poelmans, S. A. Y., & Payne, J. (2012). The healthy mind platter. *NeuroLeadership Journal, 4*(1), 1–23.

Russo, G. M., Balkin, R. S., & Lenz, A. S. (2022). A meta-analysis of neurofeedback for treating anxiety-spectrum disorders. *Journal of Counseling & Development, 100*(3), 236–251. https://doi.org/10.1002/jcad.12424

Russo, G. M., & Stevens, S. (2016, February). Counseling students gaining early exposure to neurofeedback. *Counseling Today, 58*(8), 14–16.

Watson, J. C., Lenz, A. S., Schmit, M. K., & Schmit, E. L. (2016). Calculating and reporting estimates of effect size in counseling outcome research. *Counseling Outcome Research and Evaluation, 7*(2), 111–123. https://doi.org/10.1177/2150137816660584

Weaver, Z. (n.d.). *Our history*. Society for Psychical Research. https://www.spr.ac.uk/about/our-history

Wheaton, L. A. (2021). Racial equity and inclusion still lacking in neuroscience meetings. *Nature Neuroscience, 24*(12), 1645–1647. https://doi.org/10.1038/s41593-021-00964-9

Williamson, A. M., & Feyer, A. (2000). Moderate sleep deprivation produces impairments in cognitive and motor performance equivalent to legally prescribed levels of alcohol intoxication. *Occupational and Environmental Medicine, 57*(10), 649–655. https://doi.org/10.1136/oem.57.10.649

Zalaquett, C. P., & Ivey, A. E. (2018). The role of neuroscience in advancing social justice counseling. In C. C. Lee (Ed.), *Counseling for social justice* (3rd ed., pp. 191–204). Wiley.

Chapter 15

Neuroscience-Informed Clinical Supervision: An Emerging Transtheoretical Approach

Theodore J. Chapin, Lori A. Russell-Chapin,
and Raissa Miller

The purpose of this chapter is to explore how neurocounseling can enhance supervisory relationships, expand critical issues in case conceptualization, and guide more effective treatment planning. We present basic concepts and principles of neuroanatomy and neurophysiology and apply them using an emerging transtheoretical approach to the supervisory relationship, case conceptualization, and treatment planning.

2024 CACREP Standards

This chapter addresses 2024 Council for Accreditation of Counseling and Related Educational Programs (CACREP) Standards pertinent to the Foundational Counseling Curriculum (Section 3) area of Professional Counseling Orientation and Ethical Practice (Standard A):

- The purpose of and roles within counseling supervision in the profession (Standard A.12.)

This chapter also addresses the following Doctoral Standards for Counselor Education and Supervision (Section 6) in the area of Doctoral Curriculum (Standard B):

- Purposes of counseling supervision (Standard B.2.a.)
- Roles and relationships related to counseling supervision (Standard B.2.c.)

- Skills of counseling supervision across multiple settings and across service delivery modalities (Standard B.2.d.)
- The use of technology in counseling supervision (Standard B.2.h.)
- Legal and ethical issues and responsibilities in counseling supervision (Standard B.2.k.)

Clinical Case Study: Addison

Addison sought clinical supervision to obtain hours for her second tier of counseling licensure. Addison identifies as a 24-year-old White female. She actively follows the Catholic faith system and lives alone with her cat in a midwestern city in the United States. Six months ago, Addison's boyfriend of 5 years broke up with her; however, her parents have been very supportive of this recent loss. Addison's only brother, Ben, is 22 years old. Ben calls weekly to check in with his sister. The two are planning to take a weekend biking trip in the summer. Addison works full time at a local agency in town where the majority of the caseload is made up of clients who are in a residential foster-care treatment home. Lori began supervising Addison 6 months ago. Lori asks all of her supervisees to prepare a supervision question for each session to articulate their supervision needs. Addison's questions for the case in this chapter related to her work with a client for whom a traditional cognitive behavioral approach seemed to be insufficient in promoting meaningful change. She described her client, Trystan, as a 14-year-old female who presented to counseling with depressive symptoms and suicidal ideation. Trystan was removed from two schools before being admitted into the hospital for a suicide attempt. After a week in the hospital, Trystan was admitted into a 30-day residential treatment program. During this time, Trystan's biological parents terminated parental rights, stating that this time was "just too much!" Addison noted that she was Trystan's fourth counselor in 5 months. Addison described her efforts to build the therapeutic relationship through active listening. She promoted symptom reduction by helping Trystan identify unhelpful thinking patterns and thoughts that may be contributing to her depressed mood. Addison said, however, that she had made little progress with Trystan and felt stuck. She apologetically stated, "I have no idea where to go from here."

> **Reflective Questions**
>
> As you read the case study, what aspects of Addison's supervision case stood out to you from the beginning?
>
> What thoughts or feelings came to mind when you imagined incorporating neurocounseling into the supervision process?
>
> What model(s) of supervision currently guide your approach to supervision, and does your approach include any neurocounseling elements?

Supervisory Relationship

The effectiveness of supervision depends largely on the nature of the supervisory relationship. From the beginning and throughout the supervisory relationship, supervisors must foster safety and connection; manage ruptures; and navigate feedback, confrontation, and evaluation. Neurocounseling offers several valuable insights and strategies for addressing these challenges.

Connection and Trust

Porges (2011) stated that individuals must feel safe in their interpersonal interaction to connect and more deeply engage with each other. He described this neurologically based process as the *social engagement system* and illustrated its function through what he described as the *polyvagal theory*. According to Porges's polyvagal theory (see Chapter 1), the vagus nerve, which extends from the brain into the body, serves as a modulator of the gut-brain axis and is a main component of the parasympathetic nervous system. It plays an important role in regulating internal organ functions such as digestion, heart rate, breathing, and vasomotor activity. This critical nerve is activated by responsive eye contact, warm facial expression, and calm prosodic or rhythmic speech. Together, these unconscious nonverbal and verbal cues can convey trust, safety, and care. In clinical supervision, intentionally demonstrating these cues helps to facilitate genuine communication, vulnerable self-exploration, and deeper levels of learning.

A connected and trusting supervisory relationship, from a neurocounseling perspective, begins with the supervisor's current level of healthy self-regulation. Once in place, the supervisor's regulated state allows for intentional activation of the social engagement system and enhances the supervisor's ability to coregulate with the supervisee. It is important for all clinical supervisors to consider and improve their personal level of self-regulation. A calm and relaxed supervisor presents the most encouraging, nonthreatening, and focused supervisory milieu. If not healthily self-regulated, a prospective supervisor is best advised to take a few deep, fully oxygenated breaths, relax their muscles, and warm their hands and feet—all well-established physiological approaches to healthy self-regulation. These skills can easily be learned by taking a short course of biofeedback training. Then, by providing empathetic, nurturing eye contact, and a warm and relaxed facial expression and speaking in a steady manner, supervisors can help supervisees feel safe in the supervisory relationship. After a brief period of time, the observant supervisor will notice the supervisee also becomes more relaxed. In this state of mutual, healthy coregulation, the supervisee will feel safe and connected and have more trust in the supervisory relationship.

Managing Ruptures

Ruptures in the supervisory relationship can occur for many reasons. These reasons can range from minor misunderstandings or breakdowns in communication to major conflicts of personality or multicultural experience. Fickling et al. (2019) specifically addressed sociocultural factors impacting the supervisory relationship, noting that elements of privilege and power can greatly impact supervisees' perceptions of safety and degree of comfort with disclosure. They identified the importance of supervisors honoring multicultural perspectives and diverse worldviews, openly discussing power dynamics with supervisees, and embodying a sense of cultural humility within the supervisory relationship. Multicultural factors are best discussed at the outset of supervision to prevent ruptures. However, when ruptures do occur, supervisors should consider the degree to which sociocultural factors are influencing the concern and attend to those dynamics as part of the repair process.

Some ruptures will have little impact on the supervisory alliance, whereas others may threaten to destroy the relationship (Watkins, 2021). Although many techniques for managing ruptures are noted in the literature, such as self-reflection, clarifying dialogue, and apology, the use of healthy self-regulation and the social engagement system will provide preventive value and be the best basis for repairing and deepening trust in the supervisory relationship. Nothing better potentially solidifies the supervisory alliance than successful, caring, and sometimes humanizing resolution of supervisory ruptures. When managed effectively, ruptures can strengthen the supervisory relationship while modeling for supervisees how they can manage ruptures in their clinical work.

Feedback, Confrontation, and Evaluation

The learning process inevitably involves feedback, confrontation, and evaluation. The evaluative roles of supervision are sometimes very uncomfortable for the supervisor to assume and can be threatening for the supervisee to experience. Although rapport, previously established clear expectations, mutually respectful and healthy assertive communication, encouragement, and constructive guidance are essential, sometimes they are not enough to prevent anxiety and sympathetic reaction in the supervisee or in the supervisor. Being firmly anchored in healthy self-regulation and safe social engagement is essential but sometimes insufficient. The literature on feedback, confrontation, and evaluation is plentiful, but neuroscience-informed practice contributes both a deeper understanding and practical suggestions to facilitate this challenging aspect of supervision.

The potential threat of feedback is that it may be experienced as an affront to one's previously held self-image and self-confidence. Constructive feedback can be especially overwhelming when the supervisor or supervisee has experienced past verbal abuse or cumulative microaggressions that have caused their sympathetic fight, flight, or freeze response to become hypervigilant, easily triggered, or overactivated. Although referral to counseling and trauma work may be beneficial, the evaluative task of supervision remains present in the moment. In this moment, skills in healthy self-regulation and the safety of healthy social engagement are paramount. In addition, it is important to note that evolutionarily speaking, the brain is neurologically wired to be more attentive and reactive to potential threat. This neurophysiological response is a self-protection mechanism. By remembering past threat and preparing to meet it, continued survival is better ensured. The challenge to the supervisor is how to respect this survival mechanism while fulfilling the essential role of evaluation in supervision. Perhaps some guidance can be gleaned from other disciplines.

The clinical supervision literature suggests that supervisors should provide a balance of support and challenge (e.g., Bernard & Goodyear, 2019). Studies within other fields, such as business, suggest that the ratio of support to challenge should favor the former (Losada & Heaphy, 2004). This ratio has been measured as 5.6 positive comments for every 1 negative comment for high-performing teams, as compared with 0.36 positive to 1 negative for low-performing teams (Losada & Heaphy, 2004). Gottman (Gottman & Silver, 2015) found a similar ratio in his research on stable and happy marriages. He found that happily married couples had 5 positive interactions for every 1 negative interaction during conflict. Examples of positive interactions included demonstrating interest, expressing affection (kindness), being intentionally appreciative, and finding opportunities for agreement (validation). The conclusion seems clear: Humans are much more able to receive and benefit from feedback when it is significantly more positive than negative. The neurological implication is that this ratio is more likely to elicit and maintain a parasympathetic response from the recipient, allowing for supervisory feedback to be more readily accepted and integrated into the supervisee's self-image.

Case Example: Addison

As noted in the clinical case study, Addison and Lori have been working together for 6 months. At the beginning of the supervisory relationship, they discussed and signed a contract that delineated weekly sessions, times, expectations, a supervision policy, a supervision professional will, and how Addison best receives feedback. Addison and Lori also explored multicultural backgrounds, including similarities and differences. They discussed their wide age difference as a potential difficulty, but Addison liked the fact that Lori had been teaching, counseling, and supervising for many decades.

All of these efforts were intended to support the development of a trusting supervisory relationship so that Addison would feel comfortable being open and honest in discussing supervision topics. Lori took Addison's statement that she felt stuck and had no idea what to do, an admission of vulnerability, as a positive sign that the supervisory relationship was working well.

Lori also noticed that as Addison shared about Trystan, her breathing became shallow and her voice shaky. Lori inquired about Addison's state in the moment, and Addison acknowledged feeling slightly anxious, specifically noticing that her heart was beating fast and that she felt an underlying sense of pressure. Lori became even more intentional about her own regulation in the moment and consciously took a deep breath and softened her facial expressions and tone. To maximize Addison's learning in the next hour, Lori had to use their nervous systems' natural coregulating potential to establish regulation within and between them. Before going further into discussing new ideas with Addison about her approach with Trystan, Lori spent a few minutes reflecting on Addison's strengths as a counselor. Lori offered examples from previous client sessions that she had watched and her own experience with Addison in supervision. Lori noted that Addison was warm and attentive in her sessions and that clients seemed to respond well to her natural style of helping. Lori said that she had noticed Addison was quick to identify underlying concerns, such as trauma history, rather than focus on surface-level symptoms or problematic behaviors. As Lori monitored her own regulation and shared authentic positive feedback, she could see Addison relaxing and becoming more regulated herself. Lori then knew it was a good time to move into deeper exploration around Trystan's case conceptualization and treatment planning.

> ### Reflective Questions
>
> What regulation strategies are most effective for you when facing perceived stress or threat?
>
> What signs or signals indicate a potential relationship rupture has occurred in supervision? Consider your own interoceptive signs, as well as signs from the supervisee. What feelings does evaluative feedback elicit in you, and how might that impact your approach to giving evaluative feedback to your supervisees?

Case Conceptualization

One of the primary tasks of supervision is to help counselors develop their case conceptualization skills. The supervisor can introduce several concepts that help supervisees develop neuroscience-informed case conceptualizations. A *neurocounseling-informed case conceptualization* considers both the more general neurological basis of client problems and the possible concomitant neurophysiological factors that may contribute to more significant unhealthy self-regulation or neurophysiological dysregulation. Understanding neurological factors that influence client problems can provide an appreciation for the depth, neurological function, and dynamics of their presenting concerns and associated symptoms. When more severe, chronic, or debilitating concerns and symptoms are present, it is likely the client has significant neurophysiological dysregulation that is best addressed first before more typical counseling interventions may be expected to be helpful. An example of a treatment approach that first addresses neurophysiological dysregulation before traditional counseling interventions are used is neuroscience-informed cognitive behavior therapy (nCBT). Field et al. (2015) proposed and demonstrated that for clients who initially struggle with traditional CBT, cognitive restructuring can become more effective after they have learned basic self-regulation skills, such as physiological awareness via mindfulness, biofeedback, and neurofeedback, or other healthy coping behaviors that activate the senses. This concept was affirmed in an experimental study, with control conditions, that assessed the autonomic dysregulation in children with social anxiety disorder (Asbrand et al., 2022). The research found that CBT alone did not result in changes in the children's psychophysiological measures of heart rate reactivity or skin conductance (measures of sympathetic arousal) or their lower levels of parasympathetic arousal.

In addition, conventional explanatory models of case conceptualization grounded in counseling theories may be ethnocentric and insensitive to a client's cultural background. Traditional CBT, for example, assumes that the client has full responsibility for and control of their thoughts and that thoughts themselves, rather than the activating event in their environment, are the primary cause of distress and dysfunction. The nCBT model integrates *culturally informed case conceptualization* to account for environmental events that cause distress, regardless of the client's appraisal of and response to those activating events (Beeson et al., 2017). For example, experiences of racism, misogyny, homophobia, transphobia, bullying, and exposure to interpersonal violence can be enduring activating events that persist over time. Culturally informed case conceptualization recognizes that it is important for counselors to attend to and address such ongoing environmental stressors directly, rather than addressing only the client's response to the chronic stressor.

Neurocounseling-informed case conceptualization therefore involves fuller attention to the nature and breadth of the client's presenting concern. When using this approach, counselors consider the possible underlying cultural and neurophysiological factors at play, including an assessment of the client's overall level of self-regulation, before commencing with diagnostic implications and treatment planning. Client concerns are often expressed in terms that capture their overall experience, such as anxiety, depression, trauma, attention, obsessive-compulsive disorder (OCD), or substance abuse. These terms are usually not nuanced or reflective of the complexity of their experience. However, the reality is that many clients' presenting concerns, especially those of clients who are in significant distress, are very complex. Complex symptoms have often evolved over a period of years and are chronic or have reoccurred. Many clients have sought previous treatment with counseling or medication with some success or disappointment. A neurologically designed symptom checklist can help the client better express the nuance and complexity of their symptoms and assist the supervisee in developing a fuller understanding of their experience (see Figure 15.1).

In addition, a neurocounseling assessment as presented by Chapin and Russell-Chapin (2014, 2022a, 2022b) and Russell-Chapin and Chapin (2022) uses (a) a thorough psychosocial medical history to identify possible factors and lifestyle behaviors that may contribute to neurophysiological dysregulation; (b) paper-and-pencil psychological screening of the relevant presence of the more common psychological problems (e.g., anxiety, depression, trauma, OCD, attention); (c) computerized testing of cognitive functioning to assess any impact on visual and auditory attention; and (d) if indicated, an assessment of baseline biofeedback and/or neurofeedback measures, such as peripheral skin temperature, heart rate variability, brain waves, and brain wave network functioning. When combined, this information provides an extensive view of the client's current neurophysiological functioning and level of healthy or unhealthy self-regulation. Other factors typically involved in a thorough counseling evaluation and

Neurofeedback Problem Rating Form

Client Name: __Trystan__ Rater: __Trystan__ Date: __10/2022__

Please place a check in the left column for any problems that may apply. Next circle the top ten and place an asterisk after the top three. Rate any changes since beginning neurofeedback in the right column, using the following scale: S = *same*, I = *improved*, or M = *much improved*.

Cz
(Check) (Rate)
- ☐ ___ Difficulty with visual recognition of objects or words
- ☐ ___ Retention of information
- ☐ ___ Short-term memory
- ☐ ___ Foggy thinking
- ☐ ___ Poor reading comprehension
- ☐ ___ Tired when reading or problem solving
- ☐ ___ Mental sluggishness
- ☐ ___ Hyperactive, restlessness, can't sit still
- ☐ ___ Unable to quiet or calm my body
- ☐ ___ Falling asleep
- ☐ ___ Headaches
- ☐ ___ Managing/coping with chronic pain
- ☐ ___ Tics, body tremors, involuntary muscle spasms
- ☐ ___ Seizures with a motor component

O1
(Check) (Rate)
- ☐ ___ Emotional trauma or traumatic stress
- ☐ ___ Preoccupation with artistic interests or skills
- ☐ ___ Easily tired or fatigued
- ☐ ___ Frequently ill
- ☐ ___ Easily frightened
- ☐ ___ Staying asleep or disturbed sleep
- ☐ ___ Lack of dreaming or nightmares
- ☐ ___ Racing thoughts or anxiety
- ☐ ___ Insufficient self-soothing
- ☐ ___ Self-medicating (alcohol, drugs, food)
- ☐ ___ Cognitive inefficiency or difficulty thinking

F3
(Check) (Rate)
- ☒ ___ Unhappy
- ☒ ___ Feeling worthless
- ☒ ___ Little to look forward to
- ☒ ___ Negative self-talk
- ☒ ___ Depressed mood
- ☐ ___ Poor retrieval of information
- ☒ ___ Lack of energy, motivation, interest

F4
(Check) (Rate)
- ☒ ___ Easily annoyed or irritated
- ☐ ___ Anxious mood
- ☐ ___ Easily angered
- ☐ ___ Impulsive
- ☒ ___ Emotionally volatile or explosive
- ☐ ___ Oppositional or defiant
- ☐ ___ Indifferent or unresponsive to others
- ☐ ___ Restricted emotional expression
- ☐ ___ Developmental delay or socially awkward

F3/F4
(Check) (Rate)
- ☐ ___ Inattention, day dreaming, distracted
- ☐ ___ Disorganized, poor planning and sequencing
- ☐ ___ Sustaining focus or staying on task
- ☐ ___ Talkativeness
- ☐ ___ Fibromyalgia or chronic fatigue
- ☒ ___ Poor-quality sleep

Fz
(Check) (Rate)
- ☐ ___ Stubborn, "my way or the highway"
- ☐ ___ Maintaining concentration
- ☐ ___ Forgetfulness
- ☐ ___ Fretting or excessive worry
- ☐ ___ Compulsive, repetitive behaviors
- ☐ ___ Obsessive, annoying thoughts
- ☐ ___ Excessive passiveness
- ☐ ___ Too pleasing, open-minded, conciliatory
- ☐ ___ Can't let things go
- ☐ ___ Stuck on the negative
- ☐ ___ Age-related cognitive or memory problems
- ☐ ___ Extremely focused interests or rigid behavior
- ☐ ___ Preoccupation with pain
- ☐ ___ Busy thoughts causing disrupted sleep
- ☐ ___ Marked learning or cognitive deficits

Total Checked: __9__ S ___ I ___ M
___ I + M

Please note below or on the back any worsening problem and/or any significant recent life changes involving: family or work environment, medication, emotional trauma, or physical health or injury.

FIGURE 15.1
Neurofeedback Problem Rating Form

Note. Adapted from Paul Swingle Clinical Q V3. Cz = Sensorimotor integration of both lower extremities; O1 = Right visual processing, pattern recognition, color, movement, black and white and edge perception; F3 = Motor planning, right upper extremities; F4 = Motor planning, left upper extremities; Fz = Motor planning of both lower extremities and midline.

case conceptualization remain important. These factors include, for example, the client's developmental level of functioning, interpersonal support system, cultural background and intersectional identities, environmental factors, personal resources, and current medications. When a counselor or supervisee considers these factors in total, they are better able to conceptualize the case, formulate diagnostic impressions, and begin to reflect upon and develop intervention strategies.

The supervisor can pose questions to the supervisee that facilitate exploration of the client's cultural/environmental stressors and physiological functioning. For example, is this client's general level of self-regulation relatively healthy, or are they struggling with significant dysregulation? What stressors exist in the client's environment that need to be addressed? Would the client readily benefit from typical counseling interventions, or might they need to first restore a healthier level of neurophysiological self-regulation? What working diagnoses appear to be impacting the client's presenting concern and symptoms? Rather than trying to narrow down the diagnostic impression into one or two predominant diagnoses, it is often pragmatically more helpful in case conceptualization to note all of the contributing factors. Some of the more common factors assessed across most cases of more severe dysregulation include sleep problems, a range of anxiety to depression, and cognitive inefficiency or attentional problems. Less frequent problems that are often compounding include OCD (both excessive overfocus and excessive passivity); trauma; substance or medication use or abuse; head injuries; and other sources or causes of neurological inflammation, such as chronic disease states, pain, excessive lifetime anesthesia, and poor diet. Although not explicit diagnostic categories, many lifestyle issues also impact neurophysiological dysregulation. These factors can include a lack of deep restorative sleep, excessive screen time, insufficient exercise, interpersonal conflict or isolation, inadequate intellectual or environmental stimulation, and a lack of existential or spiritual meaning and purpose. A comprehensive case conceptualization considers both diagnosable conditions and other factors that if addressed might help to improve healthy self-regulation. Once reestablished, healthy self-regulation enables the client to benefit from conventional counseling interventions more fully.

Case Example: Addison

Lori invited Addison to share more about her initial assessment with Trystan. Addison said Trystan was experiencing generalized symptoms of depression, including sleeping too much, low energy, low motivation, difficulty regulating emotions and behaviors, and thoughts of suicide. Addison recognized that Trystan's living situation (e.g., parental termination of rights, multiple counselors within a short period, living in a temporary housing situation) represented multiple significant persisting stressors that were so substantial in scope that the notion of managing emotions may have felt overwhelming to Trystan. Addison retrieved Trystan's completed symptom checklist (Figure 15.1). Lori quickly reviewed the checklist and began identifying neurophysiological underpinnings of the client's concern (e.g., brain location and function, current level of neurophysiological self-regulation).

Lori pointed out to Addison that the checklist indicated nine symptoms, which were located mainly in the left frontal cortex at F3 and F4 (see Figure 7.2 in Chapter 7 for a head map of functions). The low energy, low motivation, and impulsive and angry outbursts seemed consistent with suicidal statements. Trystan's lack of any self-harming or suicidal behaviors and plans allowed some immediate relief and time to develop a safer and more efficacious treatment plan.

At this point in supervision, Lori chose to use Dr. Norm Kagan's interpersonal process recall (Kagan, 1997) approach to process Addison's ultimate question of "Where do I go from here?" She selected two gentle leads and queries from Kagan's search and expectation lead (Bernard & Goodyear, 2019, p. 102), as elaborated in the following dialogue:

Reflective Questions

Think of a client for whom traditional cognitive or behavioral approaches seem less effective.

What underlying neurophysiological factors are influencing the client's presenting concerns that may need priority consideration or attention?

Which allied health professionals in your community can you refer to if you identify underlying neurophysiological concerns outside of your scope of practice?

How can you help supervisees to comply with current diagnostic practice requirements while also thinking in a more complex, transdiagnostic manner during case conceptualization?

Lori: What message did you want to give to Trystan? What prevented you from doing so?

Addison: Seeing Trystan's symptom checklist actually frightened me. She is so young and even fragile. I wanted to tell her it would all be OK, but I kept remembering a time when I was 14 and I too was in a very dark place. It didn't seem OK then. I guess I am afraid for her and maybe me too. She deserves a more seasoned and confident counselor.

Lori: So your own fear kept you from being genuine with Trystan. Is that correct? Such a powerful insight.

>   *Addison*: Yes, I'm realizing I need to work through my own fears.
>   *Lori*: I am looking over your case notes. Your entries are simple and measurable too. The one aspect of the psychosocial medical history that I don't see is any recent physical or comprehensive bloodwork. Let's get that accomplished. Please have Mom and Addison consult with their family physician for a complete physical and anything else the physician might need.

Counselors must know what they are truly dealing with physiologically in order to rule out and then customize treatment plans. Trystan's bloodwork results were all within normal ranges, except for her thyroid numbers. Her thyroid levels were suggestive of mild hypothyroidism, which can be associated with depressive states. Trystan began working with a primary care physician to address this. After 2 weeks of thyroid medication, Trystan's thyroid levels were in the normal range. Trystan began to feel physically better, and the suicidal thoughts dissipated. Trystan still struggled with overwhelming emotions at times and impulsive behaviors. Addison was relieved and appreciative of the holistic approach that helped identify underlying physiological concerns and was now ready to move forward with counseling that focused on Trystan's ongoing regulation struggles. Her initial diagnostic impression was for adjustment disorder with depressed mood, as described in the *Diagnostic and Statistical Manual of Mental Disorders, 5th Edition, Text Revision* (American Psychiatric Association, 2022).

Treatment Planning

Treatment planning is another area where supervisees need support. Treatment planning directly follows case conceptualization and diagnostic impression. The presenting concern, symptoms, assessment, diagnostic impression, and case conceptualization identify the issues to be addressed. For clients who do not present with significant neurophysiological dysregulation, treatment planning in neurocounseling supervision may focus on the brain locations associated with the identified problems. Russell-Chapin et al. (2021) in the book *Practical Neurocounseling: Connecting Brain Functions to Real Therapy Interventions* presented a strategy for identifying where in the brain certain problems tend to be located and what counseling interventions may activate those areas. By using a clinically based head map of brain functions, a counselor can locate the identified problem in the brain and the general lobe involved. The 10–20 international location system (Jasper, 1958) is a method used to ensure consistency in the placement of electroencephalogram (EEG) scalp electrodes. The clinically based head map of brain functions uses actual images of client EEG activity to indicate areas of activation for counseling interventions (Russell-Chapin et al., 2021). Brain mapping allows for more accurate and brain-based selection of counseling interventions. Such a text could be reviewed during supervision so the supervisee can develop competence.

For clients who present with more significant neurophysiological dysregulation, treatment planning and intervention involves the establishment of healthier self-regulation followed by more conventional counseling interventions targeted to address the client's concerns. These neurocounseling interventions include critical lifestyle changes intended to support neuroplasticity, alongside biofeedback and neurofeedback that directly target improved parasympathetic response and brain wave network functioning. All counselors, supervisors, and supervisees are well advised to learn basic skills in diaphragmatic breathing, peripheral skin temperature training, and relaxation training. The supervisor should also encourage professional training in meditation, mental imagery, and lifestyle changes that support brain health. This knowledge and skill base has the triple advantage of promoting professional self-care, supporting a facilitative supervisory relationship, and benefiting clients in need of healthier self-regulation (Corey et al., 2023).

Many of the topics identified above (e.g., lifestyle changes that support neuroplasticity, breathing techniques that support regulation) can be introduced to clients through a process of neuroeducation (Miller & Beeson, 2021). Chapter 10 presents neuroeducation best practices for clinical work. Related to supervision, supervisors can check in on supervisees' knowledge of neuroeducation best practices and provide more resources when needed. For example, supervisors can help supervisees identify and prioritize the neuroscience information most relevant to a client's case, brainstorm creative ways of exploring the neuroscience information with their clients (rather than relying solely on didactic information sharing), and consider the individual and contextual factors that can influence the timing and type of neuroeducation.

Although these approaches can help clients to better regulate their emotions, often clients' stressors may persist, making coping difficult (Beeson et al., 2017). When working with clients who face chronic and persisting stressors, counselors may need to address those stressors directly. Counselor advocacy and case management through working with systems (e.g., school systems) as well as supporting client self-advocacy are important interventions.

Specialized training and supervision in biofeedback and neurofeedback is ethically required for both supervisees and supervisors who want to deliver these services. Otherwise, referral of the client to appropriately qualified professionals is indicated. In addition, joining professional organizations that focus on these specialized trainings, such as the Biofeedback Certification International Alliance, is recommended (Russell-Chapin & Chapin, 2020).

Case Example: Addison

Addison and Lori worked together to develop a treatment plan for Trystan that included specific attention to teaching self-regulation skills via neuroeducation and biofeedback. This treatment plan included addressing chronic and persisting stressors in Trystan's life. With Lori's guidance, Addison helped Trystan identify key support people whom Trystan could deepen her relationship with to further strengthen her support system. Trystan felt she had strong rapport with one staff member at the therapeutic group home and with her individual counselor, Addison. Addison worked with the therapeutic group home to make sure Trystan had more frequent individual time with both Addison and the other staff member. Trystan also had a friend at the public school she attended, and the therapeutic group home staff organized supervised meetups at a local park. Trystan did not want more frequent contact with her biological family and reported that her scheduled phone calls were sufficient. Enhancing these supports helped Trystan feel more comfortable in her environment, more supported, and less isolated. Related to neuroeducation, Lori asked Addison to think about a creative way to explore lifestyle strengths and areas of needed improvement with Trystan. Addison thought that Trystan would like to keep a daily journal for practicing skin temperature regulation. Trystan took a thermometer and brought back the results each week. Lori also asked Addison what she knew about the physiological and neurological impact of lifestyle behaviors. In discussing her knowledge, Lori specifically noted the salient role of sleep and movement during the adolescent developmental period and the links between these factors and healthy brain development and functioning. For example, Lori shared some of the most recent scientific research indicating consistent sleep and wake times and exposure to morning sunlight as critical factors influencing adolescent sleep health. Addison said she would be sure to include this information in her exploration of lifestyle factors with Trystan.

Related to biofeedback, Addison and Lori talked about some specific interventions that could help Trystan attune to her physiological states and shift those states into calmer, more regulated ways of being. Addison had previous training in using an app that facilitated heart rate variability training, so they decided that would be a good activity to introduce to Trystan. Addison was familiar with basic breathing practices but not sure which one would be best to introduce to Trystan. Lori suggested practicing in supervision together so that Addison would feel comfortable implementing them in the counseling room. They practiced the basics of laying on one's back to observe diaphragmatic versus costal breathing (i.e., shallow breathing resulting from no contraction of the diaphragm), and then practicing diaphragmatic breathing in a sitting position. Addison felt prepared to implement these strategies with Trystan.

Addison engaged in neuroeducation, sharing with Trystan the importance of sleep, movement, and nutrition to her overall sense of well-being. Addison and Trystan worked together to identify lifestyle strengths and areas of needed improvement. Trystan specifically focused on establishing a consistent sleep schedule, learning how to go to bed at the same time and wake up at the same time every day. She also started adding a 20-minute morning walk to her routine to get daily morning sunlight and exercise. In addition, Addison facilitated heart rate variability training using an iPhone app and practiced diaphragmatic breathing with Trystan in sessions.

Over time, Trystan reported increased confidence in her ability to regulate her emotions and behaviors. She reported fewer angry outbursts and overall improvement in her mood and relationships with others. As Trystan responded to Addison's counseling approach, Lori noticed Addison's confidence growing as well. Addison learned that she was "good enough" to help foster change. Under Lori's supervision, Addison experienced the benefits of teaching regulation skills directly and incorporating essential lifestyle behaviors into a treatment plan.

Conclusion

Neurocounseling and neuroscience offer additive perspectives on the supervisory relationship and supervision process (Russell-Chapin & Chapin, 2020). As we demonstrated in this chapter, prioritizing the consideration of regulation and physiological functioning can promote stronger supervisory relationships and help supervisees consider factors impacting their clinical work that they may otherwise not have taken into account. We hope this chapter generated new ideas for ways you may consider integrating neurocounseling and neuroscience into your own supervision practice.

Reflective Questions

What curiosities do you have about the role brain mapping can play in formulating a treatment plan?

How do you incorporate attention to lifestyle behaviors into your supervision and counseling work?

In the case study, what components most contributed to Addison's movement from feeling stuck to feeling effectively confident in her work with Trystan?

Quiz

1. Looking through a neurocounseling lens, which of the following should clinical supervisors be aware of and prioritize?
   Ethnocentric case conceptualizations.
   Their own healthy regulation capabilities.
   Supervisees' knowledge of cognitive behavioral strategies.
   Their own client caseloads.
2. Neurocounseling is an essential supervision tool because it looks at:
   Neurobiological dysregulation in the brain that might be causing ruptures and concerns in both counseling and supervision.
   Neurobiological locations and functions where the supervision question might be best assisted.
   Neurobiological reasons why the client might be struggling.
   All of the above.

References

American Psychiatric Association. (2022). *Diagnostic and statistical manual of mental disorders* (5th ed., text rev.). https://doi.org/10.1176/appi.books.9780890425787

Asbrand, J., Vogele, C., Heinrichs, N., Nitschke, K., & Tuschen-Caffier, B. (2022). Autonomic dysregulation in child social anxiety disorder: An experimental design using CBT treatment. *Applied Psychophysiology and Biofeedback, 47*(3), 199–212. https://doi.org/10.1007/s10484-022-09548-0

Beeson, E. T., Field, T. A., Jones, L. K., & Miller, R. A. (2017). *nCBT: A semi-structured multiphasic treatment manual v1.0* [Unpublished manuscript]. https://www.n-cbt.com/uploads/7/8/1/8/7818585/ncbt-manual_boisefinal.pdf

Bernard, J. M., & Goodyear, R. K. (2019). *Fundamentals of clinical supervision* (6th ed.). Pearson.

Chapin, T., & Russell-Chapin, L. (2014). *Neurotherapy and neurofeedback: Brain-based treatment for psychological and behavioral problems.* Routledge.

Chapin, T., & Russell-Chapin, L. (2022a, April). The tenets of neurocounseling: Part 1. *Counseling Today, 64*(10), 13–16.

Chapin, T., & Russell-Chapin, L. (2022b, May). Brain-based assessment in neurocounseling: Part 2. *Counseling Today, 64*(11), 19–23.

Corey, G., Muratori, M., Austin, J. T., & Austin, J. A. (2023). *Counselor self-care.* Wiley.

Council for Accreditation of Counseling and Related Educational Programs. (2024). *2024 CACREP standards.* https://www.cacrep.org/wp-content/uploads/2023/06/2024-Standards-Combined-Version-6.27.23.pdf

Fickling, M. J., Tangen, J. L., Graden, M. W., & Grays, D. (2019). Multicultural and social justice competence in clinical supervision. *Counselor Education and Supervision, 58*(4), 309–316. https://doi.org/10.1002/ceas.12159

Field, T. A., Beeson, E. T., & Jones, L. K. (2015). The new ABCs: A practitioner's guide to neuroscience-informed cognitive-behavior therapy. *Journal of Mental Health Counseling, 37*(3), 206–220. https://doi.org/10.17744/1040-2861-37.3.206

Gottman, J. M., & Silver, N. (2015). *The seven principles for making marriage work.* Harmon.

Jasper, H. H. (1958). Report of the committee on methods of clinical examination in electroencephalography. *Electroencephalography and Clinical Neurophysiology, 10*, 370–375.

Kagan, N. I. (1997). *Interpersonal process recall: Influencing human interaction.* Wiley.

Losada, M., & Heaphy, E. (2004). The role of positivity and connectivity in the performance of business teams: A nonlinear dynamics model. *American Behavioral Scientist, 47*(6), 740–765. https://doi.org/10.1177/0002764203260208

Miller, R., & Beeson, E. T. (2021). *The neuroeducation toolbox: Practical translations of neuroscience in counseling and psychotherapy.* Cognella.

Porges, S. W. (2011). *The polyvagal theory: Neurophysiological foundations of emotions, attachment, communication and self-regulation.* Norton.

Russell-Chapin, L., & Chapin, T. (2020). *Integrating neurocounseling in clinical supervision: Strategies for success.* Routledge.

Russell-Chapin, L., & Chapin, T. (2022, June). Neurotherapy interventions in neurocounseling: Part 3. *Counseling Today, 64*(12), 15–19.

Russell-Chapin, L., Pacheco, N., & DeFord, J. (Eds.). (2021). *Practical neurocounseling: Connecting brain functions to real therapy interventions.* Routledge.

Watkins, C. E., Jr. (2021). Rupture and rupture repair in clinical supervision: Some thoughts and steps along the way. *The Clinical Supervisor, 40*(2), 321–344. https://doi.org/10.1080/07325223.2021.1890657

Chapter 16

Ten Guidelines for Integrating Neuroscience Into Your Practice

*Lori A. Russell-Chapin, Thomas A. Field,
and Laura K. Jones*

In this concluding chapter, we use the case study of Muna (introduced in the Preface) to provide an example of how neuroscience can be integrated into counseling practice. Our discussion of Muna's journey, which began in the first edition of this text, has been updated to describe her return to counseling after the coronavirus pandemic. We have also updated our 10 practical guidelines to encompass more ways to help clients like Muna improve their emotional and physiological self-regulation.

Clinical Case Study: Muna

Muna is a 42-year-old Iraqi woman living in a metropolitan area of a large U.S. city. Seven years ago, she sought counseling to address anxiety that she was experiencing at her new job in an accounting firm. Muna was also struggling with feelings of inadequacy related to her long-standing dating relationship of nearly a decade. Her family lives in Iraq, and she emigrated to attend a U.S. college in her early 20s. She lived in constant dread of her family finding out that she was living with her boyfriend outside of marriage. She drank alcohol to cope, mostly at night (4 to 5 units). Muna also struggled with sleep, usually only getting 3 to 5 hours per night. She had a past diagnosis of attention-deficit/hyperactivity disorder and twice a day took 20 milligrams of Adderall, a stimulant. Muna had experienced psychological abuse from her father throughout her childhood. She was warm and engaging during the initial interview, although her nonverbal fidgeting suggested she was somewhat anxious. Muna acknowledged holding deep-seated fears that something was deeply wrong with her.

During the first 2 years of counseling, Muna made significant progress. At the counselor's request, Muna scheduled a full physical exam with a primary care physician. Her bloodwork assessment indicated that she was anemic and vitamin D deficient, with estrogen levels below the normal range, and that her thyroid functioning was outside the normal range. These medical concerns were addressed, and Muna's anxiety abated somewhat. Muna explored her past and present relationships and realized her attachment pattern was to end relationships before the other person might call it quits. Treatment for Muna included discussing the effects of early trauma on her physiological hyperarousal, her development, and her overall functioning. Knowing about these connections reduced Muna's presenting symptoms of anxiety and sleeplessness. She then proceeded to achieve treatment goals of reducing her alcohol consumption through attending Alcoholics Anonymous meetings. She began attending a mosque. She also received neurofeedback and eventually was weaned off the stimulant medication. When treatment concluded, Muna reported a great deal of self-efficacy from achieving her goals.

Muna's Return to Counseling

Five years later, Muna contacts your office to arrange a new intake appointment. You learn during the first meeting that the coronavirus pandemic was particularly hard on Muna. Shaking as she talks, she discloses that she has broken up with her partner and moved into her own apartment. The close proximity and "cabin fever" of quarantine had exacerbated disputes between them to the point that she feared for her safety. During the pandemic, Muna grew uncomfortable with socializing in large groups and currently has limited social activity and few friends. She also was diagnosed with ulcerative colitis and is having difficulty managing her symptoms. Muna wonders if these conditions developed following lingering symptoms of coronavirus infection. In addition, a year before the pandemic, Muna was injured in a serious car accident that left her with chronic pain she manages daily with opioids. Muna is still employed at the accounting firm but now works from home. Her goals for counseling are to better manage her medical problems, reduce her reliance on opioids, and become more socially active again.

Conceptualizing Muna's Treatment Plan in Class

On the first day of the course Neurocounseling: Brain and Behavior, Lori gave her students this case study and asked them to conceptualize and create a treatment plan. Their discussion was lively, and they developed excellent plans. The case study was put away, and the students were informed that it would be discussed at the end of the course. As the course unfolded, the class worked on neuroanatomy concepts and definitions, reasons for neurological dysregulation, self-regulation skills, a head map of functions, and brain-based interventions.

During the final class, the students reviewed the case study and treatment plans they had previously created. The class was divided into small groups. Each group read through the case study again and created a new or edited treatment plan. Something remarkable occurred. All the groups discarded their old treatment plans and created new ones that were neurocounseling based. In that moment Lori learned that once one truly understands the brain and its physiology and functions, one can never go back to seeing the world from their former vantage point. The traditional counseling viewpoint is not lost, but an additive and adjunctive neurocounseling viewpoint is gained. The major goal of this entire text is to share the additive value of integrating neuroscience with the essential and instrumental aspects of traditional counseling.

> **Reflective Question**
>
> Think back to the initial thoughts you had about Muna when you were reading the preface to this text and consider the following question:
>
> Because neuroscience integration is now a part of your counseling foundation, how would your treatment of Muna be different than you first thought?

Ten Guidelines for Integrating Neuroscience Into Your Practice

In the following discussion, we apply 10 guidelines for integrating neuroscience into counseling practice (Chapin & Russell-Chapin, 2022; Russell-Chapin, 2016) to the clinical case study of Muna.

Guideline 1

Integrating neuroscience and related physiology into counseling offers clients a fuller opportunity for personal wellness and intrinsic locus of control through the practice of emotional, behavioral, and physiological self-regulation.

Muna has a diagnosis of ulcerative colitis, a condition that she manages with medication and ongoing care by a gastroenterologist. After completing a wellness assessment, Muna came to understand that her lifestyle habits (e.g., stress, sleep, diet, alcohol use; Russell-Chapin, 2016) were potentially exacerbating her symptoms. The counselor collaboratively discussed with Muna treatment options for addressing *therapeutic lifestyle changes* (e.g., sleep hygiene, exercise, nutrition, social relationships; Ivey et al., 2022, as cited in Russell-Chapin, 2016). Together, they developed a plan. Muna reached out to her gastroenterologist regarding a dietary consult and met with a dietitian. She also started planning an exercise and stress-reduction routine, described below. Muna had previously received psychoeducation about adults needing 7 to 9 hours of sleep per night for memory consolidation, immune system functioning, and toxin removal in the brain. Muna admitted she was back to sleeping 3 to 5 hours per night and knew that even one night of sleep deprivation had deleterious effects (Stickgold & Walker, 2015). Muna committed to better sleep hygiene patterns every night by returning to previous routines of going to bed at the same time, cutting down on alcohol consumption in the evenings, and avoiding screen time 1 hour before bed. For Muna, knowing about the brain and its functions empowered her to make more informed decisions about managing her ulcerative colitis from a complete wellness perspective.

- For more on inflammation and the role of exercise and sleep in overall neurophysiological awareness, review Chapter 1 (Anatomy and Brain Development), Chapter 2 (Neurophysiological Development Across the Life Span), and Chapter 8 (Neurocounseling Approaches to Wellness and Optimal Performance).

Guideline 2

Honoring each client's cultural identities, heritages, beliefs, and values is a core pillar of our professional identity and crucial to consider when integrating neuroscience-informed counseling or neurocounseling into counseling practice.

Muna's series of recent losses (the end of a significant dating relationship, medical issues, chronic memory difficulties) left her with spiritual and existential questions about the meaning of her suffering. As a practicing Muslim who identified as Sunni, she experienced deep inner conflict and guilt about such questions. She considered such questions as forms of doubt and lack of trust in Allah. The counselor encouraged Muna to connect with her imam at the mosque she had attended in the past to discuss these concerns. The counselor also gave Muna space to process these feelings. Helping Muna process these fears was essential to the counseling process, as Muna had felt unable to talk to her friends and family for fear of being shamed.

- To understand the importance of providing a safe space for clients who come from diverse cultural backgrounds, especially when they are marginalized and oppressed, review Chapter 3 (Biology of Marginality) and Chapter 9 (Neuroscience of Attention).

Guideline 3

Locating available licensed professionals and resources from other disciplines provides clients with more comprehensive and thorough care.

Many clients have physical health challenges that impact mental health and vice versa. Counselors often work interprofessionally to meet the needs of their clients. Professionals who could be part of a client's resource team include physicians, psychiatrists, psychiatric nurse practitioners, and dieticians. When Muna returned to counseling, it was clear that she was receiving multiple forms of care from providers who might not have been in conversation. For example, Muna did not believe her primary care physician (who was prescribing her opioids) was in contact with her gastroenterologist. Muna facilitated releases of information between doctors to allow for an integrated approach to her medical care. With Muna's approval and consent, the counselor also spoke with the primary care physician. This conversation was crucial to identifying strategies for managing chronic pain and reducing Muna's reliance on opioids. Muna also requested a referral through her gastroenterologist to a dietician, who helped Muna develop a diet plan to reduce the potential for gastrointestinal inflammation caused or worsened by diet.

- For more on hormonal functioning and an interprofessional approach to client care, review Chapter 1 (Anatomy and Brain Development), Chapter 2 (Neurophysiological Development Across the Life Span), Chapter 4 (Neurophysiology of Traumatic Stress), and Chapter 6 (Psychopharmacology Basics).

Guideline 4

Incorporating neurocounseling into counseling and becoming neuro-informed provides additional insight into the physiological basis of interpersonal and therapeutic relationships.

One of Muna's greatest regrets was the breakdown of her 15-year relationship. Not only was this loss of attachment a significant source of grief, but the ending of the relationship was disturbing to Muna and brought up past childhood memories. Just as with her father,

Advanced Applications

Muna felt vulnerable with her former partner and worried he would physically attack her in the heat of an argument. Although this did not occur, Muna's daily experience of trepidation and fear shook her confidence in the trustworthiness of relationships. Healing from this experience took many months and required traditional talk therapy interventions that included narrative retelling and reauthoring. Grounded in the safety of the counseling relationship, Muna arrived at an understanding that the stress of the pandemic had soured her relationship and that her fears for her safety reflected her hypervigilance from childhood more than an active threat from her partner. She eventually felt ready to start dating again in the knowledge that her counselor would guide her through initial anxieties pertaining to trusting another partner in a new dating relationship.

- For more on the influence of attachment relationships on brain development, review Chapter 2 (Neurophysiological Development Across the Life Span). For information on the impact of trauma, review Chapter 4 (Neurophysiology of Traumatic Stress). For the physiological basis of helping relationships, review Chapter 9 (Neuroscience of Attention).

Guideline 5
Neuro-informed group counseling can add another layer of depth.

Two of Muna's stated counseling goals were to manage her medical problems and reduce her reliance on opioids. Previously, Muna had attended Alcoholics Anonymous meetings to reduce her alcohol use and found those meetings helpful. She had learned about the biochemical aspects of addiction and its impact on the brain and body, and this information was necessary and helpful to Muna's recovery process. Currently, Muna did not feel comfortable joining large groups related to fears of contracting further viral infection and worsening her medical symptoms. She instead started attending virtual meetings and found these to be just as beneficial. She reconnected with her previous sponsor, and the two began meeting in person. Muna gradually began to cut down her prescription opioid use in consultation with her primary care provider. Six months later, Muna stopped using opioids. She felt better able to manage her pain in concert with counseling interventions described in the step below.

- For more on substance use and neuro-integrated group counseling, review Chapter 5 (Clinical Neuroscience of Substance Use Disorders) and Chapter 13 (Neuro-Informed Group Work).

Guideline 6

Becoming neuroscience informed and integrating neurocounseling into talk therapy significantly widens the array of treatment strategies available for more focused physical, emotional, and behavioral change.

Counselors adopting a neurocounseling approach can not only incorporate traditional conceptualization and treatment strategies but add basic biofeedback and neurofeedback skills to the counseling paradigm. During her first round of counseling, Muna had gained a sense of mastery and self-efficacy from witnessing her ability to regulate body temperature through visualization and breathing exercises. In her second round of counseling, Muna started attending weekly yoga classes and wanted some additional training on meditation to facilitate further stress reduction. Like many clients, Muna had preferences regarding meditation practices. Muna preferred *mantram* repetition, a practice that involves vocalized repetition (whereas mantra repetitions are silent). She appreciated receiving skin temperature feedback while repeating her mantram phrases. The counselor facilitated skin temperature feedback by giving Muna an inexpensive thermometer to hold between two fingers that assessed her skin temperature at baseline and after the mantram repetitions.

- For more on the wider array of treatment strategies available for more focused physical, emotional, and behavioral change, review Chapter 7 (Neurocounseling Assessment), Chapter 8 (Neurocounseling Approaches to Wellness and Optimal Performance), and Chapter 9 (Neuroscience of Attention).

Guideline 7

When addressing the client's world of work with career-focused counseling, counselors can also incorporate a neuro-informed approach.

Often, counselors may glide over career issues when considering a neuro-informed counseling or neurocounseling approach. Considering work-life balance and stress from a neuroscience point of view may be enlightening. More than 80% of all symptoms reported to physicians are stress related—this is a powerful statistic to share with clients (Ivey et al., 2022).

Muna already knew from her first round of counseling that excessive sympathetic activation had the potential to impair her attention and concentration and, thus, her work performance. During the workday, when she could not use mantram repetition, Muna used sensory-based stress-reduction strategies (e.g., soothing music, aromatherapy) to regulate her baseline stress level, which in turn enhanced her work performance. Working from home enabled Muna to create a low-stress and relaxing work environment.

- For more on incorporating a neuro-informed approach when addressing the client's world of work with career-focused counseling, review Chapter 12 (Neuro-Informed Career-Focused Counseling).

Guideline 8

Counselors evaluate client physiological, psychological, and behavioral struggles and plan treatment to effectively intervene.

One of Muna's lingering questions was whether her extended symptoms from coronavirus infection had brought on her autoimmune disease. The counselor encouraged Muna to discuss this causal question with her gastroenterologist and delved into the deeper existential themes that were prime discussions for counseling. Muna was given space to explore her existential shattering experiences. She would never find a final and clear answer to her question about whether her coronavirus infection had caused her autoimmune disease. She also processed that even if her prior infection had caused her current symptoms, this awareness did little to help her manage her symptoms in the present moment.

Muna had previously received neurofeedback and knew of its benefits. She was surprised to learn that neurofeedback can be used to manage pain and was excited to get started. Muna received 30 sessions of 19-channel neurofeedback that followed a pain reduction protocol. From pre- to posttest, her subjective experiences of chronic pain improved. Although her medical condition was not resolved, and she continued to experience gut inflammation flare-ups, Muna noticed improvements and felt the treatments were beneficial.

- For more on evaluating client physiological, psychological, and behavioral struggles and fitting treatment planning to this assessment, review Chapter 1 (Anatomy and Brain Development), Chapter 6 (Psychopharmacology Basics), and Chapter 7 (Neurocounseling Assessment). For more information about the theoretical basis of neurofeedback and the neuroscience basis of counseling theories, review Chapter 11 (Neuroscience-Informed Counseling Theory).

Guideline 9

Counselors attend to ethical issues when integrating neuroscience-informed counseling or neurocounseling into counseling practice.

Muna had many complicated and complex variables interacting with her presenting symptoms. Synthesizing and translating complex scientific information into something that was discernible and comprehensible to Muna required its own skill set. Counselors are encouraged to receive training in learning how to distill neuroscientific concepts without diluting critical information (Field & Ghoston, 2020). Otherwise, counselors risk alienating clients and weakening the therapeutic relationship. Furthermore, counselors are careful to deliver information in a manner that destigmatizes rather than stigmatizes the client. Counselors (a) take into consideration the client's language and cultural familiarity with neuroscience concepts and related metaphors and (b) explain the limitations of neuroscience research, particularly any issues with underrepresentation of cultural groups in neuroscience studies (Burt & Pankey, 2022). Supervision can be useful to facilitate the development of competencies pertinent to translating neuroscientific knowledge to clients, and the field needs supervisors who understand how to translate neuroscience information.

Counselors must also remember to practice within their scope of competence, and appropriate training is vital for counselors who are integrating neuroscience in any way, especially if they are considering using any brain-based assessment instruments and interventions. Muna's medical issues, such as ulcerative colitis, vitamin deficiencies, and hormonal imbalances, are outside the scope of practice for counselors and require an integrated approach of working collaboratively with medical providers.

Counselors require extensive training before using assessment tools such as the quantitative electroencephalogram. Counselors also need training and supervision in the use of interventions such as neurofeedback and biofeedback before using these approaches independently with clients. These assessments and interventions require an understanding of neurobiology in addition to familiarity with the technological components (both hardware and software) of the interventions. Counselors who are interested in integrating neurofeedback and biofeedback into their work are encouraged to consider specialized credentialing, such as from the Biofeedback Certification International Alliance, to ensure the highest levels of training. All of Muna's professionals were licensed, and her counselor was board certified in neurofeedback.

- For more on ethical considerations with neuro-informed counseling, review Chapter 10 (Leveraging the Neuroeducation Process to Enhance Outcomes). For information on supervising counselors who are providing neuroeducation and using technical tools such as neurofeedback with clients, review Chapter 15 (Neuroscience-Informed Clinical Supervision).

Guideline 10

Counselors are critical consumers of research and use research to guide their integration of neuroscience-informed counseling or neurocounseling into counseling practice.

Counselors should carefully study the research base for interventions they provide, such as neurofeedback. Prior to the use of neurofeedback, Muna was exposed to the evidence base for this approach. Research studies and results provide clients with the scientific theories and facts that can validate and support their work in counseling. This offers strength to daily practice and credibility to not only neuroscience-informed counseling or neurocounseling but counseling in general. Some clients may not want to read articles, but most enjoy listening to the facts presented. Muna was very interested. The article that impressed Muna and gave her personal motivation concerned the effect of mantram repetition on stress reduction.

- For more on becoming a critical consumer of outcome research, review Chapter 14 (Enhancing Counseling Practice With Neuroscience-Informed Research).

Conclusion

After reentering the counseling process and engaging in treatment for 7 months, Muna again reported feeling a sense of accomplishment and mastery in her life. She was regularly implementing meditative techniques to manage ongoing chronic pain and was no longer reliant on prescription opioids. She was working in concert with her providers, who now included a dietitian. Her sleep and diet improved. She processed the end of her long-standing relationship and felt ready to start dating again. She worked through her guilt and shame at the existential and spiritual questions she wrestled with following these losses. Muna started meeting with her imam and returned to regular mosque attendance. Another important point of progress for Muna was her eventual willingness to attend counseling sessions in person. Muna acknowledged the reclusiveness she had developed during the pandemic and wanted to become more socially engaged. Attending counseling sessions was the first step in becoming more comfortable in social settings, because at the time she even did most of her grocery shopping online. She decided to join a local walking group and made several friends. By the final session, Muna reported feeling "back to myself" and more optimistic about the direction of her life.

Our goal as the editors of this text has been to assist your development into counselors, educators, and supervisors who integrate neuroscience into your work. You do not have to become an expert in medicine, biochemistry, or neurofeedback. As we have discussed, some of the more important aspects of being neuro-informed include having a foundation in neuroscience and related physiology, understanding the benefits of this knowledge, being aware of the ethical implications of this work, and referring to other helping professionals with a different knowledge base and skill set when needed. As you have seen throughout this text, once you understand how knowledge of the brain, mind, and body informs and enriches case conceptualizations and clinical practice, it is difficult, if not impossible, to go back to traditional counseling approaches.

Quiz

1. Which is not a key feature of neuroscience-informed counseling or neurocounseling?
   a. Bridging brain and behavior.
   b. Using other disciplines to add to wellness.
   c. Focusing on the medical model.
   d. Realizing that many mental health concerns have a physiological basis.
2. Neuroscience-informed counseling or neurocounseling allows the client to:
   a. Focus on building an intrinsic locus of control.
   b. Rely heavily on prescribed medications.
   c. Primarily change their cognitions.
   d. Receive referrals to electroconvulsive therapy.
3. Integration of neuroscience into counseling does which of the following?
   a. Adds value to the counseling profession.
   b. Teaches personal accountability to clients.
   c. Builds client self-regulation skills.
   d. All of the above.

References

Burt, I., & Pankey, B. (2022). Critically analyzing the field of neuroscience and its therapeutic application with Black populations. *Journal of Multicultural Counseling and Development, 51*(3), 174–182. https://doi.org/10.1002/jmcd.12274

Chapin, T., & Russell-Chapin, L. (2022, April). The tenets of neurocounseling: Part 1. *Counseling Today, 64*(10), 13–16.

Field, T. A., & Ghoston, M. R. (2020). *Neuroscience-informed counseling with children and adolescents.* American Counseling Association.

Ivey, A. E., Ivey, M., & Zalaquett, C. (2022). *Intentional interviewing and counseling: Facilitating client development in a multicultural society* (10th ed.). Brooks/Cole.

Russell-Chapin, L. A. (Ed.). (2016). Integrating neurocounseling into the counseling profession: An introduction. *Journal of Mental Health Counseling, 38*(2), 93–102. https://doi.org/10.17744/mehc.38.2.01

Stickgold, R., & Walker, M. (2015). Sleep and memory: The ongoing debate. *Sleep, 28*(10), 1225–1227.

GLOSSARY

Action Potential. The basic electrical signal of a neuron and the likelihood that a neuron will transmit a signal. Important in the neurotransmission of messages from one neuron to another.

Agonists and Antagonists. An agonist is a chemical agent that binds to and activates a receptor. An antagonist is a chemical agent that impedes or blocks the physiological functioning of a receptor or another substance.

Allostasis. A person's ability to physiologically adapt in the face of changing psychological, physical, or environmental pressures and in anticipation of future stressors.

Allostatic Load. The wear and tear on the body and brain from a persistent allostatic response in the face of chronic or extreme stress or environmental pressures.

Amygdala. A central subcortical brain structure that is responsible for detecting and responding to both innate and learned threats in the environment.

Autonomic Nervous System. A part of the peripheral nervous system that consists of two different branches: the sympathetic and parasympathetic nervous systems. The enteric nervous system, which innervates the gastrointestinal tract, is also considered a branch of the autonomic nervous system.

Axon. A long, hairlike projection that carries chemical messages away from the soma or cell body to the neighboring neurons, glands, or muscles on the other side of the neuron.

Axon Terminal. The area at the end of the axon where the axon branches and extends to come in close proximity to the neighboring structure (e.g., other neurons, muscles, glands).

Basal Ganglia. A coordinated set of subcortical nuclei consisting of the caudate nucleus and putamen (together known as the striatum), nucleus accumbens, globus pallidus, substantia nigra, and subthalamic nucleus. These structures play a considerable role in learning and memory, particularly implicit learning of automatized responses. The basal ganglia also inhibit and motivate movement.

Bilateral Integration. The integration of information, by way of the corpus callosum, coming from the left and right hemispheres of the brain.

Binding Potential Values. A crucial measure in positron emission tomography studies to establish the density of available receptors.

Biofeedback. An assessment and intervention approach that provides direct feedback to clients about certain markers of their physiological functioning, such as breathing patterns, heart rate variability, skin temperature, galvanic skin responses, and muscle tension. The goal of biofeedback is to help clients modify and enhance their neurophysiological functioning and performance, a process known as self-regulation.

Bottom-Up and Top-Down Processing. Two neurophysiological sequences in which sensory information is processed. In bottom-up processing, sensory information is received from sensory neuronal receptors (e.g., pain receptors) and sent to the brain for processing. In top-down processing, this same sensory information is interpreted through the lens of past experiences and knowledge as well as current thoughts and expectations. Bottom-up processing typically involves sensory and motor neurons in the peripheral nervous system sending messages to the thalamus for processing. Top-down processing involves the thalamus sending this information to the limbic regions (amygdala, hippocampus) and prefrontal cortex for interpretation and appraisal. Note that top-down processing thus involves structures and regions of the brain associated with executive functioning and reasoning (e.g., prefrontal cortex). These processes dictate how a person might respond to a stimulus. For example, a person who touches a hot stove would first experience bottom-up processing, whereby messages are sent quickly from sensory neurons in their hand via the peripheral nervous system (PNS) to the thalamus and then back to PNS motor neurons to remove their hand. After the hand is removed, the person might then interpret this burning pain sensation through prior experience and current thoughts via top-down processing as the PNS sends messages first to the thalamus and then to the amygdala, hippocampus, prefrontal cortex, and so on. In this instance, top-down processing occurs later because of the immediate reflexive need to remove one's hand from the source of danger/threat.

Brain-Derived Neurotrophic Factor (BDNF). A class of proteins in the central nervous system that are integral to neuronal and synaptic development and survival and that play a central role in neuroplasticity.

Brain Stem. The brain stem connects the brain to the spinal cord and the rest of the body and is vital for survival, regulating such integral processes as breathing, heart rate, blood pressure, and circadian rhythms.

Brain Wave. An electrical impulse of the brain produced by neurons firing at certain characteristic frequencies. The five basic brain wave categories are alpha, beta, delta, gamma, and theta.

Brain Wave Alpha. Alpha waves are categorized as 8 to 12 hertz and are needed for idling and transitioning from one brain wave state to another.

Brain Wave Beta. Beta waves go from 13 to 30 hertz. They are considered busy waves. Low beta waves from 12 to 15 hertz assist in focused attention and problem-solving. Anxiety may be present when beta waves are too high.

Brain Wave Delta. Delta waves are often 0 to 3 hertz or 0 to 3 cycles per second. These waves are slow, low waves and are associated with sleep and, sometimes, trauma-related problems.

Brain Wave Gamma. Gamma waves are 30 or more hertz and are often found in bursts of insight.

Brain Wave Theta. Theta waves are typically 4 to 7 hertz and are associated with drowsiness and meditation. These are also slower waves.

Broca's Area. An area of the cortex in the frontal lobe, located at the base of the primary motor cortex, that is associated with language production.

Cell Body. The soma or cell body is the core part of the neuron containing the nucleus and other cellular structures such as mitochondria used in energy production.

Cerebellum. A cauliflower-shaped structure at the back base of the brain, the cerebellum (meaning "little brain") is thought to be responsible for a range of functions related to cognition, emotion, sensory perception, attention, threat, and pleasure.

Cingulate Cortex. Lies underneath the outer surface of the cortex following the line of the corpus callosum. The job of the cingulate cortex is varied, given its role in learning, memory, reward, and social and emotional processing, with the anterior (frontal) and posterior (back) sections controlling diverse functions.

Cognitive Decline. The process of neuronal loss and/or loss of synaptic connections that occurs with age, resulting in challenges with memory loss and attention. Mild cognitive decline is common in older adults and differentiated from neurocognitive disorders.

Corpus Callosum. A thick band of nerve fibers that connects the two hemispheres of the brain.

Cortisol. A hormone that helps to restore homeostasis after stress and is essential to life.

Cranial Nerves. Twelve pairs of nerves that function in sensory, motor, and parasympathetic control, with most of the nerves controlling muscles of the face and neck or regulating visual, olfactory, and auditory sensations.

Dedifferentiation. A process by which brain structures for given functions lose their specialization and become simplified and less distinct. As a result, multiple different tasks can then activate similar brain regions.

Default Mode Network. A system of functionally connected brain regions that become engaged when the brain is in a resting state and is not involved in a specific attention-demanding, goal-oriented task.

DeltaFosB (ΔFosB). A protein thought to influence behavior change (via alterations in gene expression) and to play a role in drug addiction.

Dendrites. Branches extending out from the cell body that bring chemical messages or information into the cell body from neighboring neurons.

Diaphragmatic Breathing. An assessment and intervention tool by which a client's number of breaths per minute can be compared with norms to offer clues about anxiety, body tension, and muscle rigidity. It can also regulate autonomic functioning and enhance absorption of glucose and oxygen.

Differentiation. The process by which stem cells develop into mature specialized cells. During brain development, it is the process of stem cells becoming either neurons or glial cells.

DNA. Molecules that carry the instructions or codes that direct the synthesis of proteins needed for growth, development, and other life-sustaining functions.

Dopamine. A neurotransmitter thought to play a role in movement, goal-directed behavior, cognition, attention, and reward.

Electroencephalography (EEG). A method for recording the brain's electrical activity, or brain waves, through electrodes placed on the client's scalp.

Epigenetics. An array of mechanisms in which aspects of the environment are able to manipulate how genes are expressed. Several epigenetic mechanisms are linked to the harmful psychological outcomes of chronic stress.

Epinephrine. A hormone, also called adrenaline, that is primarily produced by the adrenal glands and prepares a person for action during times of threat as a form of self-protection. This fight-or-flight response involves changes in heart rate and breathing, among other physiological effects.

Functional MRI (fMRI). A brain imaging technique that measures changes in levels of oxygenated blood across various brain regions. The amount of blood flow into a particular area is said to denote the activity level of that brain region.

Gamma-Aminobutyric Acid (GABA). The body's primary inhibitory neurotransmitter. GABA plays a role in stress responses and the regulation of anxiety.

Gene Expression. The process by which the genetic information from a gene becomes a functional gene product, most often a protein; the activation of genes in a person's DNA sequence via epigenetic influences.

Glial Cells. The nervous system has two major forms of cells: neurons and glial cells. Glial cells provide insulation, nourishment, repair, and structural support to neurons. They also remove waste from the nervous system.

Glutamate. An amino acid that neurons use to send signals to other cells.

Gyri, Sulci, and Fissures. The cerebral cortex structurally contains ridges of tissues known as gyri, shallower grooves between the gyri known as sulci, and fissures, which are similar to sulci but are deeper grooves within the tissue that more clearly divide brain regions.

Half-Life. The half-life of a drug refers to the rate at which the drug is metabolized; it is the amount of time it takes for the plasma concentration of the blood to be reduced by one half.

Healthy Mind Platter. A wellness model developed to explain how optimal brain functioning can be maintained through daily healthy mental habits. Those mental habits include connecting time, downtime, focus time, physical time, playtime, sleep time, and time-in.

Heart Rate Variability. A measure of the variation in time between heart beats that can be used to determine the amount of stress the body is under. A lower heart rate variability signifies a predominant activation of the sympathetic nervous system. This measure can also be used as a biofeedback technique to aid in self-regulation.

Hippocampus. A subcortical brain region that is located in the medial (i.e., interior) region of the temporal lobes and is responsible for the formation (i.e., consolidation) of long-term declarative memories.

Hypothalamic-Pituitary-Adrenal (HPA) Axis. The functional connection between three endocrine glands that allows a person to adapt to both emotional and physical stress. When the HPA axis is activated, a chain of events occurs that eventually results in cortisol being released.

Hypothalamus. An almond-sized structure of the diencephalon that links the endocrine and nervous systems in the body. It assists the body in maintaining a state of internal balance or equilibrium known as homeostasis, such as by controlling body temperature, food intake, and water intake. The hypothalamus is also responsible for the release of various key hormones in the body, sexual development, and the ability to respond to stress.

Immediate Early Genes (IEGs). IEGs are activated very quickly in response to cellular stimuli as part of the process of gene expression and have an important role in such functions as learning, brain development, and drug use and addiction.

Immunity. Immune function is generally divided into two types: acquired and innate. Acquired immunity refers to the ability to target and destroy specific disease-producing microorganisms such as specific viruses. Innate immunity is naturally occurring, does not require previous exposure to a pathogen, and mounts a defense that is typically generalized and nonspecific.

Inflammation. Inflammation, part of humans' innate immune functioning, involves a generalized rallying of cells (i.e., leukocytes or white blood cells) that travel to a site of tissue damage. Regardless of how the tissue became injured, the damage signals a generalized response that attempts to destroy and clear any invading microorganisms and foreign debris and initiates tissue repair.

Insula. An area of the cortex, located in the frontal lobe near the confluence of the frontal, parietal, and temporal lobes, that has a range of functions, such as homeostatic regulation, self-awareness, motor control, interoception, empathy, and emotional processing. It also helps with translating the emotions that people feel in their bodies into their cognitive understanding of those emotions, or what are known as feelings.

Interoception. The cognitive awareness of internal bodily states.

Lobe, Frontal. The largest brain region, located at the front of the skull, the frontal lobe is involved in problem-solving, decision-making, planning, moral reasoning, attention, emotion regulation, and even priming in memory. The primary motor cortex is located in the frontal lobe, along with Broca's area, which is responsible for language production.

Lobe, Occipital. The smallest of the four lobes, the occipital lobe sits at the very back of the brain and is the primary visual center of the brain.

Lobe, Parietal. Located approximately halfway between the frontal lobe and the occipital lobe, the parietal lobe contains the primary somatosensory cortex, regulating the sensations that are perceived by the physical body, such as touch and the awareness of bodily movement and the orientation of the body in space, known as proprioception.

Lobe, Temporal. Located just behind the ears, below the parietal lobe and between the frontal lobe and occipital lobe, the temporal lobe has multiple functions, including hearing and language comprehension. Wernicke's area is located in the temporal lobe.

Memory. Memory is composed of several different theoretical components. Short-term and working memory are defined as the capacity for a person to remember a certain amount of information that was received in recent history. Long-term memory refers to the consolidation of events into long-term storage for retrieval later. These memories are either consciously retrieved (known as declarative or explicit memory) or preconsciously experienced (known as nondeclarative or implicit memory). Different memory systems are regulated by different areas of the brain.

Microbiota-Gut-Brain Axis. The bidirectional connection network between the central nervous system and the gut microbiota in the gastrointestinal tract. The functioning of the microbiota-gut-brain axis has implications for depression, posttraumatic stress disorder, memory, pain, concentrations of brain-derived neurotrophic factor helpful in the development of brain cells and intercellular connections and of immune functioning.

Myelin Sheath. The fatty sheath, made out of fatty glial cells, that covers the axons of neurons and helps speed the transmission of electrochemical messages.

Necrosis. The death of a nerve cell (neuron).

Neural Plate, Groove, and Tube. During its development in the womb, the human embryo develops a neural plate, or the structure that will eventually become the nervous system, including the brain and spinal cord. A neural groove later begins to form on the neural plate, which later becomes the brain. The two sides of the groove begin to curl, folding in on themselves and becoming a tubelike structure known as the neural tube.

Neurocounseling. The integration of neuroscience into the practice of counseling by teaching and illustrating the physiological underpinnings of many mental health concerns.

Neuroeducation. A process that is grounded in the therapeutic relationship whereby neuroscience information is used not to explain but to explore past, current, and future conceptualizations of the human experience, expectations for change, and actualized changes in counseling and psychotherapy.

Neurofeedback. A treatment method designed to empower clients to alter their brain functioning through the use of auditory and/or visual feedback that represents changes in their real-time electroencephalogram. Neurofeedback is grounded in the behavioral learning theory of operant conditioning and can be used by mental health providers who have a wide range of theoretical orientations.

Neurons. Cells of the nervous system that carry electrochemical messages. A neuron has four primary parts: dendrites, cell body (soma), axon, and presynaptic or axon terminal.

Neuroplasticity. The ability of the brain to alter its structure and function in response to external or internal changes in the environment, including development, learning, memory, brain injury, and disease.

Neuroscience-Informed Counseling. The integration of neuroscience into the practice of counseling with a focus on the brain-mind-body and neurological-physiological factors that influence cognition, behavior, and emotion. Commonly understood to mean the same as *neurocounseling*, but these terms may mean different things to different practitioners.

Neurotransmission. Otherwise known as synaptic transmission, neurotransmission is the communication of neurotransmitters from one neuron to another.

Neurovegetative Symptoms. Symptoms of depression that appear to have a neurophysiological basis, such as changes in sleep, appetite, concentration, anhedonia, hypoactivity, and loss of energy.

Norepinephrine. Like epinephrine (i.e., adrenaline), norepinephrine is a chemical released by the adrenal medulla that prepares a person for action. Unlike epinephrine, noradrenaline is also a neurotransmitter in the sympathetic nervous system.

Olfactory Bulbs. The sensory organs that are responsible for the sense of smell.

Oxytocin. A neuropeptide that plays an instrumental role in social connectedness and bonding.

Pharmacokinetics. The study of how a drug moves through the body, which consists of four processes: absorption, distribution, metabolism, and excretion. Every drug has a unique kinetic profile composed of these four factors.

Pituitary Gland. A pea-sized structure that is the master gland of the body, both producing and regulating the functioning of numerous hormones. The pituitary gland is highly active in the production of sex hormones and is vital to the body's ability to respond and adapt to stress.

Polyvagal Theory. A theory proposed by Stephen Porges that suggests that evolution has led to a functional neural organization of the brain that regulates autonomic states to best support social behavior. Porges proposed that the vagus nerve consists of two branches—namely, the ventral and dorsal vagal complex—that represent diverse states of autonomic regulation and functioning.

Psychoneuroendocrineimmunology. The field of scientific inquiry that studies the complex associations among the psychological state of a given individual, the neurological and hormonal processes that respond to that state, and the immunological mechanisms that communicate with those neurological and hormonal processes.

Quantitative Electroencephalography (qEEG). The comparison of a client's electrical activity in the brain with a norm group to determine whether electrical activity is within the normal range or may represent difficulties in certain areas of functioning.

Research Domain Criteria (RDoC). An organizing system created by the National Institute of Mental Health for brain-based research aimed at detecting subgroups of mental disorders, informing treatment selection, and facilitating more direct links from research to practice. It includes numerous functional constructs (e.g., response to acute fear) that are described through their related genes, molecules, cells, circuits or networks, physiology, behavior, self-report, and research paradigms.

Reuptake. The process by which neurotransmitters that are released into the synaptic cleft and do not bind to the postsynaptic neuron are recycled back into the presynaptic neuron.

RNA. RNA is essentially a photocopy of certain portions of DNA genetic codes. RNA is involved in protein synthesis and regulating cellular processes. It can also serve as an enzyme to speed chemical reactions. Molecules are involved in the transmission of genetic information and expression of genes.

Self-Regulation. The ability of a person to intentionally regulate their emotions, cognitions, behavior, and related physiology.

Serotonin. A neurotransmitter that has an important role in regulating mood, appetite, and sleep.

Siegel Hand Model. A psychoeducational tool for explaining how the brain developed evolutionarily and developmentally.

Signal Transduction. The transmission of molecular signals from the exterior of cells to the interior that leads to a functional change within the cell.

Skin Temperature Control. A biofeedback technique that assists clients in self-regulating skin temperature to gain a state of peak performance (relaxed and attentive). A temperature of 90° is considered peak performance.

Steady State. When a person takes medication on a regular basis, there is an ongoing process of drug absorption from each dose and, at the same time, an ongoing process of drug removal by metabolism and elimination. Steady state occurs when the amount of drug taken is equivalent to the amount of drug eliminated.

Subcortical. Underneath the cortex. Subcortical brain structures are those that lie beneath the outer surface of the cortex.

Sympathetic-Parasympathetic Shift. A shift away from a state in which the sympathetic nervous system is dominant (resulting in increased heart rate, a rise in blood pressure, sleep disturbance, and fueling of anxiety) to one in which the parasympathetic nervous system prevails (resulting in lowered heart rate and blood pressure, normal sleep, and a sense of calm).

Synapse. The region between two neurons that consists of the axon terminal, the synaptic cleft, and the dendrites.

Synaptic Cleft. The small space between two neurons.

Synaptic Pruning. The process by which the brain trims back unused synapses between neurons (and thus an example of neuroplasticity).

Thalamus. The primary relay station of the brain.

Theory of Mind or Mentalization. The ability to attribute mental states to others and to grasp the perspective of another, considered one dimension of empathy.

Therapeutic Lifestyle Changes (TLCs). Seventeen stress management strategies identified by Ivey, Ivey, and Zalaquett that enhance wellness.

Tonic and Phasic Firing. During neurotransmission, neurons fire in either a tonic or a phasic manner. Low-frequency (1–8 hertz [Hz]) tonic firing represents background activity of neurons that release neurotransmitters into the extracellular space. In contrast, phasic or bursting firing refers to a series of high-frequency bursts (15 Hz; induced by glutamatergic stimulation) that flood the synaptic cleft.

Volumetric Decline. Reductions in the number of neurons within a brain structure or area, resulting in greater/lesser proliferation of neurons and resulting impairments.

Voxel. A series of cubes that are used to conceptualize brain-based activity in response to a stimulus through the analysis of oxygenation levels.

Wernicke's Area. A subcortical structure in the temporal lobe associated with language comprehension.

White Matter and Gray Matter. Visually distinct areas of the brain as viewed on brain imaging scans. The gray areas are the neuronal cell bodies, and the white areas represent the axonal tracts, in particular myelinated axons. The myelin gives the white matter its characteristic white color.

Appendix

Explanation of Quiz Answers

Chapter 1
Anatomy and Brain Development

1. Which of the following lobes of the brain is in charge of executive functioning?
   a. Frontal.
   　The frontal lobe of the brain is responsible for executive functioning.

2. Which functional system of the brain is primarily known for helping people to respond to emotionally salient cues and threats in their environment but also plays a role in memory, social processing, motivation, addiction, and sexual behavior?
   c. Limbic system.
   　The limbic system is made up of brain structures such as the hippocampus, the amygdala, and the basal ganglia, which play a role in memory, social processing, motivation, addiction, and social behavior.

3. The study of the interconnectedness of various systems of the body is called:
   c. Psychoneuroendocrineimmunology.
   　Psychoneuroendocrineimmunology is a field of scientific inquiry that studies the complex associations among the psychological state of a given individual, the neurological and hormonal processes that respond to that state, and the immunological mechanisms that communicate with those neurological and hormonal processes.

Appendix

Chapter 2
Neurophysiological Development Across the Life Span

1. During which stage of development do synapses begin developing at a rate of approximately 40,000 per minute?
   c. Third trimester.
   During the third trimester, synapses begin developing at the amazing rate of approximately 40,000 per minute.

2. During adolescent brain development, which of the following is true?
   a. Subcortical limbic regions of the brain develop before the prefrontal cortical areas.
   Because subcortical limbic regions of the brain develop before the prefrontal cortex, adolescents can struggle with cognitive control and tend to make emotionally driven decisions.

3. In older adults, which of the following *does not* seem to be impaired by healthy aging?
   b. Emotion regulation.
   Healthy aging leads to declines in memory and processing speed but not in emotion regulation. Changes in emotion regulation can be characteristic of a neurocognitive disorder.

Chapter 3
Biology of Marginality: A Neurophysiological Exploration of the Social and Cultural Foundations of Psychological Health

1. Which of the following statements does not accurately characterize what is known about psychoneuroimmunology?
   b. Chronic stress causes inflammatory shutdown.
   Research in psychoneuroimmunology has shown that in response to chronic stress, acquired immunity is suppressed, and inflammation increases in intensity. Excessive inflammation can lead to psychological symptoms such as depressed mood, anxiety, and social withdrawal, and it can also lead to numerous debilitating chronic illnesses, including cancers, arthritis, and heart disease.

2. Which of the following statements is not a characteristic of chronic stress?
   d. It mobilizes negative feedback systems that help to regulate the stress response.
   The damage caused by chronic stress is related in part

to the failure of negative feedback systems to function properly, which results in a failure to control blood cortisol levels. Over time, the abnormal cortisol levels result in physical and psychological challenges.

3. Epigenetic changes can explain which of the following?
a. A decrease in neuroplasticity.
   Under conditions of chronic stress, epigenetic mechanisms result in a decrease in brain-derived neurotrophic factor, which causes decreases in neuroplasticity that result in cognitive and emotional challenges.

4. Which of the following best describes the notion of predisposing vulnerability?
c. It can change over time as a result of the impact of environmental forces.
   Although genetic predisposition is a component of predispositional vulnerability, environmental forces can raise or lower vulnerability, creating either more resilience in the face of stressors in the case of the former or increased vulnerability to stress in the case of the latter.

Chapter 4
Neurophysiology of Traumatic Stress

1. In the face of extreme or chronic stress, which of the following statements regarding cortisol is true?
d. A negative feedback loop for the HPA axis is disrupted, which impairs the body's ability to regulate levels of cortisol.
   Cortisol disrupts the HPA axis, impairing communication to the hypothalamus for corticotropin-releasing factor to be reduced. This results in chronically high levels of cortisol.

2. Which of the following structures is not implicated in impaired traumatic memories?
b. Pineal gland.
   The pineal gland is not implicated in impaired traumatic memories, whereas the amygdala, thalamus, and hippocampus are all affected.

Chapter 5
Clinical Neuroscience of Substance Use Disorders

1. Dopaminergic neurons in the mesolimbic pathway are largely associated with:

a. Reward prediction.

Dopaminergic neurons in the mesolimbic pathway have been linked to reward prediction. In contrast, dopamine neurons in the mesocortical and nigrostriatal pathways are associated with executive function and movement disorders. Last, the insular cortex has been linked to attention and decision-making.

2. Which of the following is *not* an evidence-based counseling intervention for addiction?
   d. Interpersonal and social rhythm therapy.
   Interpersonal and social rhythm therapy is a variant of interpersonal psychotherapy designed to target the pathogenic factors underlying bipolar disorder. However, this intervention has not been applied to addiction.

3. Which of the following elements make up the addiction cycle?
   c. Binge/intoxication, withdrawal/negative affect, and preoccupation/anticipation.
   Koob and Volkow (2010) proposed that the transition from voluntary to compulsive drug use may be understood through the lens of positive and negative reinforcement. Their framework consists of three stages of the addiction cycle: binge and intoxication (positive reinforcement), withdrawal and negative affect (negative reinforcement), and preoccupation and anticipation (craving).

Chapter 6
Psychopharmacology Basics

1. When the amount of a drug available to the body is the same as the amount being eliminated, then _____ has been achieved.
   c. Steady state.
   Steady state is achieved when the amount of drug going in is the same as the amount of drug getting taken out.

2. Which of the following is not a pharmacokinetic process?
   d. Accumulation.
   Absorption, metabolism, distribution, and excretion are the four pharmacokinetic processes.

3. When a serotonin neurotransmitter binds to a receptor, it has:
   b. An inhibitory effect.
   Serotonin reduces the likelihood of action potential by

opening ion channels to negatively charged ions. This is known as an inhibitory effect.

Chapter 7
Neurocounseling Assessment

1. Which statement about validity and reliability is most true?
   a. The test needs to be valid and reliable.
   Both reliability and validity are equally important to test selection. Tests that are reliable and not valid will produce the same results over and over again, but the results will not be relevant or applicable to the client's problem.

2. The main goal of a neurocounseling assessment is:
   b. Physiological, emotional, and behavioral self-regulation.
   Neurocounseling assessment aims to help clients achieve improved self-regulation. Although neurocounseling often includes a comprehensive evaluation, this is not the main goal of neurocounseling practice.

3. Which of the following are possible sources of brain dysregulation?
   d. All of the above.
   Genetic predisposition, substance abuse, and high fever are all possible sources of brain dysregulation.

Chapter 8
Neurocounseling Approaches to Wellness and Optimal Performance

1. Which of the following is not one of the six dimensions of Hettler's (1984) model of wellness?
   d. Health.
   The six dimensions are occupational, physical, social, intellectual, spiritual, and emotional.

2. Which of the following is not a healthy lifestyle strategy?
   a. Moderate use of alcohol.
   Healthy lifestyle strategies are diet, dietary supplements, exercise, sleep, and screen time.

3. Which of the following mental activities does not help clients to regulate their emotions?
   b. Focus time.
   Focus time is not necessarily linked to emotion regulation. In contrast, time-in and mindfulness, playtime,

and adequate sleep time all help clients to regulate their emotions.

4. Which of the following functions is not typically measured by wearable devices?
   d. Cortisol secretion.
   Cortisol secretion is typically measured through saliva (e.g., spitting into a sample bottle) or blood tests. Therefore, it cannot be measured through wearable devices. In contrast, physical activity (e.g., steps walked per day), heart rate (pulse), and sleep quality (duration, frequency of physical movement as a proxy for deep sleep) can be measured through wearable devices.

Chapter 9
Neuroscience of Attention: Empathy and Counseling Skills

1. Which of the following is *not* one of the things that Porges (2022) suggested is helpful to consider in establishing safety?
   d. Open-ended questions.
   Unlike the other three choices, open-ended questions were not identified by Porges (2022) as helpful in establishing safety.

2. Which of the following is *not* a form of empathy?
   b. Situational.
   The three components of empathy are cognitive empathy, affective empathy, and mentalization.

3. In helping a client deal with microaggressions, the central goal is to:
   b. Provide a safe environment so that the client can talk easily.
   Although listening carefully and fully is important to all relationships, clients who experience microaggressions need to feel safe so that they can process what has happened to them. In such cases, close listening is not enough—the counselor's body language and nonverbal behavior must also help the client to feel safe in the room. Although providing psychoeducation is also useful, the client must first feel safe (the central goal). The last choice does not assist the client to process their experience, nor does it address the injustice of the client's situation.

Appendix

Chapter 10
Leveraging the Neuroeducation Process to Enhance Outcomes

1. Which of the following statements best describes neuroeducation?
   b. A process of exploring neuroscience information with a client.

   Delivery matters here. Clients are likely to develop negative self-cognitions if counselors attempt to explain why their brain is not working properly ("I'm damaged"). Clients who feel overwhelmed by technical information may also develop negative self-cognitions ("I can't even understand my counselor"). Referral to a neuropsychologist is also not a formal step in the neuroeducation process.

2. When exploring neuroscience information with clients, it is important to:
   a. Select only one or two concepts and deliver them as concisely as possible.

   Generally speaking, clear and concise delivery assists the client to understand the information provided without becoming overwhelmed and to begin integrating the information into their current schematic understanding. Selecting only a few concepts also enhances the likelihood that the client will not become overwhelmed. Using primary sources is important when necessary, and skillful counselors always attend to the process of delivery in addition to the content. Attending to delivery and to the client's response helps to ensure the neuroeducation process is an exploratory one rather than a lecture. Counselors also explain to clients that neurophysiology is just one part of a person's biopsychosocial holistic experience.

3. Current research on providing biogenic or neurobiological explanations suggests:
   e. There are potential benefits to sharing biogenic and neurological explanations with clients but also certain risks related to stigma, prognostic pessimism, and neuroessentialism.

   Counselors should be aware of both benefits and risks of biogenetic and neurological explanations. For some clients, this knowledge could be empowering and free them from self-blame and shame. For other clients, this knowledge could be perceived as proof that they are damaged and that they cannot change or improve because their problems are biological.

Appendix

Chapter 11
Neuroscience-Informed Counseling Theory

1. Researchers began studying the neurobiological outcomes of the use of therapeutic interventions by investigating:
   c. Phobias.
   Phobias were first studied when researchers began examining the neurobiological outcomes of counseling interventions.

2. The primary brain structure affected during the application of therapeutic interventions such as cognitive behavior therapy (CBT) and interpersonal psychotherapy (IPT) are:
   d. b and c.
   The brain structures primarily affected by counseling interventions such as CBT and IPT are the prefrontal cortex and the limbic system.

Chapter 12
Neuro-Informed Career-Focused Counseling

1. According to Donald Super (1980), vocational identity is:
   a. A projection of the self into the world of work.
   Super maintained that vocational identity is a projection of the self into the world of work. This is important because it implies that successful career development is predicated on self-identity.

2. Which system is activated during times of stress, including work stress?
   d. HPA axis.
   Stress and work stress are experienced similarly in the brain because they are governed by the HPA axis. Physiological response to stress does not vary by type of stress.

Chapter 13
Neuro-Informed Group Work

1. What factors make group so complex?
   a. The numerous points of interactions (dyads and triads) between and among members to which a facilitator must attend.
   It is not the number of members that makes group challenging; rather, it is the number of exchanges and interchanges to which a facilitator must attend that takes

so much energy.

2. According to the authors of this chapter, what is the problem with storytelling from a neuroscience perspective?
   d. Memory is not recall as much as it is reconstitution, so storytelling in the wrong context can bias the story and distort perception.

   Memory retrieval is context dependent and infused with present emotion that distorts perception of the memory.

Chapter 14
Enhancing Counseling Practice With Neuroscience-Informed Research

1. You find an article that reports mean differences between two samples. What kind of significance is being discussed with this metric?
   b. Practical significance.

   Practical significance differs from statistical significance in that it explores the strength of the relationship between variables, known as *effect size*. The standardized mean difference is a form of effect size.

2. Client outcome research can be viewed as a form of _____.
   d. Both a and b.

   Findings from client outcome research are powerful and can be used to advocate for systemic changes (e.g., changes to insurance reimbursement practices) in addition to advocating for clients (e.g., requesting more reimbursed sessions for a particular client). Although we often respect those researchers who conduct outcome studies, the value of research extends far beyond mere self-promotion!

Chapter 15
Neuroscience-Informed Clinical Supervision: An Emerging Transtheoretical Approach

1. Looking through a neurocounseling lens, which of the following should clinical supervisors be aware of and prioritize?
   b. Their own healthy regulation capabilities.

   Supervisors who use a neuro-informed approach to supervision recognize the importance of counselor emotional, mental, and physical regulation in order to provide the most helpful counseling relationship possible for the client.

Appendix

> While knowledge of techniques and skills (e.g., CBT strategies) can also be a focus of supervision, it should not be prioritized over counselor self-regulation.

2. Neurocounseling is an essential supervision tool because it looks at:
   d. All of the above.
   > Neurocounseling-informed supervision addresses several critical areas of counseling practice, including client neurophysiological regulation, neurophysiological factors that contribute to symptoms, and the use of technologies such as electroencephalograms to determine treatment targets.

Chapter 16
Ten Guidelines for Integrating Neuroscience Into Your Practice

1. Which is *not* a key feature of neuroscience-informed counseling or neurocounseling?
   c. Focusing on the medical model.
   > Counselors holding a neuroscience-informed counseling or neurocounseling lens incorporate wellness and approach clinical work from a strengths-based holistic orientation. They do not emphasize the medical model, which highlights deficits over strengths.

2. Neuroscience-informed counseling or neurocounseling allows the client to:
   a. Focus on building an intrinsic locus of control.
   > Counselors integrating neuroscience into their work empower clients and build an intrinsic locus of control by building on successful experiences. Counselors using this approach do not rely heavily on prescribed medications, nor do they refer clients regularly to electroconvulsive therapy. Neurocounseling is not restricted to cognitive behavior therapy.

3. Integration of neuroscience into counseling does which of the following?
   d. All of the above.
   > This one was easy! Integrating neuroscience into counseling practice adds value to the profession, teaches personal accountability to clients, and builds client self-regulation skills.

INDEX

Figures and tables are indicated by "f" and "t" following page numbers.

A
ABC model of CBT, 228
Abilify (aripiprazole), 133
Abraham, W. C., 170
Absorption (drugs), 128
Abuse
  child abuse, xx, 81, 88–89, 305, 309–310
  of drugs. *See* Substance use and addiction
  interpersonal violence, 81, 294
ACA (American Counseling Association), xi, 272
ACC. *See* Anterior cingulate cortex
Acetylcholine, 126
Acquired immunity, 62–63
ACTH (adrenocorticotropic hormone), 14, 110, 243
Action potential, 18–19, 317
Acute stress, 41, 243. *See also* Fight-or-flight response
Adaptive calibration model, 61
ADD/ADHD. *See* Attention deficit/hyperactivity disorder
Adderall, 133, 305
Addiction. *See* Substance use and addiction
ADDRESSING model, 200
Adjustment disorder, 298
Adolescence
  brain development during, 40–42
  case study on neuroscience-informed clinical supervision, 286–288, 292, 300–301
  case study on preadolescent, 4, 24–25, 46–47
  developmentally informed interventions for, 46–47
  sleep and, 168, 300
Adra, N., 167
Adrenal glands, 14–15, 83–84, 242, 243, 244, 245
Adrenocorticotropic hormone (ACTH), 14, 110, 243
Affective empathy, 187, 190
Affective prosody, 87
Affect regulation/dysregulation, 81, 85, 89, 90
African Americans. *See also* Marginality, biology of
  environmental stressors and systemic barriers for, 208
  group counseling and, 252
  neurofeedback research and, 281
  racism and social inequalities in research, 276–277
  underrepresentation in research, 276
Agency, 89, 208
Aging adults. *See also age-related diseases*
  brain changes in, 44–45, 181
  cognitive decline and, 46, 171
  developmentally informed interventions for, 46
  exercise and, 164, 165
  meditation and, 166
  neurofeedback and cognitive decline in, 171
  psychotropic medication and, 129
Agonists, 114–115, 127, 133, 317

337

Index

Ainslie, P. N., 170
Aksoy, Ravza N., 181, 219
Alcohol use, 114, 115, 305–306, 308, 310. *See also* Substance use and addiction
Alerting network, 184–185
Alexithymia, 108, 110
Allen, N., 276
Allostasis, 80, 84, 104, 317
Allostatic load, 80, 242, 243, 317
Alpha brain waves, 153, 156, 171, 225, 319
Alprazolam (Xanax), 131
Altruism in group work, 259*t*
Álvarez-Pérez, Y., 222
Alves, H., 165
Alzheimer's disease
  aging brain and, 44, 45
  benzodiazepines and risk of, 132
  dietary supplements and, 168
  focused attention and, 164
  neuroplasticity and, 21
  stimulants for, 134
Amblyopia (lazy eye), 245
Amen Clinic ADD Type Questionnaire, 152
American Counseling Association (ACA), xi, 272
American Mental Health Counselors Association (AMHCA), xi, xvii, 272
American Psychological Association, 277
Amgalan, A., 167
Amphetamines, 133
Amygdala
  adolescent brain and, 40–41, 47
  in alerting network, 184
  child trauma and, 88–89
  defined, 317
  depression and, 223
  in drug addiction, 109–111
  in fight-or-flight response, 59, 83–85
  functions of, 12
  infant attachment and, 38
  lithium and, 132
  memory and, 38, 240, 255
  oxytocin and, 86
  phobia therapy and, 222
  PTSD and, 223
  in sensory processing, 243
  working memory and, 193
Anabolic steroids, 16
Anatomy. *See* Brain anatomy and development
Androgens, 35

Anesthesia, 296
Angular gyrus, 226
Anhedonia, 64, 108, 110
Animal studies, 275
Antabuse, 114
Antagonists, 127, 317
Antal, B., 167
Anterior cingulate cortex (ACC)
  in adolescent brain, 40, 41
  in empathy experience, 187, 188
  functions of, 9
  neurofeedback and, 225
  in substance abuse disorders, 111
  traumatic stress and, 85
Anterior insula, 41, 85–86, 188
Anticonvulsant medication, 132
Antidepressants, 130*t*, 131, 135
Antipsychotics, 130*t*, 133
Antisocial personality disorder, 60
Anxiety and anxiety disorders
  biofeedback and, 224
  in case studies, 57, 162, 238, 239, 245–247, 305–306
  CBT and, 222
  childhood programming of stress response and, 60
  chronic stress and, 60, 64, 66
  diaphragmatic breathing and, 157
  gut-brain connection in, 15, 41
  immune functioning and, 16
  in marginalized populations, 58
  meditation and, 166
  neurofeedback and, 157, 171, 274–275
  psychotropic drugs for, 131–132
  race and, 59
  screening for, 150
  self-regulation assessment and case conceptualization, 296
  sex hormones and, 88
  thyroid hormones and, 16
  workplace and, 245, 246–247
Anxiolytics, 130*t*, 131–132
Apple Watch, 172
Applied behavior analysis treatment, 137
Aricept (donepezil), 134
Aripiprazole (Abilify), 133
Arthritis, 63, 168
Artificial intelligence, 182
Assessments, 143–179
  anxiety screening, 150
  assessment-based treatments, 156–158
  biofeedback, 157
  body perception screening, 150
  CACREP 2024 Standards and, 143–144

case study, 144–145
  brain-based approach to, 158
for client progress, 280–281
defined, 146
depression screening, 150
diaphragmatic breathing, 156–157
insomnia screening, 149
for learning difficulties
  attention deficit, 152
  continuous performance tests, 152–153
neurocounseling observational notes, 149
neurofeedback, 157–158
neurological risk assessment, 147, 148*f*
overview, 143
personality inventories, 151–152
psychosocial–medical history interview, 147
quantitative EEG, 153–156, 154–155*f*, 325
  five-channel qEEG, 153, 313
  nineteen-channel qEEG, 156
screening inventories, 149
training in, 313
trauma screening, 150–151
Association for Counselor Education and Supervision, xi
Asthma, 60
Attachment, 10, 37–38
  in case studies, 229, 309–310
Attending behavior, 182, 188
Attention, 181–198
  alerting, 184–185
  CACREP 2024 Standards and, 182–183
  case study, 183
    brain-based approach to, 194–195
  counseling skills, calming, and activating, 190–191
  empathic understanding, microskills and, 186–191
  executive functioning, 185–186
  neuroscience of, 183–186
  orienting, 185
  overview, 181–182
  polyvagal theory and safety, 188–189, 189*f*
  rewiring of memories into positive resilient action, 193–194
  self-regulation assessment and case conceptualization, 296
  sympathetic nervous system and, 311
Attention deficit/hyperactivity disorder (ADD/ADHD)

assessment for, 152
brain anatomy and, 13
in case studies, xx–xxi, 124–125, 136–137, 305–306
childhood programming of stress response and, 60
childhood trauma and, 66
neurofeedback and, 157, 171, 225
risperidone prescribed for, 124–125, 136–137
stimulants for, 133–134, 135
Auditory system, 38
Augustine, G. J., 241
Autism spectrum disorder
  antipsychotics for, 133
  brain anatomy and, 13
  diagnosis of child, 136
  neuroplasticity and, 21
  thyroid hormones and, 16
Autobiographical memory (AM), 239–241
Autoimmune diseases, 63
Automatized responses, 11
Autonomic nervous system
  anxiety disorder and, 293
  attachment and, 38
  attending, role in, 184
  cranial nerves and, 10
  defined, 317
  depression and, 223
  functions of, 14
  polyvagal theory and, 24
  PTSD and, 85, 86
  workplace stress and, 243
Autonomy, 208
Avoidance coping, 90, 247
Axon and axon terminal, 17–18, 18*f*, 317

B
Baca-Garcia, E., 42
Balkin, R. S., 274–275
Barbiturates, 132
Basak, C., 165
Basal ganglia, 11, 12, 107, 317–318
Basic listening sequence, 191
Bayro-Kaiser, E., 172
BCIA (Biofeedback Certification International Alliance), 281, 299, 313
BDMA (brain disease model of addiction), 101–102, 112–113
BDNF. *See* Brain-derived neurotrophic factor
Beblo, T., 87
Beck Depression Inventory, Second Edition (BDI-II), 150

339

Index

Beeson, Eric T., 199, 201–202, 269, 293
Bégaud, B., 132
Bekelman, T., 169
Bender Visual-Motor Gestalt Test, Second Edition, 151
Benzodiazepines, 131–132
Bergouignan, L., 239
Berkman, E. T., 242, 244
Bertelli, M., 168
Beta-blockers, 132
Beta brain waves, 153, 171, 319
Betancort, M., 222
Betancourt, R., 165–166
Bilateral integration, 45, 318
Billioti de Gage, S., 132
Binding potential values, 226, 318
Binge/intoxication stage of addiction cycle, 107–109
Biofeedback, xv, 47. *See also* Neurofeedback
   as assessment, 157, 294
   in case studies, 173, 229, 311
   as CBT component, 224
   defined, 318
   ethics and, 313
   in n-CBT, 228
   physiological awareness and, 293
   sympathetic–parasympathetic shift and, 67
   training in, 299
   wearable electronic devices for, 172, 225
   in wellness counseling, 67, 170–171, 300–301
Biofeedback Certification International Alliance (BCIA), 281, 299, 313
Biogenic explanations of mental health, 206–207
Bipolar disorder
   chronic stress and, 60
   dietary supplements and, 168
   mood stabilizers for, 132
   neuroplasticity and, 21
   thyroid hormones and, 16
Birth history, 144, 147
Black Americans. *See* African Americans
Blakemore, S. J., 41
Blame, 202, 206–207
Blind spot in group work, 260, 260*f*
Blood pressure, 10, 67, 83, 157, 165, 243
Body Perception Questionnaire–Short Form, 150
Bonding. *See* Attachment
Bottom-up processing, 89, 113, 170–171, 261, 318
Brain anatomy and development, 3–31. *See also* Neurophysiological development; *specific parts of brain*
   autonomic nervous system, 10, 14. *See also* Autonomic nervous system
   brain communication, 17–21
      neurons, 17–18, 18*f*, 125–126
      neuroplasticity, 21. *See also* Neuroplasticity
      neurotransmitters and hormones, 19, 20*t*, 41, 42. *See also* Hormones; eurotransmitters
   CACREP 2024 Standards and, 4
   case study on preadolescent isolation, 4
      brain-based approach to, 24–25, 46–47
   coordinated systems of brain and body, 13–15
   cranial nerves, 10, 34, 320
   default mode network, 9, 13–14. *See also* Default mode network
   evolution of brain, 22–24, 23*f*
   external structures, 5–10
   hemispheres, 5–6
   HPA, 14–15. *See also* Hypothalamic-pituitary-adrenal axis
   internal/subcortical structures, 10–13, 11*f*
   limbic system, 13. *See also* Limbic system
   lobes, 6–10, 6*f*
   microbiota–gut–brain axis, 15, 41, 42, 289, 323
   overview, 3–4
   polyvagal theory, 14, 24. *See also* Polyvagal theory
   psychoneuroendocrinology and immunology, 15–16
   sleep and, 300
   structure, function, and systems of brain, 4–13
Brain-based approaches case studies
   in assessment case, 158
   in career-focused counseling, 246–247
   in group work, 261
   in marginalization case, 67
   in microaggression against student case, 194–195
   in neuroeducation, 212–214
   in neuroscience-informed counseling theory, 228–229
   in preadolescent isolation case, 24–25, 46–47

in psychopharmacological medications case, 136–137
in substance use and addiction, 115–116
in traumatic stress case, 91–93
in wellness and optimal performance, 173
Brain brightening, 171
Brain-derived neurotrophic factor (BDNF), 35, 47, 64, 66, 165, 170, 318–319
Brain disease model of addiction (BDMA), 101–102, 112–113
Brain dysregulation, 147, 156
Brain mapping. *See* Head map of brain functions
Brain Research through Advancing Innovative Neurotechnologies (BRAIN of NIH), 3
Brain stem, 6f, 9–10, 319
Brain volume, 35–36, 44. *See also* Gray matter; White matter
Brain waves, 153–156, 171, 225, 273, 294, 299, 319
Breathing exercises, 47, 67, 157, 261, 311. *See also* Diaphragmatic (deep) breathing
Briere, J., 90, 91
Broca's area, 7–8, 85, 92, 111, 319
Bullying, 294
Burgdorf, J., 274
Burns Anxiety Inventory, 150
Burt, Isaac, 199, 269
Buspar (buspirone), 132

C
CACREP 2024 Standards, xi–xii, xv–xvi
Assessment and Diagnostic Processes, 102, 143–144
Career Development, 238
case study, 200–201
Counseling Practice and Relationships, 78, 102, 124, 161–162, 182, 199, 219–220
Cultural Identities and Experiences, 161–162
Group Counseling and Group Work, 252
Lifespan Development, 4, 33–34, 56, 78, 102, 124, 161–162
Professional Counseling Orientation and Ethical Practice, xii, 287
Research and Program Evaluation, 269
Social and Cultural Diversity, 56
Specialization Standards

on career counseling, 238
on counseling effectiveness, 270
on counselor advocacy, xii
on crisis and trauma on marriage, couples, and families, 78
on integration of counseling theories, 220
on legal and ethical issues, 200
on mental health and behavioral disorders, 144
on psychopharmacological medications, 124
on substance abuse and addictions, 102
on supervision, 287–288
Calming, 184, 190–191, 194
Campbell, H. A., 170
Cancer, 58, 60, 63, 168
Cardiovascular function, 166, 168
Career-focused counseling, 237–250
brain-based metaphors as normalizing and demystifying, 244–245
CACREP 2024 Standards and, 237–238
case study, 238–239
brain-based approach to, 246–247
decision-making and, 241–242
importance of integrating into neurocounseling, 311–312
metaphoric approach to understanding brain function, 245–246
neuro-based metaphors in, 244
occupational knowledge and, 240–241
overview, 237
self-knowledge and, 239–240
stress and work, 242–243
Caruso, P., 168
Case conceptualization, 293–296, 295f
Case studies
assessments, 144–145, 158
career-focused counseling, 238–239, 246–247
group work, 253, 261
marginalization, 57–59, 67
microaggression against student, 183, 194–195
neurocounseling-based approach to immigrant woman with anxiety, xx–xxi, 305–315
neuroeducation, 200–201, 212–214
neuroscience-informed clinical supervision, 286–288, 292, 300–301
neuroscience-informed counseling theory, 220, 228–229
neuroscience-informed research, 270–271

341

preadolescent isolation, 4, 24–25, 46–47
risperidone for child's ADHD, 124–125, 136–137
substance use and addiction, 103–104, 115–116
traumatic stress, 78–79, 91–93
wellness and optimal performance, 162, 173
Casey, B. J., 41
Catharsis in group work, 259*t*
CBT. *See* Cognitive behavior therapy
Cell body (soma), 17, 319
Cell death, 245
Centers for Disease Control and Prevention (CDC), 123
Central nervous system, 21, 35, 183–184, 244
Cera, N., 226–227
Cerebellum, 6*f*, 9, 35, 39, 319
Cerebral blood flow, 42, 47
Cerebral cortex, 5–10, 22, 35
Ceverino, A., 42
Chaddock, L., 165
Chapin, Theodore J., 143, 158, 287, 294
Chatters, SeriaShia, 219
Children. *See also* Adolescence
anxiety disorder and, 293
attachment patterns. *See* Attachment
brain-derived neurotrophic factor and, 66
brain development in, 36–39
child abuse, xx, 81–82, 88–89, 305, 309–310
chronic stress and, 66
developmentally informed interventions for, 45–46
early interventions targeting parent–child interaction, 69
grandmother raising children in inner city, case study of, 57–59
language development, 38–39
memory development, 39
neuroeducation for, 209
psychotropic medication and, 129
risperidone prescribed for ADHD, case study of, 124–125, 136–137
screen time limits, 169
sleep recommendations for, 168
stress response and, 60–61
trauma in childhood, 66, 88–89, 309–310
Chiurazzi, P., 168
Choanoflagellates, 22

Chromatin, 104
Chronic disease, 296, 306, 308
Chronic fatigue, 63
Chronic stress, 41, 59–65, 64*f*, 242
Cingulate cortex, 9, 13, 319
Claesson, M. J., 41
Clark, K., 167
Clarke, G., 41
Client progress, tracking, 280–281
Clinical significance, 280–281
Clonazepam (Klonopin), 131
Coan, J. A., 37
Cocaine, 111
Coconstruction, 227
Cognition and learning
in aging adults, 44–45
brain anatomy and, 9, 11
brain evolution and, 22
cognitive decline in aging, 46
neurofeedback for, 171
oxytocin and, 38
self-regulation assessment and case conceptualization, 296
Cognitive behavior therapy (CBT), 221–225
in ADHD treatment, 134
in anxiety treatment case study, 228–229
depression and, 223–224
learning theory in neurofeedback, 224–225
for marginalized populations, 68
neuroscience-informed CBT, 228, 293–294
oxytocin and, 86
phobias and, 222–223
substance abuse disorders and, 113–114, 116
trauma and, 89, 223
Cognitive decline, 46, 171, 319
Cognitive empathy, 187, 190
Cognitive processing therapy, 89
Cohen, B., 16
Cohen's *d*, 279
Cohesion in group work, 259*t*, 260
Communication skills, 191, 192*f*, 256–257
Compassion, 189, 202
self-compassion, 90, 207
Complex symptoms, 294
Compulsive behavior, 106–107, 111
Conduct disorder, 60, 61
Conduct resolution, 241
Connecting time, 165–166
Consciousness, 8–9, 193

Consultations on medications, 135
Contingency management, 113–114, 115
Continuing education, 204, 281
Continuous performance tests, 143, 149, 152–153
Coping techniques, 228, 246, 293, 311
Cordes, D., 226–227
Coregulation, 212–213, 289
Corpus callosum
  aging effects on, 44
  child trauma and, 89
  defined, 319–320
  functions of, 5, 9
Corrective emotional experience, 226
Corticolimbic connections, 92
Corticotropin-releasing hormones, 14, 110
Cortisol
  adolescents and, 41
  child trauma and, 89
  defined, 320
  depression and, 223
  drug withdrawal and, 110
  in fight-or-flight response, 59–60, 83–84
  function of, 14
  in stress response, 243, 245
  workplace stress and, 242
Cotter, J. D., 170
Council for the Accreditation of Counseling and Related Educational Programs Standards. *See* CACREP 2024 Standards
Counseling, defined, xii–xiii. *See also* Neuroscience-informed counseling
Counseling and Related Educational Programs Standards. *See* CACREP 2024 Standards
Counseling theory, neuroscience-informed, 219–236
  CACREP 2024 Standards and, 219–220
  case study, 220
    brain-based approach to, 228–229
    interpersonal therapy and depression, 226–227
    learning theory in neurofeedback, 224–225
  theoretical orientation, 221–228
    cognitive behavior therapy, 221–225
    future directions in, 228
    narrative therapy, 227–228
    psychodynamic therapy, 226–227
COVID-19 pandemic, 123, 169, 276, 306

Cozolino, L., 256
Cranial nerves, 10, 34, 320
Craving (preoccupation-anticipation), 111–112
Creative arts, 67
Cryan, J. F., 41
Crystallized tasks, 44
Csikszentmihalyi, M., 164
Culturally informed, multitargeted counseling, 68–69
Culturally informed case conceptualization, 293–294
Cultural pride, 194
Culture
  cultural competency, xix, 208
  cultural diversity, 55–57, 305, 309. *See also* Marginality, biology of
  multicultural understanding, 191, 192f, 208, 277, 290
Curcumin, 168, 173
Curiosity, 203, 211–212, 274–275
Cytokines, 63–64

D
DA. *See* Dopamine
Dabelea, D., 169
DAergic (dopaminergic), 107–109
Dahl, R. E., 41
Daniels, J., 87
Daniels, M. H., 244
Daniels, Thomas, 181
Decety, J., 187
Decision making and planning, 8, 9, 111–112, 241–242
Deconstruction, 227
Dedifferentiation, 45
Deep breathing. *See* Diaphragmatic breathing
Default mode network (DMN)
  aging and, 44
  defined, 320
  functions of, 9, 13–14
  meditation and, 166
  relaxation and activation of, 164
DeFord, J., 298
De Greck, M., 187
Delaveau, L., 239
De Leon, J., 42
DeLorenzo, C., 224
Delta brain waves, 153, 273, 319
ΔFosB, 320
Dementia, 16, 45, 58, 60, 63. *See also* Alzheimer's disease
Demos, N. J., 147
Dendrites, 17, 18f, 245, 320

Densmore, M., 87
Depakote (valporic acid), 132
Depression
    assessing aging clients for, 46
    brain anatomy and, 13
    CBT and, 223–224
    childhood programming of stress response and, 60
    chronic stress and, 60, 66
    exercise and, 165
    gut-brain connection in, 15
    hypothyroidism and, 298
    immune functioning and, 16
    inflammation and, 5, 63
    interpersonal psychotherapy for, 226–227
    in marginalized populations, 58
    meditation and, 166
    neurofeedback and, 157
    neuroplasticity and, 21
    psychodynamic therapy for, 226–227
    psychotropic drugs for, 133, 134
    race and, 59
    referrals and, 134–135
    screening for, 150
    self-regulation assessment and case conceptualization, 296
    sleep and, 165
    thyroid hormones and, 16
    workplace stress and, 243
De Quervain, D., 228
Dhuli, K., 168
Diabetes, 58, 60, 133, 168
Diagnostic and Statistical Manual of Mental Disorders
    Fifth Edition, Text Revision (DSM-5-TR), 81–82, 298
    PTSD Checklist for DSM-5 (PCL-5), 150–151
Dialectic behavior therapy, 221
Diambra, Joel F., 251
Diaphragmatic (deep) breathing
    as assessment, 156–157
    in counseling relationship, 189
    defined, 320
    heart rate variability and, 171
    as neurocounseling interventions, 299, 301
    as relaxation techniques, 246
    in trauma therapy, 90, 92
Diathesis–stress model (D-SM), 63–65
Diaz, F., 46
Diazepam (Valium), 131
Diaz-Sastre, C., 42
Diet and nutrition
    brain, effect on, 46, 47
    case conceptualization and, 296
    food security and, 69
    neuroeducation on, 301
    in ulcerative colitis case study, 309
    wellness and optimal performance, 167–168, 173
Dietary supplements, 168, 173, 200, 203–204, 213
Differentiation, 34–35, 320
Di Francesco, G., 226–227
Digestive system, 10, 245
Dill, K. A., 167
Dinan, T. G., 41
Ding, A., 172
Ding, E. L., 172
Discrimination. *See* Racial discrimination
Dissociation, 86, 87, 88
Distress, 79. *See also* Stress and stress response
Distress tolerance, 90
Distribution (drugs), 128
Disulfiram, 114
Diversity. *See* Culture; Marginality, biology of
DMN. *See* Default mode network
DNA (deoxyribonucleic acid), 43, 65, 104, 228, 320
Donepezil (Aricept), 134
Dong, Y., 169
Donovan, D. M., 114
Dopamine (DA)
    in adolescence, 41
    counselor-client relationship and, 189
    defined, 320
    estrogen and, 42
    excitatory effect of, 126
    in reward system, 38, 106–109, 108f
    ventral tegmental area and, 11–12, 107
Dopaminergic (DAergic), 107–109
Dorsal anterior cingulate cortex, 9, 85, 223
Dorsal striatum, 107
Dorsolateral prefrontal cortex, 85, 222, 226
Dorsomedial prefrontal cortex, 187
Douthit, Kathryn Z., 55, 67
Downtime, 164
Drevets, W. C., 41
Drug classifications, 129–134
Drugs. *See* Psychopharmacology
Drugs of abuse. *See* Substance use and addiction
DSM. *See Diagnostic and Statistical Manual of Mental Disorders*

D-SM (diathesis–stress model), 63–65
Ducruet, T., 132
Duncan, N., 187
Dunn, Halbert, 162
Dysphoric mood, 110, 111

E
Eating disorders, 133–134
  anorexia case study, 144–145, 158
Eccard, C. H., 41
EEG (electroencephalography), 143, 224–225, 272–273, 276, 280, 298, 320. *See also* Quantitative EEG
Effect size metrics, 274, 278–280, 279*t*
Elbert, T., 228
Elderly clients. *See* Aging adults
Elliot, A., 169
Ellis, Albert, 228
EMDR (eye movement desensitization and reprocessing), 89, 91, 274, 281
Emic vs. etic approach, 208
Emotional contagion, 39, 255
Emotional cues, 42, 87
Emotional numbing, 86, 87
Emotional regulation. *See also* Executive functioning and self-regulation
  adolescent brain and, 40–41, 47
  aging adults and, 44
  anterior cingulate cortex in, 9
  brain development and, 36
  executive functioning and, 185–186
  frontal lobe in, 8
  infants and, 38
  labeling of emotions, 47, 90–92
  memory recall and, 255
  oxytocin and, 86
  PTSD and, 87, 90, 92
Emotional wellness, 166–167
Empathy
  addiction treatment and, 113
  anterior cingulate cortex and, 9
  attention and, 186–191
  biogenic explanations and reduction in, 207
  career-focused counseling and, 244
  clinical supervision and, 290
  insula and, 8–9
  mirror neurons and, 39
  neuroeducation and, 202
  technology use and, 42
Employment. *See* Career-focused counseling
Empowerment, 68, 90, 172, 208, 279–282
emWave (software), 157, 158

Endocrine system, 12, 14, 15–16, 19. *See also* Hormones
Endogenous opioids, 83–84
Enteric nervous system, 15
Epigenetics
  addiction, epigenetic factors in, 104
  assessments and, 146
  defined, 320
  environment, influence of, 42–43
  neuroscience-informed research and race, 276
  overview, 65–66
  psychotherapy and, 228
Epinephrine
  beta-blockers and, 132
  defined, 320
  in fight-or-flight response, 83, 86, 243
Episodic prospection, 241
Epston, David, 227
Eres, R., 187
Erickson, K., 165
Eriksson, P. S., 21
Esposito, R., 226–227
Estrogen, 42, 46, 88, 91, 306
Ethics
  CACREP 2024 Standards on, xii, 200, 287
  importance in counseling practices, 312–313
  in microskills hierarchy, 191, 192*f*
  neuroeducation delivery and, 205–206, 211–212
  training in neurofeedback and biofeedback, 299
Ethnicity, 55. *See also* Marginality, biology of
Eustress, 79. *See also* Stress and stress response
Evidence-based research, 314
Excitatory neurotransmitters, 19
Excretion (drugs), 128
Executive functioning and self-regulation, 185–186
  aging brain and, 44
  in case studies, 92, 194–195, 306
  clinical supervision and, 289–291
  decision-making and, 241
  defined, 325
  drug addiction and, 111
  mesocortical pathway and, 107
  neurocounseling, offering clients approaches to, 300–301
  neurofeedback/biofeedback and, 173
  neurophysiological dysregulation and, 293–299, 295*f*

345

Index

prefrontal cortex and, 8, 185–186
Exercise
  brain, effect on, 46, 47, 165, 300–301
  neuroeducation on, 301
  wellness and, 165, 170, 173, 296, 308
Existential factors in group work, 259*t*
Explicit/declarative memory, 12, 39, 44, 85, 91, 92
Exposure therapy, 86, 89, 92, 222
Externalization, 227
Eye movement desensitization and reprocessing (EMDR), 89, 91, 274, 281

F
Facial expressions, 24, 41, 87
Family dynamics, 259*t*
  in case study, 305, 308
Fan, Y., 187
Fasting, 170
Fear/threat response, 9, 12, 222. *See also* Fight-or-flight response
Feedback in clinical supervision, 290–291
Feelings charts, 92
Ferrara, A., 169
Fesmire, S. A., 244
Fibromyalgia, 63
Fickling, M. J., 290
Field, Thomas A., 199, 237, 293, 305
Fight-or-flight response, 14, 59, 83, 84*f*, 243. *See also* PTSD; Traumatic stress
Fishbowl technique, 114
Fisher, P. A., 242, 244
Fisher, Ronald, 277–278
Fish oil supplements, 168, 173
Fissures, 5, 321
Fitbit, 172
Fitzpatrick, D., 241
Flashbacks, 82, 92, 223
Flow state, 164
Fluoxetine (Prozac), 131, 226
fMRI. *See* Functional MRI
Focus time, 164
Food insecurity, 69
Fossati, P., 239
Fox, R., 21
Freeze response, 87
Freton, M., 240
Freud, Sigmund, 3
Frewen, P., 87
Friedman, C., 169
Frontal cortex, 13, 40, 44
Frontal lobe, 6*f*, 7–8, 39, 44, 322
Fu, C. H. Y., 224

Fumero, A., 222
Functional medicine physicians, 168
Functional MRI (fMRI), 186–187, 225, 272, 321

G
Gage, Phineas, 8
Galarce, M., 169
Gamma-aminobutyric acid (GABA), 86, 126–127, 321
Gamma brain waves, 153, 273, 319
Garcia, D., 42
Gardus, J., 224
Garvey, T. S., 165–166
Gaskill, R. L., 46
Gene expression, 104, 105, 206, 321. *See also* Epigenetics
Generalizability, 275–276
Gestalt therapy, 257
Ghahremani, A., 188
Gibbons, T. D., 170, 237
Gilbert-Diamond, D., 169
Glial cells, 17, 44, 165, 321
Glucocorticoids, 14, 228, 245
Glucose metabolism, 226
Glueck, D., 169
Glutamate, 126, 165, 321
Glutamatergic, 108, 109
Glutathione, 168
Goggle Scholar, 204–205
Goldstein, S. J., 224
Gonadal hormones, 16
Gonadal steroids, 42
Gossip in group work, 257
Gottman, J. M., 291
Govindarajan, S. T., 167
G protein–coupled receptors, 108
Graden, M. W., 290
Grady, C., 45
Gray matter
  in adolescent brain development, 40
  in aging brain, 44, 181
  defined, 326
  in early brain development, 36
  meditation and, 166
  white matter and, 18
Grays, D., 290
Greenfield, P. M., 42
Gregory, H. M., 274
Grounding exercises, 92
Group work, 251–265
  CACREP 2024 Standards and, 252
  case study, 253
    brain-based approach to, 261
    facilitation fundamentals for, 253–254

importance in neurocounseling, 310
norms, establishing, 254–257
   respect and communication, 256–257
     safety and, 256
     storytelling and memory, 254–255
overview, 251–252
therapeutic factors of, 258–260, 259t, 260f
Guided imagery, 67
Guidelines for neurocounseling, 305–315
Gut-brain connection, 15, 35, 41, 42, 289
Gyri, 5, 7, 36, 227, 321

H
Half-life (drugs), 128, 321
Hall, Sean B., 101
Hall, W. C., 241
Head map of brain functions, 153, 154–155f, 157, 297, 298–299
Head trauma, 35, 153, 296
Health monitoring devices, 171–172
Healthy mind platter, 163–167, 163f, 273–274, 321
Heart disease and hypertension, 58, 60, 63
Heart rate variability (HRV)
  app for, 300–301
  as assessment, 294
  bottom-up strategies, 170–171
  in case studies, 158, 173
  defined, 321
  overview, 157
  wearable devices for, 171, 225
Hedderson, M., 169
Hedges's *g*, 274, 279
Hedonic reactions, 106, 107, 112
Heo, S., 165
Herman, J. L., 90
Herrero, M., 222
Hettler, B., 163
Hippocampus
  aging and, 44, 45
  child trauma and, 89
  cortisol and, 245
  defined, 321
  depression and, 223
  development of, 39, 40
  in drug addiction, 109, 111
  exercise and effects on, 165
  fight-or-flight response and, 83
  lithium and, 132
  memory and, 7, 12, 39, 165, 193, 240, 254–255
  PTSD and, 85
Hite, L., 243
Hoban, A. E., 41
Hockett, C., 169
Holism, 257
Holland, J. L., 239
Holt-Lunstad, J., 165–166
Homeostasis, 12, 104
Homophobia, 294
Hope in group work, 259t
Hormones. *See also specific hormones*
  in adolescence, 41, 46
  brain anatomy and, 12
  corticotropin-releasing, 14, 110
  hypothalamic–pituitary–adrenal axis and, 14–15
  neurotransmitters and, 19, 20t, 41, 42
  psychoneuroendocrinology and, 15–16
  sex hormones, 35, 42, 88, 306
Household Pulse Survey, 123
HPA. *See* Hypothalamic–pituitary–adrenal axis
HRV. *See* Heart rate variability
Humility, 211–212
Hypervigilance, 81–82, 87, 214, 291
Hypoarousal/hyperarousal, 86, 87, 91–92, 225
Hypoglycemia, 35
Hypothalamic–pituitary–adrenal axis (HPA)
  in adolescence, 40, 42, 46
  attachment and, 38
  in childhood, stress response and, 60–61
  cortisol and dysregulation of, 89
  defined, 321
  depression and, 223
  drug withdrawal and, 110
  in fight-or-flight response, 83–84, 84f, 86
  functions of, 14–15
  memory and, 193
  neuroeducation on, 203
  in stress response, 242, 244–245
Hypothalamus
  in alerting network, 184
  defined, 321–322
  in fight-or-flight response, 83
  functions of, 11
  in sensory processing, 243
Hypothyroidism, 298

I

Imitative behavior in group work, 259*t*
Immediate early genes, 322
Immigrant woman with anxiety, case study of, xx–xxi, 305–315
Immunity/immune system, 15–16, 61–64, 168, 322
Immunology, 15–16
Imparting information in group work, 259*t*, 260
Implicit/nondeclarative memory, 12, 39, 46, 91, 92
Imprinting, 37–38
Impulsivity and impulse regulation
 in adolescence, 40, 46
 substance use and, 106–107
 traumatic stress and, 81, 85
Incentive salience, 106, 109
Incentive-sensitization theory of addiction, 106
Indivisible self, 163
Inerle, S., 228
Infancy brain development, 36–39
Inferior frontal gyrus, 41
Inferior temporal gyrus, 226
Inflammation
 defined, 322
 depression and, 5, 64
 immunity and, 62–63
 nutrition and, 168
Influencing skills, 191, 194
Inhibitory control, 111, 188
Inhibitory neurotransmitters, 19
Innate immunity, 62–63
Insomnia Severity Index, 149
Insula (insular cortex)
 defined, 322
 in drug addiction, 112
 functions of, 8–9, 187–188
 psychodynamic therapy and, 227
 serotonin and, 226
 traumatic stress and, 85–86
Interdisciplinary treatment teams, 123–124
Intermittent fasting, 170
International Society for Traumatic Stress Studies, 90
Interoception, 8, 322
Interoceptive signals, 112
Interpersonal autonomic synchrony, 260
Interpersonal communication and functioning, 85, 90
Interpersonal conflict and isolation, 296
Interpersonal domain, 239, 242
Interpersonal learning, 259*t*

Interpersonal process recall approach (Kagan), 297
Interpersonal psychotherapy (IPT)
 in anxiety treatment case study, 228–229
 depression and, 226–227
 for marginalized populations, 68
Interpersonal violence, 81, 294
Intrapersonal domain, 239, 242
Intrinsic locus of control, 307
IPT. *See* Interpersonal psychotherapy
Irritability, 82, 85, 110
Irritable bowel syndrome, 67, 243
Ivey, Allen, 167, 181–182, 219, 228
Ivey, Mary Bradford, 181, 228

J

Jenkins, E. J., 170
Jones, E. M. W., 170
Jones, J. S., 224
Jones, Laura K., 3, 33, 42, 77, 101, 199, 293, 305
Journal of Counseling & Development, 272
Journal of Mental Health Counseling, xii

K

Kagan, Norm, 297
Karagas, M., 169
Karlsson, H., 226
Kawamichi, H., 187
Keane, M. M., 241
Kim, J., 165
Kivlighan, D. M., Jr., 252
Klonopin (clonazepam), 131
Knapp, E., 169
Kolassa, I.-T., 228
Koob, G. F., 107
Kramer, A., 165
Kurth, T., 132

L

Lam, S., 188
Lamantia, A. S., 241
Lamictal (lamotrigine), 134–135
Lamm, C., 187
Language, 7–8, 38–39
Lanius, R., 87
Lazy eye (amblyopia), 245
Learning. *See* Cognition and learning
Learning difficulties
 assessments, 152–153
 childhood trauma and, 66
Learning element of reward, 106, 109
LeDoux, J. E., 12, 228

Left brained/right brained, 5–6
Left frontal cortex, 297
Left temporal cortex, 226
Lehéricy, S., 239
Lemonge, C., 239
Lenz, A. S., 274–275
Leshner, Alan, 101
Leszcz, M., 261
Li, M., 169
Lifestyle habits. *See* Wellness and optimal performance
Liking element of reward, 106, 110–111
Limbic system
 depression and, 223
 emotional regulation and dysregulation, 40, 85
 functions of, 13
 memory and, 240
 in sensory processing, 243
Lin, J., 21
Listening skills
 in assessments, 149
 attention and empathetic understanding, 187–191
 in microaggression case, 193–194
 microskills, 187, 189–190
 for neuroeducation, 202
Lithen, A., 167
Lithium, 132
Liver in drug metabolism, 128, 129
Long-term depression (LTD), 21
Long-term memory, 193–194, 245
Long-term potentiation (LTP), 21
Lopez-Castroman, J., 42
LORETA brain maps, xii
Lucchini, M., 169
Lucke, C. M., 252
Luke, Chad, 199, 201, 205, 206, 237, 242, 251, 252, 254, 260

M
MacLean, Paul, 13, 23
Mailey, E., 165
Malinow, R., 21
Manganotti, P., 168
Mania, 133, 134
Mantram repetition, 246, 311
MAOIs (monoamine oxidase inhibitors), 131
Marginality, biology of, 55–76
 CACREP 2024 Standards and, 56
 case study, 57–59
  brain-based approach to, 67
  through "social determinants of health" lens, 58–59
 chronic stress and, 59
 culturally informed multitargeted counseling, 68–69
 epigenetics and, 65–66
 overview, 55–56
 preventive wellness counseling and early interventions, 69
 psychoneuroimmunology, 60–65, 64*f*
 recognizing diversity of clients, 309
 sympathetic–parasympathetic shift, 67–68
Marlatt, G. A., 114
Marmot, Michael, 58–59
Martin, S., 165
Maslow's Hierarchy of Needs, 188–189, 189*f*
Masten, A. S., 252
MAT (medication-assisted treatment), 114–116
Maze trials, 245–246
McAuley, E., 165
McDonald, J., M., 169
McDonald, K., 243
McEwen, Craig A., 242
MCMI-IV (personality inventory), 151
Meador, K. J., 273
Meaning-making process, 202, 211, 227
Medial orbitofrontal cortex, 111
Medial prefrontal cortex, 41, 222, 240
Medial temporal lobe, 240
Medical history, 147
Medication-assisted treatment (MAT), 114–116
Medications. *See* Psychopharmacology; *specific medications*
Meditation, 166–167, 299, 311
Medium spiny neurons (MSNs), 108–109
Medori, M. C., 168
Medulla oblongata, 9
Melatonin, 168
Memory. *See also specific types*
 aging adults and, 44, 45
 autobiographical memory, 239–241
 brain anatomy and, 7, 8, 9, 11
 brain evolution and, 22
 carbohydrates' effect on, 167
 children, development of, 39
 cortisol and deficits in, 245
 decision-making in career counseling, 241
 defined, 323
 exercise and, 165

explicit/declarative, 12, 39, 44, 85, 91, 92
  meditation and, 166
  narrative therapy and reauthoring, 227–228
  neuroeducation on, 203
  neuroscience of, 254–255
  oxytocin and, 38
  play therapy and, 45–46
  PTSD and, 85, 91
  recall in career counseling, 241
  rewiring of memories in therapy, 193–194
  sleep and, 165, 308
Men, PTSD in, 87–88
Menstrual cycle, 42, 46, 88
Mental health disorders. *See also specific disorders*
  adolescence and, 40, 41
  in aging adults, 44
  gender differences in, 42
  neuroplasticity impairments in, 21
Mental imagery, 299
Mentalizing. *See* Theory of mind
Mental time travel, 239
Mesocortical pathway, 107, 112
Mesocorticolimbic dopamine system, 107–109, 108*f*
Mesolimbic pathway, 106
Meta-analyses, 274, 279
Metabolism
  depression and, 223, 226
  of drugs, 128
  physical wellness and, 165
  psychiatric drugs and, 129
  stress and, 83, 245
Metabolites, 128
Metaphoric approach, 244–246, 260
Methylation, 43, 65–66, 228
Methyl groups, 65
Methylphenidate, 133
Michikyan, M., 42
Microaggressions, case study of, 183, 194–195
Microbiota–gut–brain axis, 15, 41, 42, 289, 323
Microskills
  for addiction treatment, 113, 115
  attention and empathic understanding, 187–194
  neuroeducation and, 201–202
  neuroscience-based model and, 228
  power differentials and, 208
Midbrain/mesencephalon, 9
Middle cingulate cortex, 9, 187

Miller, Raissa, 199, 201–202, 287
Millon, Theodore, 151
Mills, K. L., 41
Mind–body modalities, xv, 67, 90
Mind-brain-body conceptualization, 182
Mindfulness
  behavioral therapy and, 114
  in case studies, 47, 92, 116
  emotional and spiritual wellness, 166
  interpersonal psychotherapy and, 227
  neuroscience-informed theory and, 228
  physiological awareness and, 293
  sympathetic-parasympathetic shift and, 67
Mini-Mental State Examination, 258
Minorities. *See also* Marginality, biology of
  alerting network and, 184–185
  culturally informed case conceptualization and, 293–294
  group work and, 252
  neuroeducation and, 208, 211
  research studies and representation of, 276–277
  social determinants of health and, 58–59
Mirror neurons, 38–39, 186, 255
Mockett, B. G., 170
Molenberghs, P., 187
Monoamine oxidase inhibitors (MAOIs), 131
Monteiro, J., 226–227
Mood disorders, 41, 66, 221
Mood stabilizers, 132
Mooney, R. D., 241
Moride, Y., 132
Morris, J., 42
Motivation, 11–12, 13, 240–241. *See also* Reward system
Motivational interviewing, 113, 115
Motor functions, 9, 10, 11
Motor reflexes, 38
MSNs (medium spiny neurons), 108–109
Mujica-Parodi, L. R., 167
Multicultural and Social Justice Counseling Competencies, 208
Multicultural understanding, 191, 192*f*, 208, 277, 290
Multiple sclerosis, 17
Murthy, V., 169
Myelin/myelination, 17, 18, 36, 37, 41, 245, 323
Myers, J. E., 163

N

Nabavi, S., 21
N-acetylcysteine, 168, 173
Naltrexone (Narcan), 114, 115
Narcolepsy, 133
Narrative therapy, 68, 227–228, 229
National Institutes of Health (NIH), 3
National Sleep Foundation, 168
Naureen, Z., 168
Navarro-Jimenez, R., 42
nCBT (neuroscience-informed CBT), 228
Necrosis, 44, 45, 323
Needlework, 68
Negative affective bias, 221
Negative-feedback loops, 14, 42, 59–60, 84
Nelson, K. M., 252
Nervous system. *See* Neurophysiological development; *specific parts of nervous system*
Neural networks/pathways, 164, 227
Neural plastic responses, 242
Neural plate and groove, 34, 323
Neural tube, 34
Neuroadaptation in addiction, 106, 110
Neuroanatomy, 272
Neuroception, 256
Neurocounseling, 323. *See also* Neuroscience-informed counseling
Neurocounseling-informed case conceptualization, 293–296, 295*f*
Neuroeducation, 199–217
  CACREP 2024 Standards and, 199–200
  case study, 200–201
    brain-based approach, 212–214
  defined, 201, 323
  overview, 201–202
  supervisory relationship to facilitate, 299, 300–301
  10-step process for providing, 201–212
    assess client knowledge of neuroconcepts, 209
    attend to client and relationship, 201–202
    consider influence of social positions and power differentials, 207–208
    construct plan for next steps, 212
    deliver information ethically, 211–212
    determine method for introduction and exploration, 209–211
    explore theory and motivations of client, 202–203
    identify neuroconcepts in client's story, 204–205
    reflect on theory and motivations of counselor, 205–207
    use information for exploration, 212
Neuroendocrine system, 63
Neuroessentialism, 207
Neurofeedback (NFB), xv. *See also* Biofeedback
  in ADHD treatment, 134
  in anxiety treatment, 274–275
  as assessment, 154–155*f*, 157–158, 294
  in case studies, 47, 137, 145, 158, 173, 312, 313
  defined, 323–324
  EEGs and, 273
  ethics and, 313
  learning theory in, 224–225
  in n-CBT, 228
  physiological awareness and, 293
  in PTSD treatment, 274–275
  qEEGs and, 156
  sleep hygiene and, 168
  sympathetic–parasympathetic shift and, 67
  training in, 299
  as trauma therapy, 89, 171
  in wellness counseling, 170–171
Neurofeedback Problem Rating Form, 295*f*
Neurogenesis, 21, 34, 165, 245
Neurogenic explanations of mental health, 206
Neuroimaging. *See* EEG; Functional MRI; Quantitative EEG
Neurological risk assessment, 147–149, 148*f*
Neuromyths, xv
Neurons
  aging brain and, 44
  brain anatomy, 17–21, 18*f*, 125–126. *See also* Neurophysiological development
  defined, 324
  in prenatal development, 34–35
Neuropeptides, 20*t*
Neurophysiological development, 33–54. *See also* Brain anatomy and development
  in adolescence, 40–42
  aging brain and, 44–45, 181
  Alzheimer's disease and, 45
  attachment and, 37–38

351

Index

CACREP 2024 Standards and, 33–34
developmentally informed interventions, 45–46
increased bilateral brain activity in aging brain, 45
in infancy and childhood, 36–39
language development and, 38–39
memory development and, 39
overview, 33
in prenatal development, 34–35
Neurophysiological dysregulation, 293–299, 295f
Neuroplasticity
carbohydrates' effect on, 167
CBT and, 223
cortisol and, 245
counseling and, 181, 187
defined, 324
drug use and, 111
impairments in, 21
neurocounseling interventions to support, 299
sleep and, 165
synaptogenesis and, 245
therapeutic relationship and, 187
Neuropsysiological dysregulation, 293
Neuroscience
of attention, 181–198. See also Attention
career-focused counseling based on, 237–250. See also Career-focused counseling
clinical supervision, 287–303. See also Neuroscience-informed clinical supervision
counseling theory based on, 219–236. See also Counseling theory, neuroscience-informed
group work based on, 251–265. See also Group work
guidelines for neurocounseling, 305–315
of substance use disorders, 101–122. See also Substance use and addiction
Neuroscience-informed CBT (nCBT), 228, 293–294
Neuroscience-informed clinical supervision, 287–303
CACREP 2024 Standards and, 287–288
case conceptualization, 293–298, 295f
case study, 286–288, 292, 300–301
supervisory relationship, 289–293
  connection and trust, 289–290
  feedback, confrontation, and evaluation, 290–291
  managing ruptures, 290
  treatment planning, 298–299
Neuroscience-informed counseling. See also Case studies; Neuroscience assessments. See Assessments
brain-based research, conducting, 269–285. See also Neuroscience-informed research
brain development and informed interventions, 45–46. See also Brain anatomy and development; Neurophysiological development
consulting on medications, 134–136. See also Psychopharmacology
culturally informed counseling for marginal populations, 68–69. See also Marginality, biology of
definition and functions of, xiii–xiv, 324
group work and, 251–265. See also Group work
guidelines for, 305–315
microskills framework of, 187–194. See also Attention
strategies to enhance wellness, 167–171. See also Wellness and optimal performance
Neuroscience-informed research, 269–285
CACREP 2024 Standards and, 269–270
case study, 270–271
importance of integrating into neurocounseling, 313–314
for neuroeducation, 204–205
neuroimaging in, 112
understanding, 271–281
  curiosity and, 274–275
  generalizability and statistical significance, 275–281, 279t
  identifying research in area of interest, 271–273
  integrating research into practice, 281
  sources and client voices, 273–274
Neurotransmission, 18–19, 104, 126–127, 127t, 324
Neurotransmitters
hormones and, 19, 20t, 41, 42
nutrition and, 167–168
psychotropic drugs and, 125, 129, 130t
Neurotrophic factors, 34–35
Neurovegetative symptoms, 325
NFB. See Neurofeedback

Nicholson, L. B., 16
NIH (National Institutes of Health), 3
Nonconscious body processes, 9, 10
Norepinephrine
  in alerting response, 184
  attachment and, 38
  beta-blockers and, 132
  defined, 325
  in drug addiction, 111
  excitatory effect of, 126
  in fight-or-flight response, 83, 86, 243
Northoff, G., 187
Nucleus accumbens, 11, 38, 40, 44, 106, 107, 111
Null hypothesis testing, 277–278
Nutrition. *See* Diet and nutrition
Nutritionists, 168

O
Observational notes, 149
Obsessive-compulsive disorder (OCD), 11, 131, 168, 171, 296
Occipital lobe, 6, 6*f*, 240, 245, 322
Occupational and intellectual wellness, 164
Occupational knowledge, 240–241
Older clients. *See* Aging adults
Olfactory bulbs, 13, 21, 325
Olfactory system, 38
Omega-3 fatty acids and brain function, 168, 173
O'Neal, J. H., 129
Opara, I. N., 276
Operant conditioning, 137, 224–225, 275
Opioid use, 83–84, 306
Opponent process theory of acquired motivation, 105–106, 109–110
Oquendo, M. A., 42
Orbitofrontal cortex, 85, 111
Orienting network, 185
Ortinski, P., 273
Oura Ring, 172
Oxytocin
  defined, 325
  estrogen and, 42
  functions of, 38
  PTSD and, 86

P
Pacheco, N., 298
Pain
  emotional components of, 187
  meditation and, 166
  microbiota–gut–brain axis and, 15
  myelin deterioration and, 17
  neurofeedback for, 312
  opioids and, 83–84, 306
  self-regulation assessment and case conceptualization, 296
Palmore, M., 169
Panic attacks, case study of, 78–79, 91
Panic disorder, 224
Panksepp, Jaak, 274
Pannasch, S., 87
Papassotiropoulos, A., 228
Paquette, V., 222
Parahippocampal gyrus, 222
Paraphrasing, 187, 190, 191
Parasympathetic nervous system
  career counseling and, 242, 246
  cranial nerves and, 10
  functions of, 14
  nCBT vs. CBT and, 293
  neurocounseling interventions to support, 299
  polyvagal theory and, 24
  positive vs. negative feedback, 291
  in sympathetic–parasympathetic shift, 67–68
  in trauma response, 86–87
  vagus nerve and, 184
Pariente, A., 132
Parietal lobe, 6*f*, 7, 13, 225, 240, 322–323
Parkin, J., 16
Parkinson's disease, 11, 21
Parsey, R. V., 224
Parsons, Frank, 237, 238
Pauly, M., 228
Paxil (paroxetine), 226
Payne, J., 163, 273–274
PCC (posterior cingulate cortex), 9, 13, 240
PCL–5 (PTSD Checklist for *DSM-5*), 150–151
Peñate, W., 222
Pence, B., 165
Perez-Rodriguez, M. M., 42
Peripheral nervous system, 14, 35, 184
Peripheral skin temperature training, 170–171, 299
Perry, B. D., 46
Personality and brain anatomy, 8
Personality inventories, 151–152
Petersen, S. E., 184
PET (positron emission tomography) scans, 226
PFC. *See* Prefrontal cortex
Pfeiffer, A., 228

Pharmacokinetics, 127–129, 130*f*, 325
Phobias, 222
Phylogenetic hierarchy, 86
Physical health. *See also* Diet and nutrition
   assessment by primary care practitioner, 309
   attention and, 184
   brain function and, 5
   case conceptualization and, 297–298
   exercise and, 165
   marginal populations and, 58–59. *See also* Marginality, biology of
   medical history, 147
   sleep and, 165
Physiological awareness, 293
Pituitary gland, 12, 14–15, 243, 325
Platt, M. L., 241
Play therapy, 45–46
Playtime, 164
PNI. *See* Psychoneuroimmunology
Poelmans, S. A. Y., 163, 273–274
Polyvagal theory, 14, 24, 188–189, 189*f*, 256, 289, 325–326
Pons, 9
Porges, Steven, 14, 24, 86–87, 188–190, 256, 289
Positive asset search, 191
Positive psychology/resilience, 192*f*, 193–194
Posner, M. I., 184, 185
Postcentral gyrus, 7
Posterior cingulate cortex (PCC), 9, 13, 240
Postpartum depression, 16
Postsynaptic neurons, 19, 125–126
Posttraumatic stress disorder. *See* PTSD
Poverty. *See* Marginality, biology of
Powell, B. W., 274
Power differentials, 207–208, 290
Practical Neurocounseling: Connecting Brain Functions to Real Therapy Interventions (Russell-Chapin), 298
Practical significance, 278–280, 279*t*
Prakash, R., 165
Predictive learning, 106
Predispositional vulnerability, 63–65
Prefrontal cortex (PFC)
   in aging brain, 44
   in alerting network, 184
   bilateral activation of, 45
   in brain evolution, 23
   child trauma and, 89
   cognitive empathy and, 187
   decision-making and, 241
   depression and, 223, 226
   development of, 36, 39
      in adolescence, 40–41
   executive functioning and, 8, 185–186
   in fight-or-flight response, 85
   memory and, 240
   oxytocin and, 86
   substance use and, 106, 111
   working memory and, 193
Pregnancy, 34–35, 132
Prenatal development, 34–35, 66, 69
Preoccupation/anticipation stage of addiction cycle, 111–112
Preston, J. D., 129
Presynaptic neurons, 19, 125–126
Primary motor cortex, 7–8
Primary somatosensory cortex, 7
Primary source information, 273, 275
Problem-free narratives/problem-minimized events, 227
Procedural memory, 194
Processed-based neuroeducation, 254
Progesterone, 42
Prognostic optimism or pessimism, 206–207
Progressive muscle relaxation, 67, 261
Proinflammatory cytokines, 63
Prolux, C. D., 21
Proprioception, 6, 7
Prozac (fluoxetine), 131, 226
Psychodynamic therapy, 226–227
Psychoeducation, xv. *See also* Neuroeducation
   in case study, 46–47
   group work and, 252, 253–254
Psychoneuroendocrineimmunology, 15, 325
Psychoneuroendocrinology, 15–16
Psychoneuroimmunology (PNI), 15–16, 60–65, 64*f*, 325
Psychopharmacology, 123–139
   for addiction, 114–115
   with aging adults, 129
   antidepressants, 131, 135
   antipsychotics, 133
   anxiolytics, 131–132
   CACREP 2024 Standards and, 124
   case study, 124–125
      brain-based approach to, 136–137
   with children, 129
   consultations with clients, 136
   consultation with psychopharmacologists, 135
   drug classification, 129–134, 130*t*
   medication referrals and consultation,

134–136
mood stabilizers, 132
neurobiology of psychotropic drugs, 125–129
neurotransmission and, 126–127, 127*t*
overview, 123–124
pharmacokinetics, 127–129, 130*f*, 324
stimulants, 133–134, 135
Psychosocial–medical history interview, 147, 294
PTSD (posttraumatic stress disorder). *See also* Traumatic stress
biofeedback and, 224
brain anatomy and, 13–14
brain waves and, 156
in case study, 270–271
CBT and, 223
chronic stress and, 60
complex posttraumatic stress, 88–89
gut-brain connection in, 15
interventions for, 89–91
neurofeedback and, 274–275
neurophysiology of, 82–87, 84*f*
sex differences in, 87–88
symptoms of, 82
trauma vs., 81–82
PTSD Checklist for *DSM-5* (PCL–5), 150–151
Purves, D., 241
Putamen, 11, 107, 227

Q
Qualitative research method, 273–274
Qualitative (subjective) assessment, 146
Quantitative EEG (qEEG), 153–156, 154–155*f*, 313, 325
Quantitative meta-analyses, 279
Quantitative (objective) assessment, 146
Questions, open and closed, 191
Quick Q EEG, 153

R
Race, E., 241
Race and ethnicity. *See* Marginality, biology of
Racial discrimination
alerting network and, 184
in case study, 58–59
culturally informed case conceptualization and, 293–294
microaggression case study, 183, 194–195
neuroeducation on systemic barriers, 208
neuroscience-informed research and, 276–277

Rastogi, A., 188
Ratai, E.-M., 167
Ratey, J., 165
Rational emotive behavior therapy, 221
RDoC (Research Domain Criteria), 325
Reauthoring, 227, 229, 247
Referrals, 134–136, 203, 299
Reflecting, 187, 190, 191, 212
self-reflection by counselors, 185–186, 205–207
Reflective listening, 115
Reframing, 194
Region of interest (ROI), 272
Reid, A. G., 241
Reiss, J., 87
Relapse prevention model, 114
Relational functioning, 88, 91
Relaxation
biofeedback and, 170–171, 224, 229
breathing exercises, 157
in case study, 78
clinical supervision and, 290
downtime and, 164
sympathetic-parasympathetic shift, 68
techniques for, 67, 246, 261
training in, 299
Religion and spirituality, 69, 147, 166–167, 296, 308
Repressed emotions, 226
Reprocessing, 89, 91
Research. *See* Neuroscience-informed research
Research Domain Criteria (RDoC), 325
Respect in group work, 256–257
RESPeRATE, 157
Response inhibition, 241
Restory of identity, 68
Retraumatization, 77, 81, 85, 90, 92
Reuptake, 19, 133–134, 325
Reward system
addiction and, 12, 105–111
cingulate cortex and, 9, 188
decision-making in career counseling, 241
dopamine and, 38, 106–109, 108*f*
Riddle-Jones, L., 276
Right brained/left brained, 5–6
Right supramarginal gyrus, 187
Risk-taking behaviors, 41, 81–82, 85
Risperidone, 124–125, 136–137
Ritalin, 133
Rivero, F., 222
RNA (ribonucleic acid), 325
Rock, D., 163, 273–274
Rogers, Carl, 186, 187–188

355

Rostral anterior cingulate cortex, 9, 86
Rothbart, M. K., 185
Ruptures in supervisory relationships, 290
Russell-Chapin, Lori A., xx, 143, 287, 294, 298, 305
Russo, G. Michael, 269, 274–275
Russotti, Justin, 55
Ryan, F. J., 41
Ryan, N. D., 41

S
Safety and trust
  in case studies, 145, 188–189, 194–195, 243, 310
  in clinical supervision, 289
  in group therapy, 256
  listening skills and, 189–190
  neuroeducation and, 202
  polyvagal theory and, 24, 188–189, 256
SAM. *See* Sympathetic–adrenal–medullary axis
Sample size, 278
Sand tray work, 67
Sauder, K., 169
Savickas, M. L., 239
Schimmel, Christine J., 242, 251, 252, 260
Schizophrenia
  antipsychotics for, 133
  basal ganglia in, 11
  brain anatomy and, 13
  chronic stress and, 60
  dietary supplements and, 168
  gut microbiota and, 41
  immune functioning and, 16
  neuroplasticity and, 21
  thyroid hormones and, 16
Schneider, A., 228
School environments, 68–69
Schultz, W., 108
Scientific racism, 276–277
Scope of practice, 203–204, 206, 211
Scott, C., 90, 91
Screening inventories, 149–151
Screen time, 144, 169–170, 296, 308
Selective norepinephrine reuptake inhibitors (SNRIs), 131
Selective serotonin reuptake inhibitors (SSRIs), 19, 127, 129, 131
Self-actualization, 188
Self-advocacy, 299
Self-at-work, 239
Self-care, 299

Self-compassion, 90, 207
Self-consciousness in adolescence, 41–42
Self-correction, 187
Self-knowledge, 239–240
Self-monitoring, 81, 85
Self-reflection by counselors, 185–186, 205–207
Self-regulation. *See* Executive functioning and self-regulation
Self-report, 149–152
Self-talk in group work, 258, 261
Selye, H., 79
Sensorimotor psychotherapy, 89
Sensory-based coping, 246, 311
Sensory systems and perception, 6, 10, 13, 22–23, 243
Serotonin. *See also* Selective serotonin reuptake inhibitors
  cytokines and, 64
  defined, 325
  psychotropic drugs, effect of, 127
  serotonin (5-HT) receptors, 226
  sex hormones and, 42
Serotonin syndrome, 131
Sexual assault, case study of, 91–93
Shame, 202
Shanahan, F., 41
Sherman, Nancy, 123
Short-term memory, 193, 241, 245
Siaz-Ruiz, J., 42
Side effects and contraindications
  (drugs), 128–129, 131–134
  counselor's role, 136
Siegel, Daniel J., 23, 23f, 163, 273–274
Siegel Hand Model, 23, 23f, 90, 325
Signal transduction, 325
Singer, Tania, 187
Sisk, C. L., 42
Skiena, S., 167
Skin-deep resilience, 64
Skinner, B. F., 275
Skin temperature control
  in assessment, 294
  in case studies, 158, 173, 229, 300, 311
  defined, 325
  overview, 157
  peripheral skin temperature training, 170–171, 299
Sleep, 149, 164–165, 168, 296, 301
Small, G. W., 42
Smartwatches, 225
SNRIs (selective norepinephrine reuptake inhibitors), 131

Social cognition, 41–42, 85
Social cues, 41–42, 46
Social determinants of health, 58–59
Social engagement system, 87, 90, 289–291
Socialization techniques, 259t
Social justice/injustice, xix, 55–56, 68
Social media, 240
Social processing and functioning
  brain evolution and, 22–23
  cingulate cortex in, 9
  infant development and, 38
  polyvagal theory and, 24
  vagus nerve in, 10
Social synapse, 256
Social wellness, 165–166
Social withdrawal, 64
Society for Psychical Research, 277–278
Solino-Fernandez, D., 172
Somatic distress, 88
Somatosensory experiences, 46
Somatosensory motor cortex, 187
Spinal cord, 184
Spirituality and religion, 69, 147, 166–167, 296, 308
Spunt, R., 186
Squire, L. R., 39
SSRIs. *See* Selective serotonin reuptake inhibitors
Stage-of-change approach to neuroeducation, 209
Stallworthy, I. C., 252
Statistical significance, 277–278
Steady state, 128–129, 325–326
Steuwe, C., 87
Stevens-Johnson syndrome, 135
Stiglic, N., 169–170
Stigma of mental health, xix, 206–207
Stilling, R. M., 41
Stimulants, 130t, 133–134, 135
Storytelling
  group work and, 254–255
  in narrative therapy, 227–228
  in trauma therapy, 91, 193
Strength-based perspective, 271–272
Stress and stress response. *See also* Chronic stress; PTSD; Traumatic stress
  adaptive nature of body in, 79–80, 92
  adrenaline and cortisol in, 242
  alerting and stress, 184–185
  attachment and, 38
  counselor advocacy and, 299
  defined, 79
  gut-brain connection in, 15
  hypothalamic–pituitary–adrenal axis in, 14–15
  hypothalamus in, 11
  meditation and, 166
  neuroeducation on, 208
  workplace stress, 237, 242–245, 246, 311
Stress management, 167–170, 229, 311
Strey, H. H., 167
Stroke, 58, 60
Strong, E. K., 239
Stufflebeam, S. M., 167
Subcortical, 326
Substance use and addiction, 101–122
  addiction cycle, 106–112
    binge and intoxication, 107–109, 108f
    preoccupation and anticipation, 111–112
    withdrawal and negative affect, 109–111
  adolescents and, 41
  BDMA, limitations of, 112–113
  brain anatomy and
    basal ganglia, 11
    DMN, 13–14
    neural mechanisms of, 244
    neuroplasticity, 21
  CACREP 2024 Standards and, 102
  case study, 103–104
    brain-based approach to, 115–116
  childhood programming of stress response and, 60
  clinical implications, 113–115
  dopamine levels and, 41
  foundational concepts in neurobiology of, 104–106
  during pregnancy, 35
  psychotropic drugs, abuse potential of, 131–134
  self-regulation assessment and case conceptualization, 296
  sex hormones and, 42
  workplace stress and, 243
Suicide, 42, 66
Sulci, 5, 36, 321
Sullivan, Harry Stack, 226
Sultan, S. F., 167
Summarizing, 187, 190, 191
Sun, X., 167
Super, Donald, 239
Superior temporal sulcus, 188
  posterior superior temporal sulcus, 41
Supervision. *See* Neuroscience-informed clinical supervision
Supplements. *See* Dietary supplements
Sweeney, T. J., 163

Swingle, M., 169
Sympathetic activation, 242
Sympathetic–adrenal–medullary axis (SAM), 14–15, 83–84, 84f, 86, 243
Sympathetic nervous system
  attention and concentration, 311
  autonomic nervous system and, 14, 86–87
  in fight-or-flight response, 84f, 184, 243, 246
  group work and respect, 256
  heart rate variability and, 157
Sympathetic–parasympathetic shift, 67–68, 326
Synapses, 18, 44, 45, 125, 326
Synaptic cleft, 17, 326
Synaptic pruning, 37, 40, 245, 326
Synaptogenesis, 21, 245–246
Systematic reviews, 274
Szabo, A., 165

T
Tai Chi, 68
Talaga, M. C., 129
Tangen, J. L., 290
Tardive dyskinesia, 133
Team approach to treatment, 309
Technology
  brain development and, 42
  neuroeducation and, 213
  screen time and limits on, 144, 169–170, 308
Telencephalon, 35
Temporal lobe, 6f, 7, 13, 44, 225, 240, 323
Temporal parietal junction, 41, 186
10–20 international location system, 298
Testosterone, 42, 88
Thalamus
  in aging brain, 44
  in alerting network, 184
  defined, 326
  functions of, 10–11
  in sensory processing, 243
  in trauma response, 85
Theories and motivations of clients, 203–204, 213
Theory of mind (ToM), 41–42, 46, 186–187, 190, 326
Therapeutic lifestyle changes (TLCs), 167–170, 308, 326
Therapeutic relationship, 90, 145, 187–189, 201, 202–204, 309–310
Theta brain waves, 153, 171, 319

Thomas, K. M., 41
Thomas, K. N., 170
Thyroid hormones, 15–16
TikTok, 200, 213
Time-in, 166
ToM. *See* Theory of mind
Tonic and phasic firing, 108, 111, 326
Top-down control/processing, 40, 85, 89, 111, 113, 171, 185, 261, 318
Touch, 7, 38
Tourette's syndrome, 11, 133
Tournier, M., 132
TOVA (screening test), 152–153
Toxicity (drug), 128, 132
A Training Model for the Development of Neuroscience-Informed Counseling Competencies, xvii, xviiif
Translational neuroscience, 237, 244
Transphobia, 294
Trauma
  abuse and, 310
  CBT and, 223
  defined, 81
  neuroeducation on, 208
  neurofeedback and, 89, 171
  pervasiveness of, 77
  retraumatization and, 77, 81, 85, 90, 92
  screening for, 150–151
  self-regulation assessment and case conceptualization, 296
  stored in body, 200, 213
  workplace stress and, 245
Trauma-focused cognitive behavior therapy, 68
Traumatic brain injury (TBI), 151, 271
Traumatic stress, 77–99. *See also* PTSD
  brain anatomy and, 8
  CACREP 2024 Standards and, 77–78
  case study, 78–79
    brain-based approach to, 91–93
  complex posttraumatic stress, 88–89
  nature of stress, 79
  overview, 77
  posttraumatic stress, 81–82
  during pregnancy, 35
  trauma-focused interventions, 89–91
  veterans trauma study, 223
Trichotillomania, 168
Triune theory of brain development (MacLean), 23
Trust. *See* Safety and trust
Tsien, R. Y., 21
20/20: A Vision for the Future of Counseling, xiii

Index

U
Uchino, B. N., 165–166
Uhls, Y. T., 42
Unique outcomes, 227
United Nations Department of Economic and Social Affairs, 44
Universality in group work, 259*t*
Uno, D., 165–166

V
Vagus nerve, 10, 14, 15, 184, 188, 194, 256, 289
Valium (diazepam), 131
Valporic acid (Depakote), 132
Veech, R. L., 167
Ventral anterior cingulate cortex, 111, 226
Ventral striatum, 107, 188
Ventral tegmental area, 11–12, 107
Ventral vagal system, 24, 86–87, 90
Ventrolateral prefrontal cortex, 226
Ventromedial prefrontal cortex, 240
Verdoux, H., 132
Verfaellie, M., 241
Viera, V., 165
Vietnam veterans, trauma study of, 223
Vigilance, 184
Viña, C., 222
Viner, R. M., 169–170
Visualization, 311
Visual processing, 6, 10, 38
Vitamin D, 168, 173, 306
Vocational identity, 239–240
Volkow, N. D., 107
Volumetric decline, 326
Voss, M., 165
Voucher-based reinforcement therapy, 113–114, 115
Voxels, 272, 326
Vukojevic, V., 228

W
Waddington, C. H., 43
Wagner, B. E., 169
Walker, Kiera, 101
Wang, J., 224
Wang, L., 252
Wank, A. A., 241
Wanting element of reward, 106, 110–111
Wearable electronic devices, 171–172, 225
Wei, M., 252
Weistuch, C., 167
Wellness and optimal performance, xv, 161–179
 CACREP 2024 Standards and, 161–162
 case study, 162
  brain-based approach, 173
 dimensions of, 163
 importance of biofeedback, neurotherapy, and neurofeedback skills, 311
 neurocounseling strategies to enhance, 167–171
  biofeedback and neurofeedback, 67, 170–171, 300–301
  exercise, 165, 170, 173, 308
  lifestyle changes, 167–170
  nutrition and weight management, 167–168
  screen time limits, 169–170
  sleep, 168, 308
  wearable electronic devices, 171–172
 neuroscience-informed theory and, 228
 overview, 161
 self-regulation assessment and case conceptualization, 296
 training in, 299
 wellness models, 162–167, 163*f*
  emotional and spiritual, 166–167
  occupational and intellectual, 164
  physical, 164–165
  social, 165–166
Wernicke's area, 7, 326
Whalen, P. J., 41
Wheel of wellness, 163
White, L. E., 241
White, L. J., 244
White, Michael, 227
White, S., 165
Whitehall I/II studies, 58–59
White matter, 18, 40, 44, 226, 326
Wickman, S. A., 244
Wilker, S., 228
Withdrawal/negative affect stage of addiction cycle, 109–111
Wittgenstein, L., 37
Wojcicki, T., 165
Women
 immigrant woman with anxiety, case study of, xx–xxi, 305–315
 menstrual cycle and, 42, 46, 88
 pregnancy and, 34–35, 132
 PTSD in, 87–88
Woods, J., 165
Working memory, 44, 193
Workplace, 311. *See also* Career-focused counseling

359

Workplace stress, 237, 242–245, 246, 311
Work-related information, 240–241
World Health Organization World Mental Health surveys, 77
Wright, S. L., 46

X
Xanax (alprazolam), 131

Y
Yalom, I. D., 258, 261
Yang, J., 224
Yoga, 68, 90, 261, 311

Z
Zalaquett, C., 181, 219, 228
Zehr, J. L., 42
Zgourou, E., 42
Zhu, Y., 169

www.ingramcontent.com/pod-product-compliance
Ingram Content Group UK Ltd.
Pitfield, Milton Keynes, MK11 3LW, UK
UKHW021843140426
5217IPUK00022B/1563